国家出版基金项目
NATIONAL PUBLICATION FOUNDATION

涡轮机械与推进系统出版项目

航空发动机技术出版工程

完全流程手册：

从流程建模到管理的知识体系
卷一（下）

The Complete Business Process Handbook

Body of Knowledge from Process Modeling to BPM
Volume I

〔法〕M. V. 罗辛（Mark von Rosing）

〔德〕A. W. 舍尔（August-Wilhelm Scheer）　著

〔瑞士〕H. V. 谢尔（Henrik von Scheel）

张玉金　王占学　等译

科学出版社

北　京

图字：01-2021-1780号

内 容 简 介

本书从流程概念的发展与演变出发，汇集大量来自政府机构、标准组织、企业、大学、研究机构以及行业专家的杰出贡献，将涉及流程管理的相关知识概念与理论加以系统整合，构建了流程管理领域全面的知识体系。

本书采取理论与实践分析相结合的方法，通过对流程管理知识、模板、专家建议以及最佳实践的介绍，从什么是流程管理、流程管理的思维方式、工作方式、建模方式、实施与治理方式、培训与指导方式六部分进行编著，将每一个复杂主题拆解成便于理解的知识点，提供了有关实现流程管理的流程、框架、方法的所有内容，为商用航空发动机研制企业构建基于流程管理的自主研发体系提供了一份全面的实用指南。

本书可为企业中高级管理人员、流程管理业务人员、管理咨询行业从业人员、大学管理类专业相关人员提供全面的学习参考，也可作为高校管理类教师和研究生的参考书。

图书在版编目(CIP)数据

完全流程手册：从流程建模到管理的知识体系. 卷一
= The Complete Business Process Handbook: Body
of Knowledge from Process Modeling to BPM Volume Ⅰ:
汉、英 / (法) 马克・冯・罗辛 (Mark von Rosing) 等
著; 张玉金等译. — 北京：科学出版社，2021.12
　(航空发动机技术出版工程)
　国家出版基金项目　涡轮机械与推进系统出版项目
　ISBN 978-7-03-068886-6

　Ⅰ. ①完… Ⅱ. ①马… ②张… Ⅲ. ①航空发动机－
系统工程－流程－手册－汉、英 Ⅳ. ①V23-65

中国版本图书馆CIP数据核字(2021)第101405号

责任编辑：徐杨峰 / 责任校对：谭宏宇
责任印制：黄晓鸣 / 封面设计：殷　靓

科学出版社 出版
北京东黄城根北街16号
邮政编码：100717
http://www.sciencep.com
南京展望文化发展有限公司排版
广东虎彩云印刷有限公司印刷
科学出版社发行　各地新华书店经销
*
2021年12月第　一　版　开本：B5(720×1000)
2025年 2 月第五次印刷　总印张：91 3/4
总字数：1687 000
定价：600.00元
(如有印装质量问题，我社负责调换)

The Complete Business Process Handbook: Body of Knowledge from Process Modeling to BPM Volume Ⅰ

Mark von Rosing, August-Wilhelm Scheer, Henrik von Scheel

ISBN: 9780127999593

《完全流程手册：从流程建模到管理的知识体系 卷一》（张玉金 王占学 等译）

ISBN: 9787030688866

注意

本书涉及领域的知识和实践标准在不断变化。新的研究和经验拓展我们的理解，因此须对研究方法、专业实践或医疗方法作出调整。从业者和研究人员必须始终依靠自身经验和知识来评估和使用本书中提到的所有信息、方法、化合物或本书中描述的实验。在使用这些信息或方法时，他们应注意自身和他人的安全，包括注意他们负有专业责任的当事人的安全。在法律允许的最大范围内，爱思唯尔、译文的原文作者、原文编辑及原文内容提供者均不对因产品责任、疏忽或其他人身或财产伤害及/或损失承担责任，亦不对由于使用或操作文中提到的方法、产品、说明或思想而导致的人身或财产伤害及/或损失承担责任。

涡轮机械与推进系统出版项目
顾问委员会

主任委员
张彦仲

委 员
（以姓名笔画为序）

尹泽勇　乐嘉陵　朱　荻　刘大响　杜善义
李应红　张　泽　张立同　张彦仲　陈十一
陈懋章　闻雪友　宣益民　徐建中

航空发动机技术出版工程
专家委员会

航空发动机技术出版工程
编写委员会

主任委员

尹泽勇

副主任委员

李应红　刘廷毅

委　员

（以姓名笔画为序）

丁水汀　王太明　王占学　王健平　尤延铖
尹泽勇　帅　永　宁　勇　朱俊强　向传国
刘　建　刘廷毅　杜朝辉　李应红　李建榕
杨　晖　杨鲁峰　吴文生　吴施志　吴联合
吴锦武　何国强　宋迎东　张　健　张玉金
张利明　陈保东　陈雪峰　叔　伟　周　明
郑　耀　夏峥嵘　徐超群　郭　昕　凌文辉
陶　智　崔海涛　曾海军　戴圣龙

秘书组

组　长　朱大明
成　员　晏武英　沙绍智

航空发动机技术出版工程
基础与综合系列
编写委员会

涡轮机械与推进系统出版项目

序

涡轮机械与推进系统涉及航空发动机、航天推进系统、燃气轮机等高端装备。其中每一种装备技术的突破都令国人激动、振奋，但是由于技术上的鸿沟，使得国人一直为之魂牵梦绕。对于所有从事该领域的工作者，如何跨越技术鸿沟，这是历史赋予的使命和挑战。

动力系统作为航空、航天、舰船和能源工业的"心脏"，是一个国家科技、工业和国防实力的重要标志。我国也从最初的跟随仿制，向着独立设计制造发展。其中有些技术已与国外先进水平相当，但由于受到基础研究和条件等种种限制，在某些领域与世界先进水平仍有一定的差距。在此背景下，出版一套反映国际先进水平、体现国内最新研究成果的丛书，既切合国家发展战略，又有益于我国涡轮机械与推进系统基础研究和学术水平的提升。"涡轮机械与推进系统出版项目"主要涉及航空发动机、航天推进系统、燃气轮机以及相应的基础研究。图书种类分为专著、译著、教材和工具书等，内容包括领域内专家目前所应用的理论方法和取得的技术成果，也包括来自一线设计人员的实践成果。

"涡轮机械与推进系统出版项目"分为四个方向：航空发动机技术、航天推进技术、燃气轮机技术和基础研究。出版项目分别由科学出版社和浙江大学出版社出版。

出版项目凝结了国内外该领域科研与教学人员的智慧和成果，具有较强的系统性、实用性、前沿性，既可作为实际工作的指导用书，也可作为相关专业人员的参考用书。希望出版项目能够促进该领域的人才培养和技术发展，特别是为航空发动机及燃气轮机的研究提供借鉴。

张彦仲

2019 年 3 月

航空发动机技术出版工程

序

  航空发动机被誉称为工业皇冠之明珠,实乃科技强国之重器。

  几十年来,我国航空发动机技术、产品及产业经历了从无到有、从小到大的艰难发展历程,取得了显著成绩。在世界新一轮科技革命、产业变革同我国转变发展方式的历史交汇期,国家决策进一步大力加强航空发动机事业发展,产学研用各界无不为之振奋。

  迄今,科学出版社于2019年、2024年两次申请国家出版基金,安排了"航空发动机技术出版工程",确为明智之举。

  本出版工程旨在总结、推广近期及之前工作中工程、科研、教学的优秀成果,侧重于满足航空发动机工程技术人员的需求,尤其是从学生到工程师过渡阶段的需求,借此也为扩大我国航空发动机卓越工程师队伍略尽绵力。本出版工程包括设计、试验、基础与综合、前沿技术、制造、运营及服务保障六个系列,2019年启动的前三个系列近五十册任务已完成;后三个系列近三十册任务则于2024年启动。对于本出版工程,各级领导十分关注,专家委员会不时指导,编委会成员尽心尽力,出版社诸君敬业把关,各位作者更是日无暇晷、研教著述。同道中人共同努力,方使本出版工程得以顺利开展、如期完成。

  希望本出版工程对我国航空发动机自主创新发展有所裨益。受能力及时间所限,当有疏误,恭请斧正。

2024 年 10 月修订

译 者 序

航空发动机的研制是一项复杂的系统工程，长久以来，我国航空发动机受跟踪研发模式的影响，尚未建立起完整、统一的自主研发体系，存在数据不共享、标准不统一、管理"两张皮"、运行效率低等情况，明显阻碍型号项目研制顺利推进。要实现国产商用航空发动机的自主研制，必须遵循航空发动机发展的客观规律，建立面向航空发动机产品全生命周期完整统一的流程体系，为独立自主研制出先进可靠的航空发动机夯实基础。

中国航发商用航空发动机有限责任公司自 2009 年成立以来，就明确提出"聚焦客户、流程主导、追求卓越、持续改进"的管理政策，以流程管理为主导，建设价值驱动型流程体系，以流程统领所有业务活动，以流程绩效评价业务结果，通过持续改进，不断推动业务高质量发展。商发公司矢志不渝地推进流程型企业的建设，围绕流程、组织和文化持续打造国际一流的企业运营系统：一是企业家面向系统求价值，管理者面向流程做改进，操作者面向作业做完善；二是企业作为价值创造系统，创新与变革是永远不变的主题，企业要始终动态地适应环境而改进，必须面对环境变化作战略取舍，面对客户需求优化产品价值链，面对资源需求优化供应链与资源保障；三是企业运营系统的核心要素：流程、人与文化，流程是主航道，必须持续地"清淤拓土筑基"，人与文化要素所赋予的知识资本、信息资本与组织资本必须嵌入流程鲜活灵动地动作。

经过十余年单通道窄体干线客机发动机验证机研制走完全过程和全面推进产品研制的实践积累，商发公司在管理运营与产品研制等方面进行了全面探索与实践，初步形成了以产品研发体系为核心，涵盖公司 20 个业务领域的流程体系。同时，在不断地面向用户质量目标、聚焦价值创造、开展流程再造与数字化转型等创新变革活动中，逐步探索建立了面向商用航空发动机公司的体系运行管理知识体系，并尝试通过对供应商的管理体系延伸，进一步打造安全、可靠、稳定的供应链体系，建立产业链良好的演进生态，最终实现国产商用航空发动机的产品交付，达成客户满意。

　　本书总结吸纳了世界优秀企业实践，提供了一套流程优化剪裁模板、一套持续改进优化的方法论、一套文化重塑与习惯再造的利器。商发公司在流程体系建设实践中，借鉴了其中的原理与方法，特别是流程全生命周期以及实施与治理等部分，得到了很多启发和帮助。该书针对流程管理进行了全面的系统论述，并构建了较为完整的知识体系；该书作为流程管理的实用指南，详细揭示了流程管理中我们思考和使用流程的方式，深入论述了业务流程的本质，以及从流程建模到治理的完整知识。为了使广大的学习者和实践者能够准确掌握本书所包罗的理论与方法精髓，本译著将以中、英文逐页对照的方式出版发行。

　　企业管理就是以流程为核心、价值为目标、自我驱动的生命之旅，企业管理运作是一门技术，又是一门艺术，同时也是一门实践的系统工程。中国企业尤其是从事复杂系统工程与高端制造业的企业更应学习如何站在巨人的肩膀上不断成就自我，打造既有东方智慧又有西方商业理念优秀基因的一流企业。翻译和实践本书就是在汲取西方优秀的企业管理最佳实践，为我国航空制造企业提供参考借鉴。在此，要特别感谢支持本书翻译与校订工作，以及在过程中提供资料案例和提出宝贵建议的相关专家、同事，他们是黄博、陈楠、黄干明、陈天彧、项飞、黄飞、汤先萍、张滟滋、杨博文、吴帆、何宛文、陈婧怡、郑冰雷。希望这本译著能够进一步促进流程管理方法在我国航空制造企业乃至中国企业的管理变革中的实践，让东方智慧与西方商业理念的深度结合转化为企业治理效能，打造更多世界一流的中国企业，为国家打造央企"市营"新范式提供有益的借鉴和参考！

张玉金

2021 年 5 月

Foreword

This book has been put together to help you explore Business Process concepts and to understand what BPM really is all about.

We wrote this book for YOU—the individual. You may be a business executive, manager, practitioner, subject matter expert, student, or researcher. Or may be an ambitious career individual who wants to know more about business process concepts and/or BPM, what it is all about and how to apply it.

This, *The Complete Business Process Handbook*, provides a comprehensive body of knowledge written as a practical guide for you—by the authorities that have shaped the way we think and work with processes today. You hold the first of three books in the series in your hand.

- The first volume endows the reader with a deep insight into the nature of business process concepts and how to work with them. From BPM Ontology, semantics, and BPM Portfolio management, to the BPM Life Cycle, it provides a unique foundation within this body of knowledge.
- The second volume bridges theory and application of BPM in an advanced modeling context by addressing the subject of extended BPM.
- The third volume explores a comprehensive collection of real-world BPM lessons learned, best practices, and leading practices examples from award-winning industry leaders and innovators.

We wish you well on your Business Process journey and that is why we also have invested years putting this Handbook series together. To share the knowledge, templates, concepts, best and leading practices and to ensure high quality and standards, we have worked and coordinated with standard development organizations like International Organization for Standardization (ISO), Object Management Group (OMG), Institute of Electrical and Electronics Engineers (IEEE), North Atlantic Treaty Organization (NATO), Council for Scientific and Industrial Research (CSIR), MITRE—a Federally Funded Research and Development Center, European Committee for Standardization (CEN), The Security Forum, World Wide Web Consortium (W3C), and LEADing Practice.

We have also identified and worked with leading organizations and with their process experts/architects, and have described their practices. Among them are Lego, Maersk Shipping, Carlsberg, FLSmidth, the US Government, AirFrance, KLM, German Government, SaxoBank, Novozymes, the Canadian Government, US Department of Defense, Danish Defense, Johnson & Johnson, Dutch Railway, Australian Government, and many more. At last but not least the Global University Alliance consisting of over 400 universities, lecturers, and researchers have analyzed and examined what works, again and again (best practice), and what are the unique practices applied by these leading organizations (leading practices). They then identified common and repeatable patterns, which provide the basis for the BPM Ontology, BPM Semantics, the BPM standards, and the process templates found in this book.

原 书 序

本书已经整合在一起,可以帮助您探索业务流程概念,并了解业务流程管理(BPM)的真正含义。

我们为您编写的这本书。您可能是:业务主管、经理、工作者、某一领域专家、学生或研究员,也或许是一个雄心勃勃的职场人士,想要更多地了解业务流程概念和BPM是什么以及如何应用它。

这本《完全流程手册》,它为您提供一个全面的知识体系,作为一本实用指南,由那些塑造了我们今天思考和使用流程的方式的权威人士编写。您手里拿着的这本是这个系列的三卷书中的第一卷。

- 第一卷让读者深入了解业务流程概念的本质以及如何使用它们。从BPM本体论、语义、BPM项目组合管理,到BPM生命周期,它在这个知识体系中提供一个独特的基础。
- 第二卷通过处理扩展BPM主题,在高级建模背景中架起BPM理论和应用之间的桥梁。
- 第三卷探讨屡获殊荣的行业领导者和创新者的全面实际BPM经验教训、最佳实践和领导实践示例。

我们祝愿您在业务流程之旅中取得成功,这就是我们花费多年时间将本系列手册整合在一起的原因,分享知识、模板、概念、最佳和领导实践。为了确保高质量和高标准,我们与国际标准化组织(ISO)、对象管理组织(OMG)、电气和电子工程师协会(IEEE)、北大西洋公约组织(NATO)、科学与工业研究理事会(CSIR)、联邦资助的研究与发展中心(Federally Funded Research and Development Center,FFRDC)MITER、欧洲标准化委员会(CEN)、安全论坛(The Security Forum)、万维网联盟(W3C)和领导实践(LEADing Practice)等组织进行了协调。

我们还选择与领先的组织及其流程专家/架构师合作,并描述他们的实践。其中包括:乐高、马士基航运、嘉士伯、艾法史密斯(FLSmidth)、美国政府、法国航空、荷兰皇家航空、德国政府、盛宝银行、诺维信、加拿大政府、美国国防部(United States Department of Defense,DOD)、丹麦国防部、强生、荷兰铁路、澳大利亚政府等。最后但同样重要的是,由400多所大学的讲师和研究人员组成的全球大学联

We have worked years on this book, and as you just read, with contributions of standard bodies, governments, defense organizations, enterprises, universities, research institutes and individual thought leaders. We put these chapters and their subjects carefully together and hope you enjoy reading it—as much as we did writing, reviewing and putting it together.

Name	Organization
Mark von Rosing	Global University Alliance
August-Wilhelm Scheer	Scheer Group GmBH
Henrik von Scheel	LEADing Practices, Google Board
Adam D.M. Svendsen	Institute of Future Studies
Alex Kokkonen	Johnson & Johnson
Andrew M. Ross	Westpac
Anette Falk Bøgebjerg	LEGO Group
Anni Olsen	Carlsberg Group
Antony Dicks	NedBank
Asif Gill	Global University Alliance
Bas Bach	NS Rail
Bob J. Storms	LEADing Practices
Callie Smit	Reserve Bank
Cay Clemmensen	LEADing Practices
Christopher K. Swierczynski	Electrolux
Clemens Utschig-Utschig	Boehringer Ingelheim Pharma
Dan Moorcroft	QMR
Daniel T. Jones	Lean UK
David Coloma	Universitat Politècnica de Catalunya, Spain
Deb Boykin	Pfizer Pharmaceuticals
Dickson Hunja Muhita	LEADing Practices
Duarte Gonçalves	CSIR—Council for Scientific and Industrial Research
Ekambareswaran Balasubramanian	General Motors
Fabrizio Maria Maggi	University of Estonia
Fan Zhao	Florida Gulf Coast University
Fatima Senghore	NASA
Fatma Dandashi	MITRE
Freek Stoffel	LEADing Practices
Fred Cummins	OMG

盟（Global University Alliance）分析和检查哪些有效，以及这些领先组织独特的实践应用（领导实践）。然后，他们确定了常见和可重复的模式，这些模式为BPM本体、BPM语义、BPM标准和本书中的流程模板提供了基础。

正如您刚才所读，我们已经为编写这本书工作了多年，您将在本书中找到标准机构、政府、国防组织、企业、大学、研究机构和个人思想领袖的贡献。我们将这些章节及其主题精心放在一起，希望您就像我们写作、复习和整理一样喜欢阅读它。

姓名	组织
Mark von Rosing	Global University Alliance
August-Wilhelm Scheer	Scheer Group GmBH
Henrik von Scheel	LEADing Practices, Google Board
Adam D.M. Svendsen	Institute of Future Studies
Alex Kokkonen	Johnson & Johnson
Andrew M. Ross	Westpac
Anette Falk Bøgebjerg	LEGO Group
Anni Olsen	Carlsberg Group
Antony Dicks	NedBank
Asif Gill	Global University Alliance
Bas Bach	NS Rail
Bob J. Storms	LEADing Practices
Callie Smit	Reserve Bank
Cay Clemmensen	LEADing Practices
Christopher K. Swierczynski	Electrolux
Clemens Utschig-Utschig	Boehringer Ingelheim Pharma
Dan Moorcroft	QMR
Daniel T. Jones	Lean UK
David Coloma	Universitat Politècnica de Catalunya, Spain
Deb Boykin	Pfizer Pharmaceuticals
Dickson Hunja Muhita	LEADing Practices
Duarte Gonçalves	CSIR—Council for Scientific and Industrial Research
Ekambareswaran Balasubramanian	General Motors
Fabrizio Maria Maggi	University of Estonia
Fan Zhao	Florida Gulf Coast University
Fatima Senghore	NASA
Fatma Dandashi	MITRE
Freek Stoffel	LEADing Practices
Fred Cummins	OMG

Name	Organization
Gabriel von Scheel	LEADing Practices
Gabriella von Rosing	LEADing Practices
Gary Doucet	Government of Canada
Gert Meiling	Tommy Hilfiger
Gert O. Jansson	LEADing Practices
Hans Scheruhn	University of Harz, Gemany
Hendrik Bohn	Nedbank
Henk de Man	OMG, VeeBee
Henk Kuil	KLM, Air France
Henrik Naundrup Vester	iGrafx
Jacob Gammelgaard	FLSchmidt
James P. Womack	Cambridge University-Massachusetts Institute of Technology (MIT)
Jeanne W. Ross	Cambridge University-Massachusetts Institute of Technology (MIT)
Jeff Greer	Cardinal Health
Jens Theodor Nielsen	Danish Defense
John A. Zachman	Zachman International
John Bertram	Government of Canada
John Golden	iGrafx
John M. Rogers	Government of Australia
Jonnro Erasmus	CSIR—Council for Scientific and Industrial Research
Joshua Michael von Scheel	LEADing Practices
Joshua Waters	LEADing Practices
Justin Tomlinson	LEADing Practices
Karin Gräslund	RheinMain University-Wiesbaden Business School
Katia Bartels	Office Depot
Keith Swenson	Fujitsu
Kenneth Dean Teske	US Government
Kevin Govender	Transnet Rail
Klaus Vitt	German Federal Employment Agency
Krzysztof Skurzak	NATO ACT
LeAnne Spurrell	QMR
Lloyd Dugan	BPM.com
Lotte Tange	Carlsberg Group

姓名	组织
Gabriel von Scheel	LEADing Practices
Gabriella von Rosing	LEADing Practices
Gary Doucet	Government of Canada
Gert Meiling	Tommy Hilfiger
Gert O. Jansson	LEADing Practices
Hans Scheruhn	University of Harz, Gemany
Hendrik Bohn	Nedbank
Henk de Man	OMG, VeeBee
Henk Kuil	KLM, Air France
Henrik Naundrup Vester	iGrafx
Jacob Gammelgaard	FLSchmidt
James P. Womack	Cambridge University-Massachusetts Institute of Technology (MIT)
Jeanne W. Ross	Cambridge University-Massachusetts Institute of Technology (MIT)
Jeff Greer	Cardinal Health
Jens Theodor Nielsen	Danish Defense
John A. Zachman	Zachman International
John Bertram	Government of Canada
John Golden	iGrafx
John M. Rogers	Government of Australia
Jonnro Erasmus	CSIR—Council for Scientific and Industrial Research
Joshua Michael von Scheel	LEADing Practices
Joshua Waters	LEADing Practices
Justin Tomlinson	LEADing Practices
Karin Gräslund	RheinMain University-Wiesbaden Business School
Katia Bartels	Office Depot
Keith Swenson	Fujitsu
Kenneth Dean Teske	US Government
Kevin Govender	Transnet Rail
Klaus Vitt	German Federal Employment Agency
Krzysztof Skurzak	NATO ACT
LeAnne Spurrell	QMR
Lloyd Dugan	BPM.com
Lotte Tange	Carlsberg Group

Name	Organization
Mads Clausager	Maersk Group
Mai Phuong	Northrop Grumman Electronic Systems
Maria Hove	LEADing Practices
Maria Rybrink	TeliaSonera
Marianne Fonseca	LEADing Practices
Mark Stanford	iGrafx
Marlon Dumas	University of Tartu
Mathias Kirchmer	BPM-d
Maxim Arzumanyan	St. Petersburg University
Michael Tisdel	US Government, DoD
Michel van den Hoven	Philips
Mikael Munck	SaxoBank
Mike A. Marin	IBM Corporation
Mona von Rosing	LEADing Practices
Nathaniel Palmer	BPM.com, Workflow Management Coalition (WfMC)
Neil Kemp	LEADing Practices
Nils Faltin	Scheer Group GmBH
Partha Chakravartti	AstraZeneca
Patricia Kemp	LEADing Practices
Peter Franz	BPM-d
Philippe Lebacq	Toyota
Régis Dumond	French Ministry of Defense, NATO, ISO
Rich Hilliard	IEEE, ISO
Richard L. Fallon	Sheffield Hallam University
Richard N. Conzo	Verizon
Rod Peacock	European Patent Office
Rogan Morrison	LEAD Enterprise Architect Professional
Ronald N. Batdorf	US Government, DoD, Joint Staff
Sarel J. Snyman	SAP Solution Design
Scott Davis	Government of Canada
Simon Polovina	Sheffield Hallam University
Stephen White	IBM Corporation
Steve Durbin	Information Security Forum
Steve Willoughby	iGrafx

姓名	组织
Mads Clausager	Maersk Group
Mai Phuong	Northrop Grumman Electronic Systems
Maria Hove	LEADing Practices
Maria Rybrink	TeliaSonera
Marianne Fonseca	LEADing Practices
Mark Stanford	iGrafx
Marlon Dumas	University of Tartu
Mathias Kirchmer	BPM-d
Maxim Arzumanyan	St. Petersburg University
Michael Tisdel	US Government, DoD
Michel van den Hoven	Philips
Mikael Munck	SaxoBank
Mike A. Marin	IBM Corporation
Mona von Rosing	LEADing Practices
Nathaniel Palmer	BPM.com, Workflow Management Coalition (WfMC)
Neil Kemp	LEADing Practices
Nils Faltin	Scheer Group GmBH
Partha Chakravartti	AstraZeneca
Patricia Kemp	LEADing Practices
Peter Franz	BPM-d
Philippe Lebacq	Toyota
Régis Dumond	French Ministry of Defense, NATO, ISO
Rich Hilliard	IEEE, ISO
Richard L. Fallon	Sheffield Hallam University
Richard N. Conzo	Verizon
Rod Peacock	European Patent Office
Rogan Morrison	LEAD Enterprise Architect Professional
Ronald N. Batdorf	US Government, DoD, Joint Staff
Sarel J. Snyman	SAP Solution Design
Scott Davis	Government of Canada
Simon Polovina	Sheffield Hallam University
Stephen White	IBM Corporation
Steve Durbin	Information Security Forum
Steve Willoughby	iGrafx

Name	Organization
Sven Vollbehr	SKF
Thomas Boosz	German Government
Thomas Christian Olsen	NovoZymes
Tim Hoebeek	SAP
Tom Preston	Booz Allen Hamilton
Ulrik Foldager	LEADing Practices
Victor Abele	Government of Canada
Vincent Snels	Nationale Nederlanden
Volker Rebhan	German Federal Employment Agency
Wim Laurier	Université Saint-Louism Bruxelles
Ýr Gunnarsdottir	Shell
Yury Orlov	Smart Architects
Zakaria Maamar	Zayed University, United Arab Emirates

姓名	组织
Sven Vollbehr	SKF
Thomas Boosz	German Government
Thomas Christian Olsen	NovoZymes
Tim Hoebeek	SAP
Tom Preston	Booz Allen Hamilton
Ulrik Foldager	LEADing Practices
Victor Abele	Government of Canada
Vincent Snels	Nationale Nederlanden
Volker Rebhan	German Federal Employment Agency
Wim Laurier	Université Saint-Louism Bruxelles
Ýr Gunnarsdottir	Shell
Yury Orlov	Smart Architects
Zakaria Maamar	Zayed University, United Arab Emirates

Abbreviation Meaning

A2A	Application to application
AAIM	Agility adoption and improvement model
ACM	Adaptive case management
ADDI	Architect design deploy improve
API	Application programming interface
APQC	American productivity and quality center
B2B	Business to business
BAM	Business activity monitoring
BCM	Business continuity management
BEP	Break even point
BI	Business intelligence
BITE	Business innovation and transformation enablement
BOM	Business object management
BPA	Business process analysis
BPaaS	Business process as a service
BPCC	Business process competency center
BPD	Business process diagram
BPE	Business process engineering
BPEL	Business process execution language
BPEL4WS	Business process execution language for web services
BPG	Business process guidance
BPI	Business process improvement
BPM	Business process management
BPM CM	Business process management change management
BPM CoE	Business process management center of excellence
BPM LC	Business process management life cycle
BPM PM	Business process management portfolio management
BPMaaS	BPM as a service
BPMI	Business process management institute
BPMN	Business process model and notation
BPMS	Business process management system
BPO	Business process outsourcing
BPPM	Business process portfolio management
BPR	Business process reengineering
BRE	Business rule engine
BRM	Business rules management
CDM	Common data model
CE-BPM	Cloud-enabled BPM
CEAP	Cloud-enabled application platform
CEN	European committee for standardization
CEP	Complex event processing
CM	Configuration management
CMS	Content management system
COBIT	Control objectives for information and related technology

缩 略 词

A2A	应用到应用
AAIM	敏捷应用和改进模型
ACM	适应性案例管理
ADDI	架构师设计部署改进
API	应用程序编程接口
APQC	美国生产力和质量中心
B2B	业务到业务
BAM	业务活动监控
BCM	业务连续性管理
BEP	盈亏平衡点
BI	商务智能
BITE	业务创新和转型支持
BOM	业务对象管理
BPA	业务流程分析
BPaaS	业务流程即服务
BPCC	业务流程能力中心
BPD	业务流程图
BPE	业务流程工程
BPEL	业务流程执行语言
BPEL4WS	Web服务的业务流程执行语言
BPG	业务流程指导
BPI	业务流程改进
BPM	业务流程管理
BPM CM	业务流程管理变更管理
BPM CoE	业务流程管理卓越中心
BPM LC	业务流程管理生命周期
BPM PM	业务流程管理组合管理
BPMaaS	BPM即服务
BPMI	业务流程管理机构
BPMN	业务流程建模标记法
BPMS	业务流程管理系统

CPO	Chief process officer
CRM	Customer relationship management
CSF	Critical success factor
CSIR	Council for Scientific and Industrial Research
CxO	Chief x officer
DB	Database
DBMS	Database management system
DMS	Document management system
DNEAF	Domain neutral enterprise architecture framework
DSDM	Dynamic systems development method
EAI	Enterprise application integration
EITE	Enterprise innovation & transformation enablement
EMR	Enterprise-wide metadata repositories
EPC	Event-driven process chain
EPSS	Electronic performance support system
ERM	Entity relationship modeling
ERP	Enterprise resource planning
ESB	Enterprise service bus
FEAF	Federal enterprise architecture framework
FI	Financial
iBPM	Intelligent business process management
IDE	Integrated development environment
IE	Information engineering
IEEE	Institute of electrical and electronics engineers
ISO	International Organization for Standardization
ITIL	Information technology infrastructure library
KPI	Key performance indicator
L&D	Learning and development
LEADP	Layered enterprise architecture development and/or LEADing big in Practice
MDM	Master data management
NATO	North Atlantic Treaty Organisation
NIST	National Institute of Standards and Technology
OCM	Organizational change management
OLAP	Online analytic processing
OLTP	Online transaction processing
OMG	Object management group
PDC	Process data collection
PIM	Process instance management
PM	Portfolio management
PM	Project management
PMBOK	Project management body of knowledge
PMO	Project management offices
POA	Process oriented architecture
PPI	Process performance indicator
PPM	Project portfolios management
PPPM	Portfolio, program and project management
PRINCE	PRojects IN Controlled Environments
QM	Quality management

BPO	业务流程外包
BPPM	业务流程组合管理
BPR	业务流程再造
BRE	业务规则引擎
BRM	业务规则管理
CDM	通用数据模型
CE-BPM	支持云端的BPM
CEAP	支持云的应用平台
CEN	欧洲标准化委员会
CEP	复杂事件处理
CM	配置管理
CMS	内容管理系统
COBIT	信息及相关技术控制目标
CPO	首席流程官
CRM	客户关系管理
CSF	关键成功因素
CSIR	科学和工业研究委员会
CxO	首席x官员
DB	数据库
DBMS	数据库管理系统
DMS	文件管理系统
DNEAF	领域中立的企业架构框架
DSDM	动态系统开发方法
EAI	企业应用程序集成
EITE	企业创新和转型支持
EMR	企业范围的元数据存储库
EPC	事件驱动的流程链
EPSS	电子绩效支持系统
ERM	实体关系建模
ERP	企业资源规划
ESB	企业服务总线
FEAF	联邦企业架构框架
FI	金融
iBPM	智能业务流程管理
IDE	集成开发环境
IE	信息工程
IEEE	电气和电子工程师协会
ISO	国际标准化组织
ITIL	信息技术基础架构库
KPI	关键绩效指标
L&D	学习和发展
LEADP	分层企业架构开发和/或领导实践
MDM	主数据管理

ROI	Return on investment
SBO	Strategic business objective
SCM	Supply chain management
SCOR	Supply chain operations reference model
SD	Sales and distribution
SNA	Social network analysis
SOA	Service oriented architecture
SPI	Service performance indicator
SRM	Supply relationship management
SW	Software
TCO	Total cost of ownership
TOGAF	The open group architecture framework
TQM	Total quality management
UI	User interface
ULM	Unified modeling language
USGAP	United States general accounting principles
VDML	Value delivery modeling language
VNA	Value network analysis
W3C	World Wide Web consortium
xBPMN	eXtended business process model and notation
XLM	Extensible markup language
XMI	Metadata interchange
XSD	XML schema definition

NATO	北大西洋公约组织
NIST	美国国家标准与技术研究院
OCM	组织变革管理
OLAP	联机分析处理
OLTP	联机事物处理
OMG	对象管理组织
PDC	过程数据收集
PIM	流程实例管理
PM	项目组合管理
PM*	项目管理
PMBOK	项目管理知识体系
PMO	项目管理办公室
POA	面向流程的体系结构
PPI	流程绩效指标
PPM	项目组合、项目集、项目管理
PPPM	投资组合、计划和项目管理
PRINCE	受控环境中的项目
QM	质量管理
ROI	投资回报率
SBO	战略业务目标
SCM	供应链管理
SCOR	供应链运作参考模型
SD	销售和分销
SNA	社会网络分析
SOA	面向服务的架构
SPI	服务绩效指标
SRM	供应关系管理
SW	软件
TCO	总拥有成本
TOGAF	开放组体系结构构框架
TQM	全面质量管理
UI	用户界面
UML	统一建模语言
USGAP	美国通用会计准则
VDML	价值交付建模语言
VNA	价值网络分析
W3C	万维网联盟
xBPMN	扩展业务流程模型和符号
XML	可扩展标记语言
XMI	元数据交换
XSD	XML 模式定义

---

* 正文中出现 PM 缩写时请对照原文。

Introduction to the Book

Prof. Mark von Rosing, Henrik von Scheel, Prof. August-Wilhelm Scheer

It is not a new phenomenon that the markets are changing; however, the business environment in which firms operate lies outside of themselves and their control. So, while it is their external environment, which is always changing, most changes on the outside affect the need for innovation and transformation on the inside of the organization. The ability to change the business and to manage their processes is symbiotic, which is, among others, one of the reasons for such a high Business Process Management (BPM) adoption rate in the market. It is, however, important to note that unlike some analysts might claim, the size of the market and its adoption is in no way an indicator of maturity. As a matter of fact, the maturity of many of the BPM concepts can have a low maturity, even though the adoption is widespread. So while the high demand for BPM as a management method and a software solution, and the maturing BPM capabilities develop and unfold, the challenge quickly develops to provide concise and widely accepted BPM definitions, taxonomies, standardized, and integrated process templates, as well as overall frameworks, methods, and approaches.

Written as the practical guide for you—by the authorities that have shaped the way we think and work with process today. This handbook series stands out as a masterpiece, representing the most comprehensive body of knowledge published on business process. The first volume endows the reader with a deep insight into the nature of business process, and a complete body of knowledge from process modeling to BPM, thereby covering what executives, managers, practitioners, students, and researchers need to know about:

- Future BPM trends that will affect business
- A clear and precise definition of what BPM is
- Historical evolution of process concepts
- Exploring a BPM Ontology
- In-depth look at the Process Semantics
- Comprehensive Frameworks, Methods, and Approaches
- Process analysis, process design, process deployment, process monitoring, and Continuous Improvement
- Practical usable process templates
- How to link Strategy to Operation with value-driven BPM
- How to build BPM competencies and establish a Center of Excellence
- Discover how to apply Social Media and BPM
- Sustainable-Oriented process Modeling
- Evidence-based BPM
- Learn how Value and Performance Measurement and Management is executed
- Explore how to enable Process Owners

本 书 介 绍

Mark von Rosing, Henrik von Scheel, August-Wilhelm Scheer

市场的变化并不是一个新的现象,但是企业的经营环境是不受自身控制的。因此,虽然外部环境一直在变化,但大多数外部变化都会影响组织内部的创新和转型需求。改变业务和管理流程的能力是共生的,这也是BPM在市场上被广泛应用的原因之一。不过,值得注意的是,与一些分析人员发表的意见不同,市场的规模和应用程度绝不是BPM成熟的主要标志。事实上,尽管许多BPM概念被广泛应用,但是成熟度依然很低,因此,对BPM作为管理方法和软件解决方案的高需求以及日益成熟的BPM能力,两者不断发展和展现的同时,在提供简洁并广泛接受的BPM定义、分类法、标准化集成流程模板以及总体框架、方法和途径等方面的挑战也在快速发展。

作为实用指南,本书作者塑造了BPM中我们思考和使用流程的方式。本手册系列作为杰作脱颖而出,代表了当前业务流程文档中最全面的知识体系。第一卷使读者深入了解业务流程的本质,以及从流程建模到BPM的完整知识体系,涵盖高管、经理、工作者、学生和研究人员需要了解的内容:

- 影响未来业务的BPM趋势;
- 一个对BPM清晰而精确的定义;
- 流程概念的历史演变;
- 探索BPM本体;
- 深入研究流程语义;
- 综合框架、方法和途径;
- 流程分析、流程设计、流程部署、流程监控以及持续改进(continuous improvement,CI);
- 实用的流程模板;
- 如何将战略与价值驱动的BPM运营联系起来;
- 如何建立BPM能力和卓越中心(center of excellence,CoE);
- 了解如何应用社交媒体和BPM;

- BPM Roles and Knowledge Workers
- Discover how to develop information models within the process models
- Uncovering Process Life cycle
- BPM Maturity
- BPM Portfolio Management and BPM Alignment
- BPM Change Management and BPM Governance
- Learning a structured way of Thinking, Working, Modeling, and Implementing processes.

This book is organized into various chapters that have been thoughtfully put together to communicate many times a complex topic into a replicable and manageable structure—that you as a reader can apply. Furthermore, the book is structured into six parts with the intention to guide you in turning business processes into real assets.

In Part I, we introduce a comprehensive "history of process concepts" from Sun Tzu's, to Taylorism, to Business Process Reengineering to Lean and BPMN, providing the reader with an in-depth understanding of the evolution of process thinking, approaches, and methods: a fundamental insight to what has shaped and what is shaping process thinking.

In Part II, we introduce the "Way of Thinking" around Business Process with focus on the value of Ontology, and a comprehensive BPM Ontology—the essential starting point that creates the guiding principles.

In Part III, we establish a "Way of Working" with Business Processes—the critical discipline of translating both strategic planning and effective execution. Exploring the current and future process trends that you need to be aware of with a detailed practical guide on how to apply them in areas such as BPM Life cycle, BPM Roles, process templates, evidence-based BPM, and many more.

In Part IV, we provide the essential guidance to help you in a "Way of Modeling" in traditional Process Modeling concepts to BPMN and Value-Oriented Process Modeling, how to work with and model Business Processes variations, as well as how to interlink information models and process models.

In Part V, we focus on the "Way of Implementation" and "Way of Governance"—the approach the practitioner follows in order to apply and steer what exists, spanning issues ranging from BPM change management, agile BPM, business process outsourcing, and holistic governance to project, program, and portfolio alignment.

In Part VI, we focus on the "Way of Training and Coaching"—to provide insight into ideal process expert, process engineer and process architecture training, from online to class-based learning and coaching.

While this book certainly can be read cover to cover, depending on where you are in your Business Process journey, you may wish to choose a different path. If you are new to Business Process concepts, you might start at the beginning, with Part I. If you are beginning a BPM project, or it has already begun its journey, or you are looking for inspiration, we recommend using the book as a reference tool to access it by the topic of interest.

But no matter how you plan on building your knowledge, the book has been designed and architected to be a guide and a handbook able to create the right way of thinking, working, modeling implementation, and governance.

- 面向可持续发展的流程建模；
- 基于证据的 BPM；
- 了解如何执行战略、绩效测量和管理；
- 探索如何启用流程责任人；
- BPM 角色和知识工作者；
- 了解如何在流程模型中开发信息模型；
- 揭示流程生命周期；
- BPM 成熟度；
- BPM 组合管理和 BPM 协调机制；
- BPM 变革管理和 BPM 治理；
- 学习思维、工作、建模和实施流程的结构化方法。

本书由不同的章节构成，这些章节经过深思熟虑被组合在一起，将一个复杂的主题分解成一个可复制和可管理的结构，便于读者应用。此外，本书分为六个部分，旨在指导您将业务流程转化为公司的宝贵资产。

第一部分，我们介绍从孙子到泰勒主义（Taylorism）、业务流程再造（BPR）、精益和业务流程建模标记法（BPMN）的流程概念的演变历史，有助于读者形成对进化流程思维、途径和方法的深入理解，建立对流程思维及形成流程的基本认识。

第二部分，我们介绍围绕业务流程的思维方法，重点介绍本体论的价值，以及一个全面的 BPM 本体论，这是创建指导原则的基本出发点。

第三部分，我们建立业务流程的工作方法，这是战略规划和有效执行的关键。我们协助您探索您需要了解的当前和未来流程趋势，并提供详细的实践指导，包括：如何将其应用于诸如 BPM 生命周期、BPM 角色、流程模板、基于证据的 BPM 等领域。

第四部分，我们为您提供基本的指导，帮助您以传统流程建模概念的建模方法为基础，实现 BPMN 和价值导向的流程建模，包括：如何应对与建模业务流程有关的变化，以及如何将信息模型和流程模型相互链接等。

第五部分，我们将重点放在实施方法和治理方法——从业者遵循的方法方面，尤其是如何面对存在的问题，包括从 BPM 变革管理、敏捷 BPM、业务流程外包（BPO）和整体治理到项目、计划和组合调整等方面的各种问题。

第六部分，我们将重点放在"培训和指导的方法"上，从在线、课堂学习到经验指导，我们将深入学习流程专家、流程工程专家和流程架构专家的培训内容与知识体系。

尽管这本书可以从头到尾详细地进行阅读，但您也可根据您在业务流程中所处的环节选择不同的章节学习。如果您对业务流程概念不熟悉，我们建议您可以从第一部分开始。如果您正在计划或者已经开始了一个 BPM 项目，我们建议您使用这本书作为参考工具，研究其中您感兴趣的主题。

但是，不管您打算如何构建您的知识体系，这本书在设计和架构方面都是一本能够创建正确的思考、工作、建模和治理方式的指南和手册。

目 录

第五部分

第七部分

第七部分

The BPM Way of Modeling

Mark von Rosing, Henrik von Scheel, August-Wilhelm Scheer

INTRODUCTION

Part IV provides practical and essential guidance to the "*Way of Modeling*" with and around business process concepts. Part IV outlines the approach the practitioner follows to apply principles for representing process and making an objective assessment of the possible. By using decomposition and composition modeling techniques within the different layers, for example, business, application, and technology, the approach provides you, the practitioner, faced with real world challenges, a uniform and structured description of the model objects and artifacts within one or more different types of models.

Identifying and classifying the different objects, for example, business information and/or data, is not always easy, relating such to the process and/or service model as well as the execution and realization of such into application software solutions is also quite complex. The ways to model one's process models both in terms of EPC-Event-Driven Process Chain from ARIS (Software AG), with examples of information and process models, BPNM, to value-oriented process modeling, and sustainability modeling, are the focus of Part IV.

The Complete Business Process Handbook. http://dx.doi.org/10.1016/B978-0-12-799959-3.00020-3

第四部分

4.1　BPM建模方法

Mark von Rosing, Henrik von Scheel, August-Wilhelm Scheer

介绍

本部分为业务流程概念的建模方法提供了实用和基本的指导,概述了从业人员采用的方法,以适用于代表流程和作出客观评估的原则。通过在不同的层(如业务、应用程序和技术)中使用分解和组合建模技术,该方法为您(从业者)提供了面对现实挑战的统一和结构化的模型对象描述,以及一种或多种不同类型的模型中的组件。

不同的对象(如业务信息和/或数据)进行识别和分类并不总是容易的,将这些对象与流程、服务模型以及它们的执行和实现联系到应用软件解决方案中也相当复杂。从ARI(Software AG)的EPC事件驱动流程链(event-driven process chain)(包括信息和流程模型、BPNM、面向价值的流程建模和可持续性建模)方面,对流程模型进行建模的方法是第四部分的重点。

Business Process Model and Notation—BPMN

Mark von Rosing, Stephen White, Fred Cummins, Henk de Man

INTRODUCTION

This chapter is intended to provide an overview and introduction to the Business Process Model and Notation (BPMN). We will describe BPMN and its historic development. In addition, we will provide the general context and usage of BPMN, layered upon the technical details defined in the BPMN 2.0 Specification. The basics of the BPMN notation will be described—that is, the types of graphical shapes, their purpose, and how they work together as part of a Business Process Model/Diagram. Also discussed will be the different uses of BPMN diagram types, including how levels of precision affect what a modeler will include in a diagram. Finally, the value in using BPMN as a standard notation will be defined.

It is vital to note that because both the main authors and the additional four authors all officially work with the Object Management Group (OMG) to develop standards, this chapter and its content be based on the official OMG BPMN specification.[1]

WHAT IS BPMN?

Business Process Model and Notation (BPMN) is a standard for business process modeling that provides graphical notation for specifying business processes in a Business Process Diagram (BPD),[2] based on traditional flowcharting techniques. The objective of BPMN is to support business process modeling for both technical users and business users, by providing notation that is intuitive to business users, yet able to represent complex process semantics. The BPMN 2.0 specification also provides execution semantics as well as mapping between the graphics of the notation and other execution languages, particularly Business Process Execution Language (BPEL).[3]

BPMN is designed to be readily understandable by all business stakeholders. These include the business analysts who create and refine the processes, the technical developers responsible for implementing them, and the business managers who monitor and manage them. Consequently, BPMN serves as a common language, bridging the communication gap that frequently occurs between business process design and implementation.

The Complete Business Process Handbook. http://dx.doi.org/10.1016/B978-0-12-799959-3.00021-5

4.2　业务流程模型和表示方法——BPMN

Mark von Rosing, Stephen White, Fred Cummins, Henk de Man

4.2.1　介绍

本节旨在提供 BPMN 的概述和简介。我们将描述 BPMN 及其历史发展。此外，我们将提供 BPMN 的一般语义和用法，并按照 BPMN 2.0 规范中定义的技术细节进行分层，将描述 BPMN 的基础知识（即图形形状的类型）、它们的目的，以及它们如何作为业务流程模型/图表的一部分一起工作。还将讨论 BPMN 图类型的不同用法，包括精度级别如何影响建模者将在图中包含的内容。最后，将阐述使用 BPMN 作为标准符号的价值。

需要注意的是，由于主要作者和其他四位作者都与 OMG 正式合作开发标准，所以本节及其内容都基于官方 OMG BPMN 规范[1]。

4.2.2　什么是 BPMN？

BPMN 是业务流程建模的标准，它提供了基于传统流程图技术在业务流程图（BPD）[2] 中指定业务流程的图形表示法。BPMN 的目标是通过为业务用户提供直观的符号能够表示复杂的流程语义的同时来支持技术用户和业务用户的业务流程建模。BPMN 2.0 规范还提供了执行语义以及符号图形和其他执行语言之间的映射，特别是业务流程执行语言（BPEL）[3]。

BPMN 旨在让所有业务利益相关者都能轻松理解，这些利益相关者包括：创建和优化流程的业务分析师、负责实施流程的技术开发人员以及监控和管理流程的业务经理。因此，BPMN 作为一种通用语言，弥合了业务流程设计和实现之间经常出现的沟通鸿沟。

THE HISTORIC DEVELOPMENT OF BPMN

In 2001, the process-modeling marketplace was fragmented with many different modeling notations and viewpoints. It was in this context that members of Business Process Management Institute (BPMI), many of whom represented companies that contributed to the fragmented market, began discussing the idea of standardizing business-oriented techniques for visually representing process components and aligning the notation with an executable process language. The BPMN 1.0 specification was released to the public in May 2004. With this, the primary goal of the BPMN specification was to provide a notation that is readily understandable by all business users, from the business analysts that create the initial drafts of the processes, to the technical developers responsible for implementing the technology that will perform those processes, and finally, to the business people who will manage and monitor those processes. BPMN 1.0 was also supported with an internal model that was mapped to executable BPEL4WS.

It was February 6, 2006, when BPMI was subsumed by the OMG, who has since maintained and developed the BPMN standard. The BPMN 1.1 version was published in January 2008[4] and a year later version 1.2[5] was published. Work on the well-known version 2.0 took another two years, and it was published in January 2011.[6] This international standard represents the amalgamation of best practices within the business modeling community to define the notation and semantics of collaboration diagrams, process diagrams, and choreography diagrams. In doing so, BPMN will provide a simple means of communicating process information to other business users, process implementers, customers, and suppliers.

Another goal, but no less important, is to ensure that the models created by BPMN are executable. BPMN 1.x provided mappings to Extensible Markup Language (XML) designed for the execution of business processes, such as Web Services Business Process Execution Language (WSBPEL). The ability to execute BPMN via BPEL (BPEL, also known as WS-BPEL) breathed life into model-driven process execution. In essence, the equation Application = Computation + Coordination has become reality with network-addressable computation being provided by Web Services and BPMN graphically depicting the coordination logic. BPMN 2.0 provided its own execution semantics in addition to an updated mapping to BPEL. Thus, new process engines can directly execute BPMN models without the potential behavioral restrictions that might result in the complex mapping of the more free-form BPMN to the more structured BPEL.

Some of the main changes that the BPMN versions 2.0 brought with them are among others:

- The addition of a Choreography diagram.
- The addition of a Conversation diagram.

4.2.3　BPMN的发展历史

2001年，流程建模市场被许多不同的建模符号和观点分割开来。正是在这种背景下，业务流程管理研究所（BPMI）的成员（其中许多代表了在各细分市场做出贡献的公司）开始讨论面向业务标准化的技术（用于可视化地表示流程组件）和将符号与可执行流程语言进行统一的想法，最终BPMN 1.0规范于2004年5月公开发布。有鉴于此，BPMN规范的主要目标是提供一个易于让所有业务用户理解的符号，从创建流程初始草稿的业务分析师到负责实现将执行这些流程的技术的技术开发人员，最后是负责管理和监控这些流程的业务人员，BPMN 1.0也受到映射到可执行BPEL4WS（是2002年8月由微软公司、IBM公司和BEA公司联合发布的"网络服务业务流程执行语言"）的内部模型的支持。

BPMI是在2006年2月6日被OMG纳入的，OMG此后一直维护和开发BPMN标准。BPMN 1.1版本于2008年1月发布[4]，一年后又发布了1.2版本[5]。著名的2.0版本又花了两年的时间，并于2011年1月发布[6]，该国际标准代表了业务建模社区中最佳实践的融合，以定义协作图、流程图和编排图的表示法和语义。通过这样做，BPMN将提供一种简单的方法，将流程信息传递给其他业务用户、流程实现人员、客户和供应商。

另一个目标（但同样重要）是确保BPMN创建的模型是可执行的。BPMN 1.x提供了到XML的映射，XML设计用于执行业务流程，如Web服务业务流程执行语言（WSBPEL，Web services business process execution language）。通过BPEL（BPEL，也称为WS-BPEL）执行BPMN的能力为模型驱动的流程执行注入了活力。实际上，随着Web服务提供的网络寻址计算和BPMN图形化地描述协调逻辑，等式Application=Computing+Coordination已经成为现实。除了更新到BPEL的映射之外，BPMN 2.0还提供了自己的执行语义。因此，新的流程引擎可以直接执行BPMN模型，而不需要潜在的行为限制，这些限制可能导致更自由的BPMN与更结构化的BPEL之间的复杂映射。

BPMN 2.0版本带来的一些主要变化包括：

- 添加编排关系图；
- 添加对话图；

- Noninterrupting Events for a Process.
- Event Subprocesses for a Process.

The major technical changes include:

- A definition of the process execution semantics.
- A formal metamodel as shown through the class diagram figures.
- Interchange formats for abstract syntax model interchange in both XML Metadata Interchange (XMI) and XML Schema Definition (XSD).
- Interchange formats for diagram interchange in both XMI and XSD.
- Extensible Stylesheet Language Transformations (XSLT) between the XMI and XSD formats.

Other technical changes include:

- Reference Tasks are removed. These provided reusability within a single diagram, as compared to Global Tasks, which are resuable across multiple diagrams. The new Call Activity can be used to reference a Global Task or another Process to be used within a Process (instead of Reference Tasks).

Because of the version 2.0 updates, the number of elements more than doubled from 55 elements to 116. Many of these new elements were applied to modeling interactions between processes and/or entities, such as the new choreography diagram.

BPMN 2.0.2, released in December 2013,[7] included only minor modifications in terms of typo corrections and a change in clause 15.

THE BPMN NOTATIONS/SHAPES

A major goal for the development of BPMN was to create a simple and understandable notation for creating Business Process models, while providing the semantics and underlying mechanisms to handle the complexity inherent in Business Processes. The approach taken to handle these two conflicting requirements was to organize the graphical aspects of the notation into specific categories. This provides a small set of notation categories so that the reader of a BPMN diagram can easily recognize the basic types of elements and understand the diagram. The various basic BPMN shapes are shown below (Tables 1–6):

Within the basic categories of elements, additional variation and information can be added to support the requirements for complexity without dramatically changing the basic look and feel of the diagram. In the following sections, we will illustrate how the BPMN shapes are used in various end-to-end BPMN models.

- 流程的非中断事件；
- 流程事件的子流程。

主要技术变化包括：

- 流程执行语义的定义；
- 通过类图所示的正式元模型；
- XML元数据交换（XMI）和XML模式定义（XSD）中抽象语法模型交换的交换格式。
- XMI和XSD中的图表交换格式。
- XMI和XSD格式之间的可扩展样式表语言转换（extensible stylesheet language transformation，XSLT）。

其他技术变更包括：

- 删除引用任务。与可以跨多个图重用的全局任务相比，这些任务在单个图中提供了可重用性。新的调用活动可用于引用一个全局任务或流程中要使用的另一个流程（而不是引用任务）。

由于2.0版的更新，元素数量从55个增加到了116个，增加了一倍多。其中许多新元素被应用于流程和/或实体之间的交互建模，如新的编排图。

2013年12月发布的BPMN 2.0.2仅包括对拼写纠正方面的微小修改和第15条的修改。

4.2.4 BPMN符号/形状

开发BPMN的一个主要目标是为创建业务流程模型创建一个简单易懂的符号，同时提供语义和底层机制来处理业务流程中固有的复杂性。处理这两个相互冲突的要求所采用的方法是将符号的图形方面组织成特定的类别。这提供了一小组符号类别，以便BPMN图的读者可以轻松识别元素的基本类型并理解图表。各种基本BPMN形状如下所示（表1～表6）。

在元素的基本类别中，可以添加其他变体和信息以支持复杂性要求，而无须显著更改图表的基本外观。在以下部分中，我们将说明如何在各种端到端BPMN模型中使用BPMN形状。

Table 1 *BPMN Task Description*

BPMN 2.0.2	Task Description
None	No special task type is indicated.
User Task	A User Task is a typical "workflow" task in which a human performer performs the task with the assistance of a software application and could be scheduled through a task list manager of some sort.
Manual Task	A Manual Task is a task that is expected to be performed without the aid of any business process execution engine or application.
Service Task	A Service Task is a task that uses some sort of service, which could be a web service or an automated application.
Receive Task	A Receive Task is a simple task that is designed to wait for a message to arrive from an external participant (relative to the process).
Send	A Send Task is a simple task that is designed to send a message to an external participant (relative to the process).
Script	A Script Task is executed by a business process engine. The modeler or implementer defines a script in a language that the engine can interpret. When the task is ready to start, the engine will execute the script. When the script is completed, the task will also be completed.

表1　BPMN任务描述

BPMN 2.0.2	任　务　描　述
无	没有指出特殊任务类型
用户任务	用户任务是一个典型的工作流任务,在此任务中,人工执行者在软件应用程序的帮助下执行任务,并且可以通过某种类型的任务列表管理器进行调度
手工任务	手工任务是预期在没有任何业务流程执行引擎或应用程序帮助的情况下执行的任务
服务任务	服务任务是使用某种服务的任务,可以是Web服务或自动化应用程序
接收任务	接收任务是一个简单的任务,用于等待来自外部参与者(相对于流程)的消息到达
发送任务	发送任务是一个简单的任务,其设计目的是将消息发送给外部参与者(相对于流程)
脚本任务	脚本任务由业务流程引擎执行。建模者或实现者用引擎可以解释的语言定义脚本。当任务准备启动时,引擎将执行脚本。当脚本完成时,任务也将完成

Table 1 BPMN *Task Description—Cont'd*

BPMN 2.0.2	Task Description
Business Rule	A Business Rule Task provides a mechanism for the process to provide input to a Business Rules Engine and to get the output of calculations that the business rules engine might provide. The input/output specification of the task will allow the process to send data to and receive data from the Business Rules Engine.
Sub-Process	A Sub-Process is a type of activity within a process, but it also can be "opened up" to show a lower-level process. This is useful for process decomposition or general process organization.
Call Activity	A Call Activity is a type of activity within a process. It provides a link to reusable activities: for example, it will call a task into the Process (see upper figure on the left) or another Process (see lower figure on the left).

Table 2 BPMN *Flow Description*

BPMN 2.0.2	Flow Description
Sequence Flow	A Sequence Flow is represented by a solid line with a solid arrowhead and is used to show the order (the sequence) in which activities will be performed in a process or choreography diagram.
Message Flow	A Message Flow is represented by a dashed line with an open arrowhead and is used to show the flow of messages between two separate process participants (business entities or business roles) that send and receive them.
Association	An Association is represented by a dotted line, which may have a line arrowhead on one or both ends, and is used to associate text and other artifacts with flow objects.
Data Association	A Data Association is represented by a dotted line with a line arrowhead and is used to associate data (electronic or nonelectronic) with flow objects. Data Associations are used to show the inputs and outputs of activities.

（续表）

BPMN 2.0.2	任 务 描 述
业务规则任务	业务规则任务为流程提供了一种机制,用于向业务规则引擎提供输入,并获取业务规则引擎可能提供的计算输出。任务的输入/输出规范将允许流程向业务规则引擎发送数据并从业务规则引擎接收数据
子流程	子流程是流程中的一种活动类型,但它也可以"打开"以显示较低级别的流程。这对于流程分解或一般流程组织非常有用
调用活动	调用活动是流程中的一种活动。它提供了一个到可重用活动的链接:例如,它将调用流程中的一个任务(参见左上方的图)或另一个流程(参见左下方的图)

表2 BPMN流程描述

BPMN 2.0.2	流 程 描 述
序列流	序列流由带有实心箭头的实线表示,用于显示将在流程或编排关系图中执行活动的顺序(序列)
消息流	消息流由带开放箭头的虚线表示,用于显示发送和接收消息的两个独立流程参与者(业务实体或业务角色)之间的消息流
关 联	关联由虚线表示,虚线的一端或两端可能有一个箭头,用于将文本和其他构件与流对象关联
数据关联	数据关联由带线箭头的虚线表示,用于将数据(电子的或非电子的)与流对象关联。数据关联用于显示活动的输入和输出

Table 3 BPMN *Marker Description*

BPMN 2.0.2	Markers Description
↻ Loop Marker	A Loop Marker is used to represent an activity that will be executed multiple times until the condition is satisfied. The condition can be validated either at the start or end of the activtiy.
‖‖‖ Parallel Multiple Instance Marker	A Parallel Multi-Instance Marker is used to represent an activity that can be executed as multiple instances performed in parallel. The number of instances will be determined through a condition expression that is evaluated at the start of the activity. All the instances will start in parallel and each instance can have different input parameters. The activity, as a whole, is completed after all the instances are completed. However, another expression, if it becomes true, will stop all instances and complete the activity.
≡ Sequential Multiple Instance Marker	A Sequential Multi-Instance Marker represents an activity that is similar to a Parallel Multi-Instance activity, but its instances will be executed in sequence. The second instance will wait until the first instance is completed and so on.
～ Adhoc Marker	The Adhoc Marker is a tilde symbol and used to mark a Sub-Process for which the normal sequence patterns are relaxed and its activities can be performed in any order at the discretion of the users. Tasks can start any time without any direct dependency on other tasks.
Text Annotation Annotation Marker	An Annotation Marker is a mechanism for a modeler to provide additional text information (i.e., notes) for the reader of a BPMN diagram. Annotations can be connected to other objects through an Association (see above).

Table 4 BPMN *Data Object Description*

BPMN 2.0.2	Data Description
▢ Data Object	A Data Object represents the data that are used as inputs and outputs to the activities of a process. Data Objects can represent singular objects or collections of objects.
▢ Data Input	A Data Input is an external data input for the entire process. It is a kind of input parameter.

表3　BPMN标记描述

BPMN 2.0.2	标　记　描　述
↻ 循环标记	循环标记用于表示将执行多次的活动,直到满足条件为止。可以在活动的开始或结束时验证条件
||| 并行多实例标记	并行多实例标记用于表示可以作为并行执行的多个实例执行的活动。实例的数量将通过在活动开始时计算的条件表达式来确定。所有实例将并行启动,每个实例可以有不同的输入参数。整个活动是在所有实例完成之后完成的。然而,如果另一个表达式为真,它将停止所有实例并完成活动
≡ 顺序多实例标记	顺序多实例标记表示类似于并行多实例活动的活动,但其实例将按顺序执行。第二个实例将一直等到第一个实例完成,以此类推
~ 特定的标志	Adhoc标记符是一个波浪符号,用于标记一个子进程,该子进程的正常序列模式是放松的,它的活动可以按照用户的任意顺序执行。任务可以在任何时间启动,而不直接依赖于其他任务
Text Annotation 注释标记	注释标记是建模者为BPMN图的读者提供附加文本信息(即,注释)的一种机制。注释可以通过关联连接到其他对象(参见上面)

表4　BPMN数据对象描述

BPMN 2.0.2	数　据　描　述
数据对象	数据对象表示用作流程活动的输入和输出的数据。数据对象可以表示单个对象或对象集合
数据输入	数据输入是整个流程的外部数据输入。它是一种输入参数

Table 4 BPMN *Data Object Description—Cont'd*

BPMN 2.0.2	Data Description
Data Output	A Data Output is the data result of the entire process. It is a kind of output parameter.
Data Store	A Data Store is a place where the process can read or write data (e.g., a database or a filing cabinet). It persists beyond the lifetime of the process instance.
Collection of Data Objects	A Collection of Data Objects represents a collection of data elements related to the same data entity (e.g., a list of order items).

Table 5 BPMN *Event Description*

BPMN 2.0.2	Event Description
Event: Start	Start Events indicate the instance or initiation of a process or an Event Sub-Process and have no incoming sequence flow. A Process can have more than one Start Event, but an Event Sub-Process only have one Start Event.
Event: Event Sub-Process non-interrupting	Non-interrupting Start Events can be used to initiate an Event Sub-Process without interfering with the main process flow.
Event: Intermediate and Boundary	Intermediate Events indicate something that occurs or may occur during the course of the process, between Start and End. Intermediate Catching Events can be used to catch the event trigger and can be in the flow or attached to the boundary of an activity. Intermediate Throwing Events can be used to throw the event trigger.
Event: Boundary non-interrupting	Non-interrupting Boundary Events can be attached to the boundary of an activity. When they are triggered, flow will be generated from them, but the source activity will continue to be performed.
Event: End	The End Event indicates where a path in the Process will end. A Process can have more than one end. The Process ends when all active paths have ended. End Events have no outgoing sequence flows.

Continued

（续表）

BPMN 2.0.2	数　据　描　述
数据输出	数据输出是整个过程的数据结果。它是一种输出参数
数据存储	数据存储是进程可以读写数据的地方（如数据库或文件柜）。它在流程实例的生存期之后仍然存在
数据对象集合	数据对象集合表示与同一数据实体（如订单项列表）相关的数据元素集合

表5　BPMN事件描述

BPMN 2.0.2	事　件　描　述
事件开始	事件开始是指流程或事件子进程的实例或启动，并且没有输入的序列流。流程可以有多个启动事件，但事件子流程只有一个启动事件
事件：事件子流程不中断	非中断启动事件可用于启动事件子流程，而不会干扰主流程
事件：中级和边界	中间事件表示在流程中、开始和结束之间发生或可能发生的事情。中间捕获事件可用于捕获事件触发器，可以在流中或附加到活动的边界。中间投掷事件可用于抛出事件触发器
事件：边界不中断	非中断边界事件可以附加到活动的边界。当它们被触发时，将从它们生成流，但源活动将继续执行
事件：结束	结束事件表示进程中的路径将结束的位置。流程可以有多个目标。当所有活动路径结束时，进程结束。结束事件没有传出的序列流

Table 5 BPMN *Event Description—Cont'd*

BPMN 2.0.2	Event Description
Message (receive)	Receive messages to start a Process or in the middle of a Process, either in the flow or attached to the boundary of an activity.
Message (send)	Send messages in the middle or at the end of a Process path.
Timer (catch)	A Timer Event is always of catch type and used to signify waiting for a specific time condition to evaluate to true, which will start a Process, start an Event Sub-Process, wait in the middle of a flow, or wait as a Boundary Event.
Escalation (catch)	An Escalation Event handles escalation conditions, triggering the start of an Event Sub-Process or a Boundary Event.
Escalation (throw)	A throw Escalation Event will cause the escalation conditions that will trigger the catch Events.
Link (throw and catch)	A Link Event has no significance related to how the Process is performed, but it facilitates the diagram-creation process. For example, you can use two associated links as an alternative to a long sequence flow. There is a throwing Link Event as the "exit point," and a catching Link Event as the "entrance point," and the two events are marked as a pair.
Error (catch)	A catch Error Event is used to capture errors and to handle them. This event can only be used as the start an Event Sub-Process or as a Boundary Event. These events can catch errors thrown by the throw Error Events or errors thrown by a BPM system or services used by the Process.

（续表）

BPMN 2.0.2	事　件　描　述
消息（接收）	接收消息以启动流程或处于流程的中间，或在业务流中，或者附加到活动的边界上
消息（发送）	在流程路径的中间或末尾发送消息
定时器（捕获）	定时器事件始终是捕获类型，用于表示等待一个特定时间条件评估为真，这将启动进程，启动事件子进程，在工作流中间等待，或等待一个完整的事件
升级（捕获）	升级事件处理升级条件，触发事件子流程或边界事件的开始
升级（抛出）	引发升级事件将导致引发捕获事件的升级条件
链接（抛出和捕获）	链接事件与执行流程的方式无关，但它有助于图表创建过程。例如，您可以使用两个关联链接作为长序列流的替代方法。有一个投掷链接事件作为"退出点"，并且捕获链接事件作为"入口点"，并且这两个事件被标记为一对
错误（捕获）	捕获错误事件用于捕获错误并对其进行处理。此事件只能用作开始事件子流程或边界事件。这些事件可以捕获由抛出错误事件引发的错误或由流程管理系统或流程使用的服务引发的错误

Table 5 BPMN *Event Description*—Cont'd

BPMN 2.0.2	Event Description
 Error (throw)	A throw Error Event is used to set an error to be handled. This event can only be used as an End Event (i.e., never as an Intermediate Event).
 Cancel (catch)	Cancel Events can only be used in the context of the transactions. The catch Cancel Events are used as Boundary Events for the transaction Sub-Process, and will trigger the roll back of the transaction (i.e., the Activities of the Sub-Process).
 Cancel (throw)	Cancel Events can only be used in the context of the transactions. The throw Cancel Events are only used within a transaction Sub-Process.
  Conditional (catch)	Conditional Events are used to determine whether to start (or continue) only if a certain condition is true. Like the Timer Event, the Conditional Event can only exist as a catching event. They can be used at the start of a Process or an Event Sub-Process, in the middle of the flow, or as a Boundary Event.
  Compensation (catch)	A Compensation Event is used to handle compensation in the process. The catching Compensation Event be triggered as an Event Sub-Process Start Event, or as a Boundary Event.
  Compensation (throw)	A Compensation Event is used to handle compensation in the process. The throwing Compensation Event can be used in the middle or end of a Process path.
  Signal (start)	Catching Signal Events are used for receiving signals. They are a generic, simple form of communication and exist within pools (same participant), across pools (different participants), and across diagrams. They can be used at the start of a Process or an Event Sub-Process, in the middle of the flow, or as a Boundary Event.
  Signal (end)	Throwing Signal Events are used for sending signals. They are a generic, simple form of communication and exist within pools (same participant), across pools (different participants), and across diagrams. They can be used in the middle or end of a Process path.

Continued

（续表）

BPMN 2.0.2	事 件 描 述
错误（抛出）	抛出错误事件用于设置要处理的错误。此事件只能用作结束事件（即从不作为中间事件）
取消（捕获）	取消事件只能在事务的上下文中使用。捕获取消事件用作事务子流程的边界事件,并将触发事务的回滚（即子流程的活动）
取消（抛出）	取消事件只能在事务的上下文中使用。抛出取消事件仅在事务子流程中使用
有条件的（捕获）	条件事件是用来决定是否开始（或继续）,只有当某个条件为真时才启动（或继续）。与定时器事件一样,条件事件只能作为一个捕获事件存在。它们可以在流程或事件子流程的开始、流程的中间或边界事件中使用
补偿（捕获）	补偿事件用于处理流程中的补偿。捕获补偿事件可以作为事件子流程启动事件或边界事件触发
补偿（抛出）	补偿事件用于处理流程中的薪酬。抛出补偿事件可以在流程路径的中间或末尾使用
信号（开始）	捕获信号事件用于接收信号。它们是通用的、简单的通信形式,存在于池（同一参与者）、跨池（不同参与者）和跨图。它们可以在流程或事件子流程的开始、流程的中间或边界事件中使用
信号（结束）	抛出信号事件用于发送信号。它们是通用的、简单的通信形式,存在于池（同一参与者）、跨池（不同参与者）和跨图。它们可以在流程路径的中间或末尾使用

Table 5 BPMN *Event Description—Cont'd*

BPMN 2.0.2	Event Description
  Multiple (catch)	The Multiple Event is used to summarize several event types with a single symbol. The event is triggered if any one of those types is satisfied. They can be used at the start of a Process or an Event Sub-Process, in the middle of the flow, or as a Boundary Event.
  Multiple (throw)	The Multiple Event is used to summarize several event types with a single symbol. When this event is reached, then all the event types are thrown. They can be used in the middle or end of a Process path.
  Parallel Multiple (catch)	The Parallel Multiple Event is used to summarize several event types with a single symbol. The difference between this event and the Multiple Event is that the Parallel Multiple is only triggered if *all* of those types are satisfied. They can be used at the start of a Process or an Event Sub-Process, in the middle of the flow, or as a Boundary Event.
  Terminate (throw)	The Terminate End Event is the "stop everything" event. When a Terminate End Event is reached, the entire process is stopped, including all parallel activities.

Table 6 BPMN *Gateway Description*

BPMN 2.0.2	Gateway Description
 Gateway	Gateways are used to control how process paths converge and diverge within a process.
  Exclusive Gateway	The Event Gateway, when splitting, routes sequence flow to only one of the outgoing branches, based on conditions. When merging, it awaits one incoming branch to complete before continuing the flow. The Gateway can be displayed with or without the "X" marker, but the behavior is the same.
 Inclusive Gateway	The Inclusive Gateway, when splitting, allows one or more branches to be activated, based on conditions. All active incoming branches must complete before merging.

Continued

（续表）

BPMN 2.0.2	事 件 描 述
多个（捕获）	多个事件用于用单个符号汇总多个事件类型。如果满足其中任何一种类型，则触发事件。它们可以在流程或事件子流程的开始、流程的中间或作为边界事件使用
多次（抛出）	终止结束事件是"停止一切"事件。到达终止结束事件时，将停止整个过程，包括所有并行活动
并行多个（捕获）	并行多事件用于用单个符号汇总多个事件类型。此事件和多个事件之间的区别在于，只有满足所有这些类型时，才会触发并行多个事件。它们可以在流程或事件子流程的开始、流程的中间或作为边界事件使用
终止（抛出）	终止结束事件是"停止一切"事件。到达终止结束事件时，将停止整个过程，包括所有并行活动

表6   BPMN网关描述

BPMN 2.0.2	网 关 描 述
网　关	网关用于控制过程路径如何在过程中收敛和分散
独家门户	分割时，事件网关根据条件将路线顺序分流到唯一一个传出分支。合并时，它会在继续流程之前等待一个传入的分支完成。可以使用或不使用"X"标记显示网关，但行为是相同的
包容性网关	包含网关在分割时允许基于条件激活一个或多个分支。所有活动的传入分支必须在合并之前完成

Table 6 BPMN *Gateway Description* —Cont'd	
BPMN 2.0.2	**Gateway Description**
Parallel Gateway	The Parallel Gateway, when splitting, will direct the flow down all the outgoing branches. When merging, it awaits all the in branches to complete before continuing the flow.
Event-based Gateway	The Event-based Gateway is always followed by catching events or receiving tasks. The flow of the Process is routed to the subsequent event/task which happens first. When merging, it behaves like an Event Gateway. This Gateway can be configured such that it can be used to start a Process, based on the first event that follows it (see the lower figure on the left).
Parallel Event-based Gateway	The Parallel Event-based Gateway is only used for starting a Process. It is configured like a regular Event Gateway, but *all* of the subsequent events must be triggered before a new process instance is created.
Complex Gateway	The Complex Gateway defines behavior that is not captured by other gateways. Expressions are used to determine the merging and splitting behavior.

BPMN DIAGRAMS

Business Process Modeling is used to communicate a wide variety of process configurations to a wide variety of audiences. Thus, BPMN was designed to cover many types of modeling and allow the creation of end-to-end Business Processes. The structural elements of BPMN allow the viewer to be able to easily differentiate between sections of a BPMN Diagram. There are three basic types of submodels within a BPMN modeling environment:

1. Processes (*Orchestration*), including:
   a. *Private non-executable* (internal) Business Processes.
   b. *Private executable* (internal) Business Processes.
   c. *Public* Processes.
2. Choreographies.
3. Collaborations, which can include Processes and/or Choreographies.
   a. A view of Conversations.

TO POINT (1) PRIVATE (INTERNAL) BUSINESS PROCESSES

Private Business Processes are those internal to a specific organization. These Processes have been generally called workflow or BPM Processes (see Figure 1). Another synonym typically used in the Web services area is the *Orchestration* of services. There are two types of *private* Processes: *executable* and *non-executable*.

（续表）

BPMN 2.0.2	网　关　描　述
✛ 并行网关	拆分时,并行网关将引导所有传出分支的流。合并时,它会等待所有分支完成,然后再继续流程
⬠ ⬠ 基于事件的网关	事件网关始终跟踪事件或接收任务。进程的流程被发送到随后发生的事件/任务。合并时,它的行为类似于事件网关。可以配置此网关,使其基于其后面的第一个事件可以用于启动进程（请参见左侧的右图）
⊕ 基于并行事件的网关	并行事件网关仅用于启动进程。它配置为常规事件网关,但必须在创建新流程实例之前触发所有后续事件
✳ 复杂网关	复杂网关定义了其他网关未捕获的行为。表达式用于确定合并和拆分行为

4.2.5　BPMN图

业务流程建模用于将各种流程配置传达给各种受众。因此,BPMN被设计为涵盖多种类型的建模,并允许创建端到端的业务流程。BPMN的结构元素使查看器能够轻松地区分BPMN图的各个部分。在BPMN建模环境中,有三种基本的子模型类型。

（1）流程（编排）,包括:

a. 私有不可执行（内部）业务流程;

b. 私有可执行（内部）业务流程;

c. 公共过程。

（2）编排。

（3）协作,包括流程和编排。

对话视图。

4.2.6　指向（1）私有（内部）业务流程

私有业务流程是特定组织内部的流程。这些流程通常被称为工作流或BPM流程（图1）。Web服务领域中通常使用的另一个同义词是服务编排。私有流程有两种类型:可执行流程和不可执行流程。

FIGURE 1

Example of private process.

An *executable* Process is a Process that has been modeled for being executed according to the defined BPMN execution semantics. Of course, during the development cycle of the Process, there will be stages in which the Process does not have enough detail to be "executable."

A non-executable Process is a *private* Process that has been modeled for documenting Process behavior at a modeler-defined level of detail. Thus, information needed for execution, such as formal condition expressions are typically not included in a *non-executable* Process.

If a swim lanes-like notation is used (e.g., a Collaboration, see below) then a *private* Business Process will be contained within a single Pool. The Process flow is therefore contained within the Pool and cannot cross the boundaries of the Pool. The flow of Messages can cross the Pool boundary to show the interactions that exist between separate *private* or *public* Business Processes.

PUBLIC PROCESSES

A *public* Process represents the interactions to and from another Process or *Participant* (see Figure 2). Only those Activities and Events that are used to communicate to the other *Participants* are included in the *public* Process. These Activities and Events can be considered the "touch-points" between the participants. All other "internal" Activities of the *private* Business Process are not shown in the *public* Process. Thus, the *public* Process shows to the outside world the Message Flows and the order of those Message Flows that is needed to interact with that Process. *Public* Processes can be modeled separately

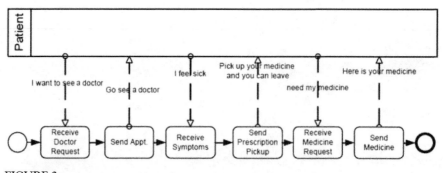

FIGURE 2

Example of public process.

图1　私有流程的示例

可执行流程是一个根据定义的BPMN执行语义建模以执行的流程。当然,在流程的开发周期中,会有一些阶段,流程没有足够的细节来"可执行"。

非可执行流程是一个私有流程,它已经经过建模,用于在建模者定义的详细级别上记录流程行为。因此,执行所需的信息(如正式条件表达式)通常不包括在不可执行的进程中。

如果使用类似"通道"的表示法(如协作,见下文),则私人业务流程将包含在单个池中。因此,流程包含在池中,不能跨越池的边界。消息流可以跨越池边界以显示单独的私有或公共业务流程之间存在的交互。

4.2.7　公共流程

公共流程表示与另一个流程或参与者之间的交互(图2)。只有那些用于与其他参与者进行通信的活动和事件才会包含在公共流程中。这些活动和事件可被视为参与者之间的"接触点"。私有业务流程的所有其他内部活动未在公共流程中显示。因此,公共流程向外界显示消息流以及与该流程交互所需的那些消息流的顺序。公共流程可以单独建模,也可以在协作中建模,以显示消息的定向流。请

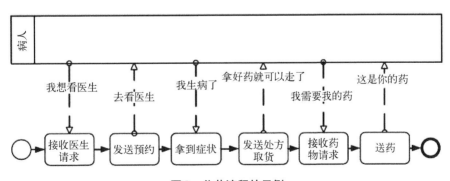

图2　公共流程的示例

or within a Collaboration to show the directional flow of Messages. Note that the *public* type of Process was named "abstract" in BPMN 1.2 (2009 release).

COLLABORATIONS

A Collaboration depicts the interactions between two or more business entities. A Collaboration usually contains two or more Pools, representing the *Participants* in the Collaboration. The Message exchange between the *Participants* is shown by a Message Flow that connects two Pools (or the objects within the Pools). The Messages associated with the Message Flows can also be shown graphically. The Collaboration can be shown as two or more *public* and/or *private* Processes communicating with each other (see Figure 3). Or a Pool MAY be empty, a "black box." Choreography elements MAY be shown "in between" the Pools as they bisect the Message Flows between the Pools. All combinations of Pools, Processes, and a Choreography are allowed in a Collaboration.

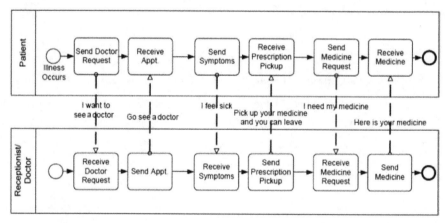

FIGURE 3

Example of a Collaboration.

TO POINT (2) CHOREOGRAPHY

A self-contained Choreography (no Pools or *Orchestration*) is a definition of the expected behavior, basically, a procedural contract between interacting *Participants*. Although a normal Process exists within a Pool, a Choreography exists between Pools (or *Participants*).

The Choreography looks similar to a *private* Business Process because it consists of a network of Activities, Events, and Gateways (see Figure 4). However, a Choreography is different in that the Activities are interactions that represent a set (one or more) of Message exchanges, which involves two or more *Participants*. In addition, unlike a normal Process, no central controller, responsible entity, or observer of the Process exists.

注意,BPMN 1.2(2009版)中公共类型的流程被命名为"抽象"。

4.2.8　协作

协作描述两个或多个业务实体之间的交互。协作通常包含两个或多个池,表示协作中的参与者。参与者之间的消息交换由连接两个池(或池内的对象)的消息流显示。与消息流关联的消息也可以图形化地显示。协作可以显示为两个或多个相互通信的公共进程和/或私有进程(图3),或者一个池可能是空的,即一个"黑盒子"。编排元素可以显示在池之间,因为它们平分池之间的消息流。协作中允许所有池、流程和编排的组合。

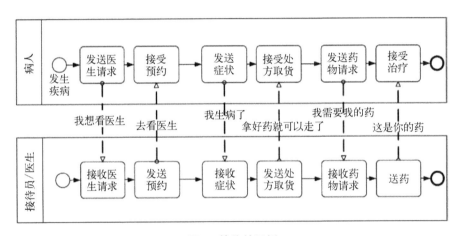

图3　协作的示例

4.2.9　要点(2)编排

自包含的编排(无池或编排)是预期行为的定义,基本上是参与者之间交互的程序合同。虽然池中存在正常的流程,但池(或参与者)之间存在编排关系。

因为它由活动、事件和网关组成(图4),编排看起来类似于私有业务流程,但是,编排的不同之处在于活动是一种表示一组(一个或多个)消息交换的交互,其涉及两个或更多参与者。此外,与正常流程不同,编排不存在负责实体或观察者的流程中央控制器。

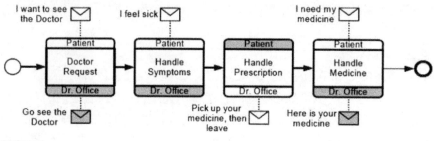

FIGURE 4

Example of a Choreography.

TO POINT (3) CONVERSATIONS

The Conversation diagram is a particular usage and an informal description of a Collaboration diagram. However, the Pools of a Conversation diagram usually do not contain a Process, and a Choreography is usually not placed between the Pools of a Conversation diagram. An individual Conversation (within the diagram) is the logical relation of Message exchanges. The logical relation, in practice, often concerns a business object(s) of interest, for example, "Order," "Shipment and Delivery," or "Invoice."

Thus, the Conversation diagram is a high-level modeling diagram that depicts a set of related Conversations that reflect a distinct business scenario Table 7. For example, in logistics, stock replenishments involve the following types of scenarios: creation of sales orders, assignment of carriers for shipments combining different sales orders, crossing customs/quarantine, processing payment, and investigating exceptions. Thus, a Conversation diagram, as shown in Figure 5, shows Conversations (as hexagons) between *Participants* (Pools). This provides a "bird's eye" perspective of the different Conversations that relate to the domain.

Table 7 *BPMN Conversation Description*	
BPMN 2.0.2	**Conversations Description**
⬡ Conversation	A Conversation defines a set of logically related Message Flows. When marked with a (+) symbol it indicates a Sub-Conversation, a compound conversation element.
⬢ Call Conversation	A Call Conversation is a wrapper for a globally defined, re-usable Conversation or Collaboration. A call to a Collaboration is marked with a (+) symbol.
══════ Conversation Link	Connects Conversations and Participants.

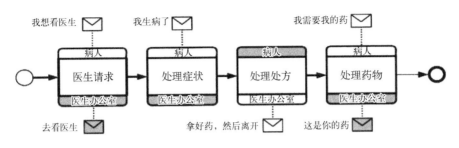

图4 编排的示例

4.2.10 要点(3)对话

对话图是协作图的一种特殊用法和非正式描述。然而,对话图的池通常不包含流程,并且编排通常不放置在会话图的池之间。单个对话(在图5中)是消息交换的逻辑关系。在实践中,逻辑关系通常与感兴趣的业务对象有关,如"订单""装运和交付"或"发票"。

因此,对话图是一个高级建模图,表7描绘了一组反映不同业务场景的相关对话。例如,在物流中,库存补充涉及以下类型的场景:创建销售订单,分配运输承运人结合不同的销售订单、过关/隔离,处理付款和调查例外情况。因此,如图5所示,对话图显示了参与者(池)之间的对话(作为六边形)。这提供了与域相关的不同对话的"鸟瞰"视角。

表7 BPMN对话描述

BPMN 2.0.2	对 话 描 述
⬡ 对　话	对话定义了一组逻辑相关的消息流。当用一个(+)符号标记时,它表示一个子对话,一个复合对话元素
⬣ 呼叫对话	呼叫会话是全局定义的、可重用的会话或协作的包装器。对协作的调用标记有一个(+)符号
══ 对话链接	连接对话和参与者

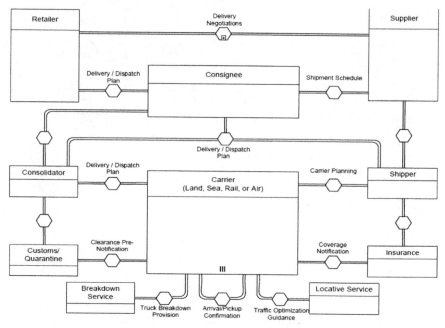

FIGURE 5

Example of a conversation diagram.

BPMN USAGE

We have just illustrated the three basic BPMN models of Processes—*private* Processes (both *executable* and *non-executable*), *public* Processes - Collaborations (including Conversations), and Choreographies. Within and between these BPMN sub-models, many types of Diagrams can be created.

The following are examples of Business Processes that can be modeled:

- High-level *non-executable* Process Activities (not functional breakdown).
- Detailed executable Business Process.
- As-is or old Business Process.
- To-be or new Business Process.
- A description of expected behavior between two (2) or more business *Participants*—a Choreography.
- Detailed *private* Business Process (either *executable* or *non-executable*) with interactions to one or more external *Entities* (or "Black Box" Processes).
- Two or more detailed *executable* Processes interacting.
- Detailed *executable* Business Process relationship to a Choreography.
- Two or more *public* Processes.
- *Public* Process relationship to Choreography.
- Two or more detailed *executable* Business Processes interacting through a Choreography.

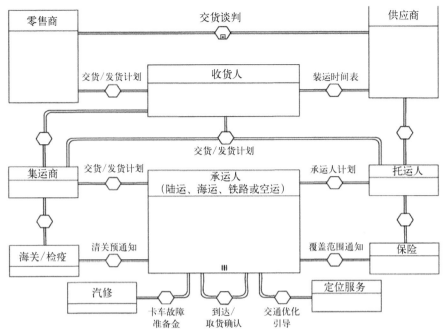

图 5　对话图的示例

4.2.11　BPMN 的使用方法

我们刚刚演示了流程的三个基本 BPMN 模型：私有流程（可执行和不可执行）、公共流程——协作（包括对话）和编排。在这些 BPMN 子模型内部和之间，可以创建许多类型的图。

以下是可以建模的业务流程示例：

- 高级非可执行流程活动（非功能分解）。
- 详细的可执行业务流程；
- 原有的或旧的业务流程；
- 未来的或新的业务流程；
- 描述两个或多个业务参与者之间的预期行为的业务流程——编排；
- 详细的私有业务流程（可执行或不可执行），与一个或多个外部实体（或"黑盒"流程）进行交互；
- 两个或更多详细的可执行流程交互；
- 与编排的详细可执行业务流程关系；
- 两个或多个公共进程；
- 与编排的公共流程关系；
- 通过编排交互两个或更多详细的可执行业务流程。

One of the benefits of BPMN, among others, is that it has the flexibility to allow the development of all the above examples of business processes. However, the ways that different submodels are combined within a specific tool is a choice of the vendors and can vary quite a bit.

DIAGRAM POINT OF VIEW

Because a BPMN diagram may depict the processes of different participants, each participant could view the diagram differently. That is, the participants have different points of view regarding how the processes will apply to them. Some of the activities will be internal to a participant (that is, they are performed by or under control of that participant) and other activities will be external to that participant. Each participant will have a different perspective as to which are internal and external. At run time, the difference between internal and external activities is important in how a participant can view the status of the activities or troubleshoot any problems. However, the diagram itself remains the same. Figure 3, above, displays a business process that has two points of view. One point of view is of a patient, the other is of the doctor's office.

The diagram may show the activities of both participants in the process, but when the process is actually being performed, each participant will only have control over their own activities. Although the diagram point of view is important for a viewer of the diagram to understand how the behavior of the process will relate to that viewer, BPMN will not currently specify any graphical mechanisms to highlight the point of view. It is open to the modeler or modeling tool vendor to provide any visual cues to emphasize this characteristic of a diagram.

UNDERSTANDING THE BEHAVIOR OF DIAGRAMS

So far, we have mentioned how sequence flows are used within a process. To facilitate the understanding of process behavior, we employ the concept of a *token* that will traverse the sequence flows and pass through the elements in the process. A *token* is a theoretical concept that is used as an aid to define the behavior of a process that is being performed. However, modeling and execution tools that implement BPMN are NOT REQUIRED to implement any form of *token*.

Process elements can be defined by describing how they interact with a *token* as it moves through the structure of the Process. A Start Event generates a *token* that MUST eventually be consumed at an End Event (which MAY be implicit if not graphically displayed). The path of a *token* should be traceable through the network of Sequence Flows, Gateways, Events, and Activities within a process.

Note: A *token* does not traverse a Message Flow since it is a Message that is passed down a Message Flow (as the name implies).

BPMN EXAMPLE

The following is an example of a manufacturing process from different perspectives (Figure 6–8).

　　BPMN的优点之一是,它具有灵活性,允许开发上述所有业务流程示例。然而,不同的子模型在特定工具中组合的方式是由供应商选择的,并且可能会有很大的不同。

4.2.12　图表的观点

　　因为BPMN图可以描绘不同参与者的流程,所以每个参与者可以不同地查看该图。也就是说,参与者对于流程如何适用于他们有不同的观点。一些活动将在参与者内部(即由该参与者执行或在该参与者的控制下),并且其他活动将在该参与者外部。每个参与者对内部和外部都有不同的看法。在运行时,内部和外部活动之间的差异对于参与者如何查看活动状态或解决任何问题非常重要。但是,图表本身保持不变。上面的图3显示了一个有两个观点的业务流程,一种是患者的观点,另一种是医生的观点。

　　图3可以显示该流程中两个参与者的活动,但是当该流程实际执行时,每个参与者将仅控制他们自己的活动。虽然图表的观点对于图表的查看者理解流程的行为将如何与该查看者相关是很重要的,但BPMN目前不会指定任何图形机制来突出观点。建模者或建模工具供应商可以提供任何视觉提示来强调图表的这一特性。

4.2.13　理解图的行为

　　到目前为止,我们已经提到了如何在流程中使用序列流。为了便于理解流程行为,我们使用令牌(token)的概念,它将遍历序列流并通过流程中的元素。令牌是一个理论概念,用于帮助定义正在执行的流程的行为。然而,实现BPMN的建模和执行工具并不需要实现任何形式的令牌。

　　可以通过描述流程元素在流程结构中移动时如何与令牌交互来定义流程元素,一个开始事件生成一个必须最终在结束事件中消耗的令牌(如果没有以图形方式显示,则可能是隐含的)。令牌的路径应该可以通过流程中的顺序流、网关、事件和活动的网络进行跟踪。

　　注意:令牌不会遍历消息流,因为它是沿消息流传递的消息(正如同其名所寓意)。

4.2.14　BPMN示例

以下是不同角度的制造流程的示例(图6～图8)。

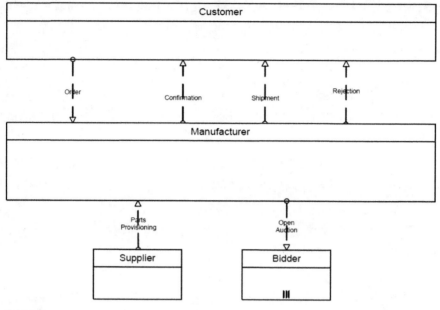

FIGURE 6

An example of a Collaboration diagram with black-box Pools.

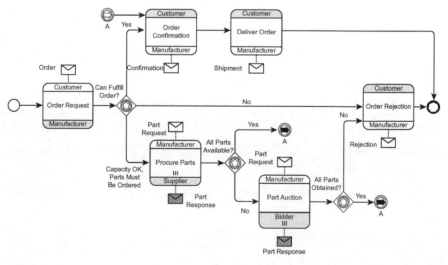

FIGURE 7

An example of a standalone Choreography diagram.

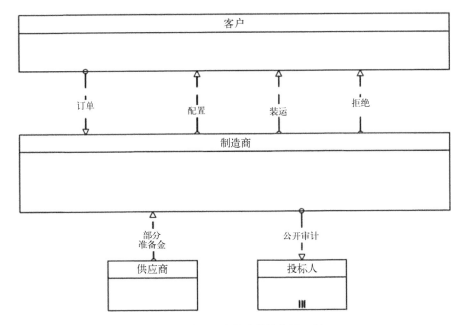

图6　带有"黑盒"池的协作图示例

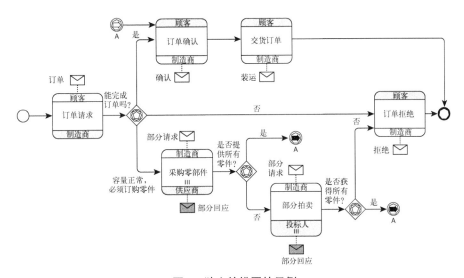

图7　独立编排图的示例

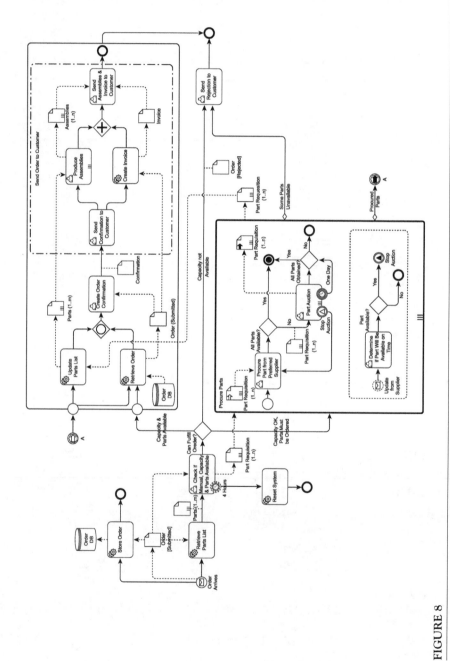

FIGURE 8

An example of a stand-alone Process (Orchestration) diagram.

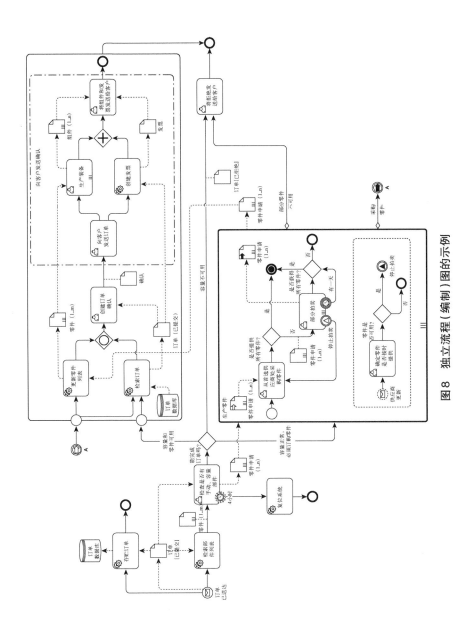

图8　独立流程(编制)图的示例

BPMN CAVEATS

The focus of BPMN is to enhance primary process modeling capabilities. It does not attempt to model other business models, such as organization, strategic direction, business functions, rules/compliance aspects, etc. Therefore, it is vital to understand that other types of modeling done by organizations outside the primary process purposes are out of scope for BPMN, but they all fit within larger BPM solutions. Below is therefore a specification of modeling principles and concepts excluded from BPMN:

- The linking of business strategies, critical success factors, and value drivers to processes.
- The relation between organizational structures, including business competencies, capabilities, and resources to processes.
- Functional breakdowns of business functions into process tasks.
- Arrangement of business objects such as product, machine, warehouse, and so on, throughout the process models.
- Specification of information objects and thereby information flow within the process models.
- The ability to illustrate or model business measurement that is, Key Performance Indicators or Process Performance Indicators (PPIs) within the process.
- Data models, whereas BPMN shows the flow of data (messages), and the association of data artifacts to activities, it is not a data model or even a data flow diagram.
- Even though the data objects are specified within the process, real-time process monitoring in terms of Scorecards, Dashboards, and/or Cockpits.
- The support for Business Rules Modeling, in terms of business rules, rule script, flow rule, decision table, report, and thereby decision-making support.
- The ability to run process ownership gap analysis, that is, to both process and processes rules or process measurements.

So although we realize that many BPM teams wish the ability to relate process models to other vital aspects of enterprise modeling, that is, business modeling, value modeling, performance management, and enterprise architecture (e.g., business architecture, allocation/information systems architecture, and technology architecture). The scope of BPMN does not provide such modeling capabilities, but a robust BPM modeling environment could provide the linkages between the various BPM modeling domains.

THE FUTURE OF BPMN

At some point, the OMG will update BPMN to version 3.0. Although some discussions have occurred on this topic, no certain timeline exists as to when this will happen. BPMN versions 1.0 and 2.0 did not cover the wide landscapes and complexities that exist in the process-modeling domain. Thus, certain topics and capabilities

4.2.15　BPMN附加说明

BPMN的重点是增强主要流程建模功能。它不会尝试对其他业务模型进行建模,如组织、战略方向、业务功能、规则/合规性方面等。因此,重要的是组织在主要流程目的之外,所做的其他类型的建模是超出BPMN范围的,但它们都适合更大的BPM解决方案。因此,下面是从BPMN中排除的建模原则和概念的规范:

- 将业务战略、关键成功因素和价值驱动因素与流程联系起来;
- 组织结构之间的关系,包括业务能力、能力和流程资源;
- 业务功能的功能细分为流程任务;
- 在整个流程模型中安排业务对象,如产品、机器、仓库等;
- 规范信息对象,从而规范过程模型中的信息流;
- 能够说明或模拟业务衡量,即流程中的KPI或PPI;
- 数据模型,而BPMN显示数据流(消息),以及数据制品与活动的关联,它不是数据模型甚至数据流图;
- 即使在流程中指定了数据对象,也可以根据记分卡、仪表板和/或驾驶舱进行实时流程监控;
- 业务规则建模的支持,包括业务规则、规则脚本、流规则、决策表、报告以及决策支持;
- 能够运行流程所有权差距分析,即运行流程和处理规则或流程测量。

因此,虽然我们意识到许多BPM团队希望能够将流程模型与企业建模的其他重要方面联系起来,即业务建模、价值建模、绩效管理和企业架构(如业务架构、分配/信息系统架构和技术架构)。BPMN的范围不提供此类建模功能,但强大的BPM建模环境可以提供各种BPM建模域之间的联系。

4.2.16　BPMN的未来

在某些时候,OMG会将BPMN更新到3.0版。尽管在这个主题上已经进行了一些讨论,但是没有确定何时会发生这种情况的时间表。BPMN版本1.0和2.0未涵盖流程建模领域中存在的广泛环境和复杂性。因此,某些主题和功能可以并且应该在BPMN 3.0中解决。请注意,本小节中提供的材料完全是本节作者的观点。

could and should be addressed in BPMN 3.0. However, note that the material presented in this section is solely the opinion of the authors of this chapter. The OMG membership, which does include the authors, will determine what will be included in the next version of BPMN.

FULFILLING THE BPMN VISION

In a presentation introducing BPMN to the Business Process Management Initiative (BPMI) in April 2002, the following statement was made: "The BPMN will provide businesses with the capability of understanding their internal and external business procedures with graphical notation and will give organizations the ability to communicate these procedures in a standard manner."

Business Process types cover a wide range that is required for normal operations of most organizations. In the first two versions of BPMN, the standard has focused on more controlled, prescriptive types of internal processes as well as external processes modeled through Collaboration, Conversation, and Choreography. Nevertheless, BPMN does not yet have the built-in capabilities to easily model the entire range of process types that organizations require to run their businesses.

To fulfill this vision, BPMN eventually must be able to cover the entire range of processes that occur in the real world. This range is bounded on one side by very structured processes and on the other side by very unstructured (ad hoc) processes (see Figure 9). Potential work is available on both ends of the spectrum.

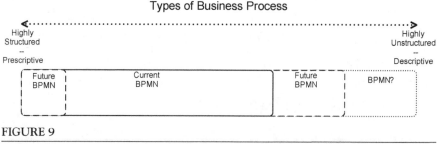

FIGURE 9

A diagram representing the range of process types that are performed by organizations.

There are different areas where future work can be applied to BPMN, including:

- Collaboration, Choreography, and Conversation.
- Metamodel changes.
- Implementation Level Modeling.
- Case Management.

OMG成员（包括作者）资格将确定将在下一版BPMN中包含的内容。

4.2.17　现BPMN愿景

在2002年4月将BPMN引入业务流程管理研究所（BPMI）的演示中，发表了以下声明："BPMN将为企业提供以图形符号理解其内部和外部业务流程的能力，并使组织具备这种能力，以标准方式传达这些程序。"

业务流程类型涵盖了大多数组织正常运营所需的广泛范围。在BPMN的前两个版本中，该标准侧重于更加受控制的、规范性的内部流程类型以及通过协作、对话和编排建模的外部流程。尽管如此，BPMN还没有内置功能来轻松模拟组织运营业务所需的所有流程类型。

为了实现这一愿景，BPMN最终必须能够涵盖现实世界中发生的所有流程。这个范围一方面是非常结构化的流程，另一方面是非结构化（临时）流程（图9）。在范围的两端都有潜在的工作。

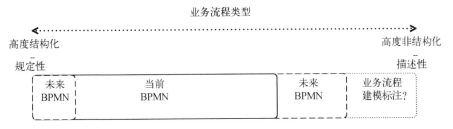

图9　表示组织执行的流程类型范围的图表

未来的工作可以应用于BPMN，包括：

- 协作、编排和对话；
- 元模型改变；
- 实施级别建模；
- 案例管理。

We don't expect much work to be done on Collaboration, Choreography, or Conversation. More vendor/customer experience and feedback is required.

In terms of metamodel work, the following could be applied:

- Various extensions could be added.
- Separate ad hoc processes for better case management support.
- Inherent support for element substitution.
- Allowing different levels of detail or local variations of detail based on single model.

The next two sections discuss the two other major topics that could be added to the BPMN standard.

IMPLEMENTATION LEVEL MODELING

This type of modeling involves highly structured diagrams and fits on the left side of Figure 9 (above). BPMN allows multiple levels of process detail through sub-processes and tasks. But tasks are the lowest level of detail that can be modeled in BPMN. However, some BPM tools provide modelers of executable BPMN models with additional modeling capabilities for modeling the execution details of tasks, which are provided by the services that implement the tasks. These details include the sequence of steps or user interface screens in a service (sometimes called screen flow).

Thus, a process-like level of modeling exists at the service or implementation level. The layout of these models looks very similar to standard BPMN processes, but they are not, at this point, BPMN processes. They have slightly different semantics and visualizations. Figure 10 shows how a BPMN user task could be broken down to an implementation flow.

Some of the characteristics of service flow models include (for example):

- No Lanes. They exist fully within the lane of their parent task.
- Only one Start Event. This Start Event does not have a trigger. Control is always passed from the parent task.
- There are no parallel paths.
- Gateways are allowed.
- They can nest lower level service flow models.
- Semantics of the user events in a service level.
    - They do not interrupt activity in normal sense.
    - They represent a normal completion of the activity.
    - For example, through the clicking of a screen button.
    - User Event notation: User (like a User Task) or a button icon.

Given that modeling tools already exist that provide modeling at the implementation level, this type of diagram could easily be built into the BPMN standard.

我们不希望在协作、编排或对话方面做太多工作,但需要更多的供应商/客户体验和反馈。

在元模型工作方面,可以应用以下内容:

- 可以添加各种扩展;
- 单独的临时流程,以获得更好的案例管理支持;
- 对元素替换的内在支持;
- 根据单一模型允许不同级别的细节或细节的局部变化。

接下来的两节我们将讨论可以添加到BPMN标准的另外两个主要主题。

4.2.18　实施级别建模

这种类型的建模涉及高度结构化的图表,并且适合图9的左侧。BPMN通过了流程和任务允许多个级别的流程细节,但任务是可以在BPMN中建模的最低细节级别。但是,一些BPM工具为可执行BPMN模型提供了建模器,并具有额外的建模功能用于对任务的执行细节进行建模,这些任务由实现任务的服务提供。这些细节包括服务中的步骤序列或用户界面屏幕(有时称为屏幕流程)。

因此,在服务或实现级别存在类似级别的流程建模。这些模型的布局看起来与标准BPMN流程非常相似,但此时它们并不是BPMN流程,它们的语义和可视化略有不同。图10显示了如何将BPMN用户任务分解为实现流程。

服务流模型的一些特征包括(例如):

- 没有通道,它们完全存在于其父任务的通道内;
- 只有一个开始事件,此启动事件没有触发器,始终从父任务传递控制;
- 没有平行路径;
- 允许网关;
- 他们可以嵌套较低级别的服务流模型;
- 服务级别中用户事件的语义。
    - 它们不会在正常意义上中断活动;
    - 它们代表活动的正常完成;
    - 例如,通过单击屏幕按钮;
    - 用户事件表示法:用户(如用户任务)或按钮图标。

鉴于已经存在可在实现级别提供建模的建模工具,这种类型的图表可以很容易地构建到BPMN标准中。

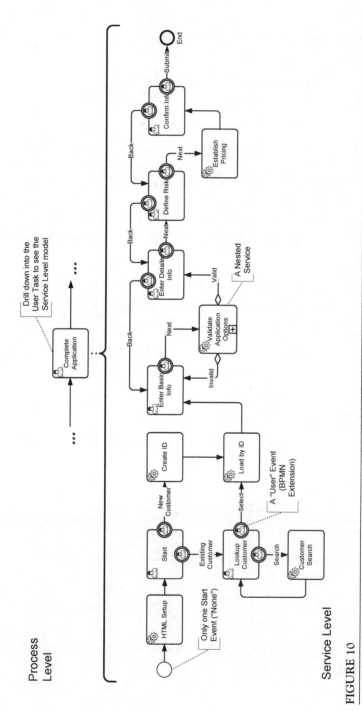

FIGURE 10

An example of an Implementation Level Diagram.

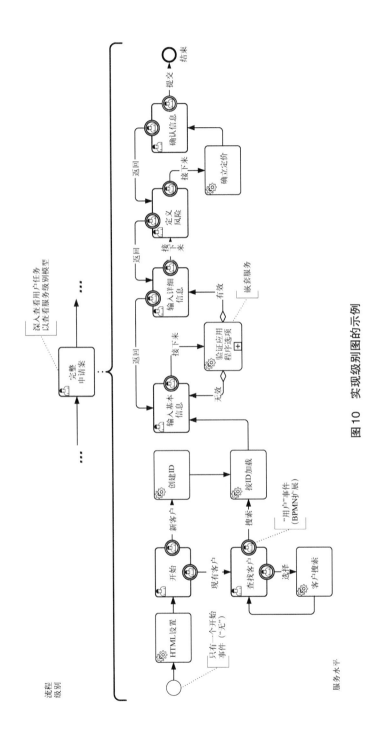

图 10　实现级别图的示例

CASE MANAGEMENT MODELING

Case Management is a hot topic in BPM. This type of modeling involves highly unstructured diagrams and fits on the right side of Figure 9 (above). However, not all businesses have the same understanding of what Case Management is or how it works. Sometimes a case involves mainly straight-through prescriptive processes, with some trouble-shooting. However, most of the time a case involves mainly free-form descriptive processes.

BPMN 2.0 has incomplete support for Case Management (unstructured) Processes. BPMN mainly defines "Structured" Processes—those processes that have a well-defined sequence flow. But BPMN does provide for "Unstructured" Processes—The Ad Hoc Sub-Process. However, additional descriptive process types and behavior are required to fully handle all the unique aspects of unstructure processes.

When BPMN 1.0 was first developed, there was an understanding that descriptive processes were an important part of the process landscape. However, the initial focus of BPMN was to create a business process modeling language for business people that could also be executed by the available BPMSs. The Ad Hoc Process was included in BPMN as a placeholder that provides many of the capabilities required for modeling descriptive processes. It is expected as BPMN evolves, the Ad Hoc Process will also evolve to handle all Case Management Process requirements, which include:

- No predefined sequence flow exists.
- Activities can occur in any order or any frequency.
- But some sequence flow and data flow can be shown.

Unstructured Processes have additional requirements, such as:

- Milestones—e.g., a Case state life cycle.
- New types of events.
- For example, the Case state (life cycle) changes, document updates, and so on.
- Preconditions, dependencies.
- Activities that can be started manually or automatically.
- Activities that are optional.
- Activities that can be repeated.

Figure 11 displays some potential notational updates to BPMN elements that would allow the standard to provide the modeling of more sophisticated unstructured processes.

The OMG has been developing a Case Management Modeling Notation (See the CMMN Chapter) standard. It is focused on a specification for tools that specialize in free-form Case Management behaviors. Because both CMMN and BPMN provide modeling, the authors believe that the best course is that the OMG consolidate the two specifications in the next update to BPMN (version 3.0).

4.2.19　案例管理建模

案例管理是BPM的热门话题。这种类型的建模涉及高度非结构化的图,并且适合于图9的右侧。但是,并非所有企业对案例管理是什么或如何运作都有相同的理解。有时一个案例主要涉及直接的规范程序,并有一些故障排除。但是,大多数情况下,案件主要涉及自由形式的描述性流程。

BPMN 2.0不完全支持案例管理(非结构化)流程,BPMN主要定义结构化流程(那些具有定义良好的序列流的流程)。但是BPMN确实提供了非结构化流程(即特别子流程)。然而,需要额外的描述性流程类型和行为来完全处理非结构化流程的所有独特方面。

当BPMN 1.0首次开发时,人们就认识到描述性流程是流程环境的重要组成部分。然而,BPMN最初的重点是为业务人员创建的业务流程建模语言,该语言也可以由可用的BPMS执行。BPMN中包含了Ad Hoc流程作为占位符,它提供了建模描述性流程所需的许多功能。预计随着BPMN的发展,Ad Hoc流程也将不断发展,以处理所有的案例管理流程需求,包括:

- 不存在预定义的序列流;
- 活动可以以任何顺序或频率发生;
- 可以显示一些序列流和数据流。

非结构化流程有额外的要求,例如:

- 里程碑,如一个案例状态生命周期;
- 新的事件类型;
- 案例状态(生命周期)更改、文档更新;
- 先决条件,依赖关系;
- 可以手动或自动启动的活动;
- 可选的活动;
- 可以重复的活动。

图11显示了BPMN元素的一些潜在的符号更新,这些更新将允许标准提供更复杂的非结构化过程的建模。

OMG一直在开发案例管理建模标记(参见CMMN章节)标准,它专注于专门针对自由形式案例管理行为的工具规范。由于CMMN(case management modeling notation)和BPMN都提供建模,作者认为最好的方法是OMG在下次更新BPMN(版本3.0)时合并两个规范。

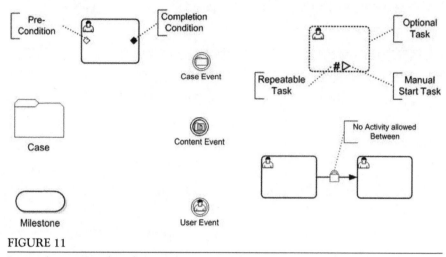

FIGURE 11

Potential notation updates for BPMN 3.0.

CONCLUSIONS

This chapter provides an overview and introduction to the Business Process Model and Notation (BPMN), what it is, and how it is used. We illustrated the primary goal of BPMN and how it provides a standard notation readily understandable by various stakeholders. Further, through its model types, BPMN provides the flexibility to integrate various views from business to technical perspectives. However, as we talk with many organizations about how BPMN can and cannot be used, we have discovered that BPMN has been, by choice, constrained to support only the concepts of modeling applicable to traditional business processes. Therefore, extended business process modeling aspects such as linking processes to business goals, the ability to do Value-Oriented Process Modeling, defining relationships between business competencies and processes, specifying measurements and reporting aspects, or defining rule sets (business, application, etc.), while all relevant, are not the focus of BPMN. It was, however, more vital to have a standard in the marketplace that enables all to have a common platform, than having the ability to do extended business process modeling. It is the start of a great journey, one that enables organizations and BPM teams around the world to analyze, design, build, and implement their processes. More will come!

End Notes

1. Object Management Group, "BMI Standard Specification 2.0.2," (2013), http://www.omg.org/spec/BPMN//2.0.2/PDF.
2. Simpson S., "An XML Representation for Crew Procedures, Final Report NASA Faculty Fellowship Program (Johnson Space Center)," (2004).

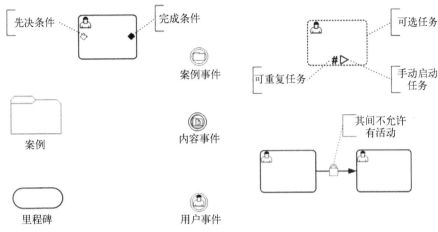

图11   BPMN 3.0潜在的符号更新

4.2.20   结论

本节概述和介绍了BPMN、它是什么以及如何使用它。我们说明了BPMN的主要目标,以及它如何提供各种利益相关者易于理解的标准符号。此外,通过其模型类型,BPMN提供了从业务到技术角度集成各种视图的灵活性。然而,当我们与许多组织讨论如何使用BPMN和不使用BPMN时,我们发现BPMN已经被限制为只支持适用于传统业务流程的建模概念。因此,扩展的业务流程建模方面,例如,将流程链接到业务目标、进行面向价值的流程建模的能力、定义业务能力和流程之间的关系、指定度量和报告方面,或者定义规则集(业务、应用程序等),虽然都是相关的,但并不是BPMN的重点。然而,在市场中拥有一个能够使所有人都拥有一个公共平台的标准比拥有进行扩展业务流程建模的能力更为重要。它是一个伟大旅程的开始,使世界各地的组织和BPM团队能够分析、设计、构建和实现他们的流程。后文将会对此详细说明。

参考文献

[1] Object Management Group, "BMI Standard Specification 2.0.2," (2013), http://www.omg.org/spec/BPMN//2.0.2/PDF.

[2] Simpson S., "An XML Representation for Crew Procedures, Final Report NASA Faculty Fellowship Program (Johnson Space Center)," (2004).

[3] White S., "Business Process Modeling Notation v1.0." for the Business Process Management Initiative

3. White S., *"Business Process Modeling Notation v1.0." for the Business Process Management Initiative (BPMI)* (May 2004).
4. Object Management Group, "BMI Standard Specification 1.1," (2008), http://www.omg.org/spec/BPMN/1.1/.
5. Object Management Group, "BMI Standard Specification 1.2," (2009), http://www.omg.org/spec/BPMN/1.2/.
6. Object Management Group, "BMI Standard Specification 2.0," (2011), http://www.omg.org/spec/BPMN/2.0/.
7. Object Management Group, "BMI Standard," http://www.omg.org/spec/BPMN/2.0.2/.

(BPMI) (May 2004).

[4] Object Management Group, "BMI Standard Specification 1.1," (2008), http://www.omg.org/spec/BPMN/1.1/.

[5] Object Management Group, "BMI Standard Specification 1.2," (2009), http://www.omg.org/spec/BPMN/1.2/.

[6] Object Management Group, "BMI Standard Specification 2.0," (2011), http://www.omg.org/spec/BPMN/2.0/.

[7] Object Management Group, "BMI Standard," http://www.omg.org/spec/BPMN/2.0.2/.

Variation in Business Processes

Mark von Rosing, Jonnro Erasmus

INTRODUCTION

During business processes analysis and development, we are often confronted with the criticism that we introduce additional administration and stifle the creativity of those who play a part in the process. As analysts and architects, we naturally oppose this notion in our quest for simplicity and optimal solutions. With good reason too, enterprises are complicated, and it is not uncommon to encounter hundreds of business processes, categorized into dozens of process groups and several process areas.[1] Such decomposition may lead to thousands of process activities and tasks, quickly becoming a managerial burden. This makes it easy for us to make the argument for standardization, because allowing every business unit or competency area to have its own variation of common processes will exponentially grow those numbers.

Yet we know that variation exists, and the goal of business process modeling is to capture things that reflect reality. For example, the way quality management is performed in manufacturing operations differs quite significantly from the way it is done in accounting. Although these business processes have similar goals and contribute towards the overall quality targets of the enterprise, the techniques applied and skills required are so different that we do not even consider these the same processes. Manufacturing applies techniques like statistical process control and is concerned with concepts such as sample sizes, whereas accounting relies more heavily on checks, balances, and audits. Clearly, these two very different practices should not be unified and standardized, as that will decrease value while cutting very little cost. In fact, the decision to allow variation in business processes is always a trade-off between value to the enterprise and the potential to decrease costs.

This chapter explores the phenomenon of variation in business processes. It explains what process variance is and how to identify it. Much of the discussion is related to the need for business process variation and how much is enough. Some examples are provided on how to model and manage process variance and how to relate it to the business model and strategy. Finally, the benefits, and typical pitfalls when modeling and managing process variance, are explained to give insight into how variance should be handled and what to avoid.

BUSINESS PROCESS VARIANCES: WHAT IS IT?

According to ISO9000:2005, a process is quite simply a set of activities that transforms inputs into outputs.[2] However, even though the activities performed may be the same, various other factors influence how those activities are actually performed. The most obvious factors that cause such variation are unequal quality of input or

4.3　业务流程的变化

Mark von Rosing, Jonnro Erasmus

4.3.1　介绍

在业务流程分析和开发过程中,我们经常面临这样的批评:由于我们引入了额外的管理,扼杀了在流程中扮演角色的人的创造力。事实上,作为分析师和架构师,我们在寻求简单和最佳解决方案的过程中自然反对这种观点,并且也有很好的理由。企业很复杂,经常会遇到数百个业务流程,这些流程分为几十个流程组和多个流程区域[1]。这种分解可能导致数千个流程活动和任务,很快成为管理负担。由于每个业务单元或能力领域允许拥有自己的流程变化并将以指数形式增长,这使得我们很容易提出标准化的想法。

然而,我们知道差异存在,业务流程建模的目标是捕获反映现实的东西,例如,质量管理在生产操作中的执行方式与在会计中的执行方式有很大的不同。尽管这些业务流程具有相似的目标,并有助于实现企业的总体质量目标,但所应用的技术和所需的技能是如此不同,以至于我们甚至不认为这些流程是相同的,制造业应用SPC等技术关注样本量等概念,而会计则更依赖于检查、平衡和审计。显然,这两种截然不同的做法难以统一和标准化,因为这样降低成本的同时会降低价值。事实上,允许业务流程发生变化的决定总是企业价值和降低成本的潜力之间的权衡。

本节探讨业务流程中的变更现象,它解释什么是流程差异以及如何识别它。讨论的大部分内容都与业务流程变更的程度以及数量有关。我们提供一些关于如何建模和管理流程差异以及如何将其与业务模型和策略相关联的示例。最后,解释建模和管理流程差异时的好处和典型缺陷,以深入了解应如何处理差异以及应避免什么。

4.3.2　业务流程差异: 它是什么?

根据ISO9000: 2005,一个流程就是一组将输入转化为输出的活动[2]。然而,即使所执行的活动可能相同,各种其他因素也会影响这些活动的实际执行方式。导致这种差异的最明显因素是不均衡的输入质量或角色参与者在此流程中的能

the ability of the role players in the process. Factors that are more difficult to detect and define are influences such as the motivation of practitioners or poorly defined interdependencies between business processes. In reality, slight differences in the way business processes are executed are always present, though it may be at a level of detail that does not lend itself to concern.

Variation in process performance prompted the pioneering research of Frederick Winslow Taylor on scientific management. He improved industrial efficiency by performing work studies and providing detailed instructions to standardize the work.[3] Yet, by studying what is normal, we will remain merely normal, and, in many cases, the outliers or nonstandard data points inspire the inquiry.[4] It is not in the interest of a modern business to remain normal, but rather to become an outlier and differentiate itself from its competitors. In fact, those differentiating capabilities offer superior value to customers that give businesses their competitive advantage.[5]

Variations of business processes occur in different ways. The example given above, showing how different quality is managed within the manufacturing and accounting domains of a business, represents the type of variation that can be found in any enterprise. Most processes are variations of the basic themes in business, such as governance, resource management, product realization and service delivery.

Systems Applications Products (SAP) define process variances the following way[6]: "A Business Process Variant is a fundamental flow variant of a Business Process which uses the same input and delivers the same measurable outcome." The flow of process steps is defined at Business Process Variant level. To keep level consistency, it is necessary that each business process has at least one Business Process Variant attached. The SAP modeling handbook proceeds by stating that a Business Process Variant should differ from another at least in one of the following ways:

- Flow of documents;
- Business objects needed;
- Life cycle schema of the business objects (status and status transitions);
- Application to Application/Business to Business (A2A/B2B) message choreography or choreography with direct interactions with other business processes.

The SAP handbook also offers the following characteristics of business process variants:

- A business process variant is not just an alternative user interface (UI);
- A process variant is not just another sequence of tasks that a user decides to perform on the UI;
- Two business process variants differ in the way the business process flows. The difference is so important that the variants are to be considered separately in a business process analysis.[7]

Thus, variance is a challenge within BPM, information management, and operations management, effecting entire end-to-end flows of processes, information, and services. To illustrate this, consider the following scenario: finance, people, and raw materials are all types of resources consumed and managed in the business, through

力。更难以检测和定义的因素是诸如从业者的动机或业务流程之间定义不明确的相互依赖性等因素。实际上,业务流程执行方式的细微差别总是存在,尽管它可能处于一个不容易引起关注的细节层面。

流程绩效的变化促使 F. W. Taylor 对科学管理进行了开创性的研究[3]。他通过开展工作研究和提供详细的指导使工作标准化,提高了工业效率[4]。然而,通过研究什么是正常的标准,我们将仅仅保持正常的内容,在许多情况下,异常值或非标准数据点成为我们研究的目标。事实上,这些差异化的能力为客户提供卓越的价值,使企业获得竞争优势[5]。

业务流程的变更以不同的方式发生。上面给出的示例展示了如何在企业的制造和会计领域中管理不同的质量,它表示了可以在任何企业中找到的变更类型。大多数流程是业务中基本主题的分支,如治理、资源管理、产品实现和服务交付。

SAP(Systems Applications Products)公司通过以下方式定义流程差异[6]:"业务流程变量是业务流程的基本流变量,它使用相同的输入并提供相同的可度量结果。"流程步骤在业务流程中定义分支级别。为了保持级别一致性,每个业务流程必须至少附加一个业务流程分支。SAP 建模手册详细说明了业务流程变更应至少包含以下方式之一:

- 文件流;
- 所需的业务对象;
- 业务对象的生命周期模式(状态和状态转换);
- 应用到应用(A2A)/业务到业务(B2B)消息编排或与其他业务流程编排的直接交互。

SAP 手册还提供了以下业务流程变体特征。

- 业务流程变体不仅仅是替代 UI。
- 流程变更不仅仅是用户决定在 UI 上执行的另一系列任务。
- 两种业务流程变更在业务流程的流动方式上有所不同。区别非常重要,因此在业务流程分析中应单独考虑变体[7]。

因此,差异是 BPM、信息管理和操作管理中的一个挑战,它影响整个流程、信息和服务的端到端流。为了说明这一点,请考虑以下场景:财务、人员和原材料是业务中通过财务管理、人力资源管理、物料管理和生产/运营的流程所消耗和管理的所有类型的资源。识别共性是非常困难的,部分原因是流程具有不同的目的、角

the processes of financial management, human resource management, material management, and production/operation. It is very difficult to identify commonality, partly because the processes have different purposes, roles, and so on. At the other extreme, however, we find those business processes serve the same purpose and seem identical, except for the finest of details. These processes are typically found in the same business area and can therefore truly be considered variances of each other. To illustrate, consider the differences in how engineers perform failure analysis for two different pieces of equipment. Failure analysis is a well-documented industry standard, but for the equipment specialists the differences are huge.

These two extremes help us distinguish between what are typically considered separate processes and what we consider business process variances. Both of these, though, are controlled variations. The differences in how the different competencies and people perform quality management and failure analysis are understood and acknowledged. Variation can also be uncontrolled, in which the state of the business has an impact on the processes, for example, emergency state as opposed to normal operation.

COMPLICATIONS AND CHALLENGES

Although it is very important to be able to define process variances, many organizations suffer under the phenomenon of having too much variation. Typically, this happens when the various business units are allowed to specify how unique they are during process development and mapping. This leaves the organization with too much variation, resulting in increased cost and complexity of operation. Even more debilitating is when such processes are used in blueprints for information systems and are implemented into customer relationship management (CRM), enterprise resource planning (ERP), supply chain management (SCM), security risk management (SRM), and/or mobile solutions. Not only are high cost and complexity built into the business processes, but now also into the information systems of the organization.

Some vendors offer standard reports on variances of work. These reports enable organizations to identify where they have too many variances and perhaps even duplication or unnecessary processes. These variances in work processes occur when the total costs charged to a job or schedule do not equal the total costs relieved from a job or schedule. Oracle provides the following usage, efficiency, and standard cost adjustment variance calculations in transactions[8]:

- Material Usage Variance;
- Resource and Outside Processing Efficiency Variance;
- Move-Based Overhead Efficiency Variance;
- Resource-Based Overhead Efficiency Variance;
- Standard Cost Adjustment Variance.

Variance is a known pitfall and issue of business process modeling and management, so much so that the ERP vendors officially ask their customers to "Define

色等。然而,在另一个极端,我们发现这些业务流程具有相同的目的,并且看起来完全相同,除了最细微的细节。这些流程通常位于相同的业务领域,因此可以真正地考虑彼此之间的差异。为了说明这一点,请考虑工程师对两种不同设备进行故障分析的不同方式。故障分析是一个文档化良好的行业标准,但对于设备专家来说,差异是巨大的。

这两个极端有助于我们区分什么是典型的独立流程,什么是我们认为的业务流程差异。不过,这两种情况都是可控的变化。理解和承认不同能力和人员如何执行质量管理和故障分析的差异。业务状态对流程有影响,因此,变更经常不受控制。例如,紧急状态是正常操作的对立面。

4.3.3　复杂性和挑战

尽管定义流程差异是非常重要的,但是许多组织都面临着变化太多的现象。通常,当允许不同的业务单元指定它们在流程开发和映射期间的唯一性时,就会发生这种情况。这给组织留下了太多的变化,从而增加了成本和操作的复杂性。在信息系统的蓝图中使用这些流程,并将其实现到客户关系管理(CRM)、ERP、供应链管理(SCM)、安全风险管理(security risk management, SRM)或移动解决方案中时,会造成更大的损害。高成本和复杂性不仅存在于业务流程中,而且还存在于组织的信息系统中。

一些供应商提供关于工作差异的标准报告。这些报告使组织能够确定它们在哪些地方有太多的差异,甚至可能出现重复或不必要的过程。当某项工作或计划的总成本不等于从某项工作或计划中节省的总成本时,就会出现工作过程中的这些差异。Oracle 在事务中提供了以下使用、效率和标准成本调整方差计算[8]:

- 材料使用差异;
- 资源与外部处理效率差异;
- 基于移动的开销效率差异;
- 基于资源的管理费用效率差异;
- 标准成本调整方差。

差异是业务流程建模和管理的一个已知陷阱和难题,以至于 ERP 供应商正式要求客户"定义差异的原因"。例如,SAP 要求客户定义发生任何差异的原因,并

Reasons for Variances." For example, SAP asks customers to define the reasons for any variance that occurs and to document this variance.[9] The official action is to "define the possible causes for variation that could occur in your company." Such variances might be any of the following[10]:

- Scrap on a quantity basis;
- Excess consumption of activities;
- Longer execution time;
- Other resources, and so on.

Anyone following that recipe will end up with multiple variations, which has also been evident of the high level of ERP system customization. Not realizing that there might be a variation because of the duplication of business function, roles, services, and so forth. Another missing aspect is the lack of ability to identify aspects in which the organization is unique and not unique. Therefore, as with most aspects of business process modeling and management, no simple recipe exists to model and manage business process variance. It depends on what is expected from the modeling and management effort. Business processes are developed, modeled, and managed for different reasons, and different stakeholders have different expectations. For example, the manager of an engineering department may be entirely aware of the differences between failure analysis for mechanical and that for electronic components, but does not consider the differences of enough significance to warrant separate processes. For the engineers who design and develop the mechanical and electronic products, those differences are of the greatest significance and may even be the unique value that a specific group or team contributes to the organization. The difference between the two variations of the failure analysis process does not necessarily have to lie in the steps carried out, but can be in the type of information used or how the results are captured and presented. Both those cases are from the perspectives within the process, however. Tsikriktsis and Heineke found that customer satisfaction is significantly affected by inconsistency in process, especially when the average service quality is low.[11] Clearly then, allowing variation is a decision best taken with care.

This example is meant to serve the purpose of showing how different expectations will inform the decision to allow business process variation and how much to allow. Again, the basis for decision returns to the business model of the organization. It has been shown that most organizations, especially highly successful organizations, have more than one business strategy concurrently in effect, in an effort to maintain its competitive advantage.[12] The model that is concerned with delivering unique value to customers will most likely embrace business process variance in the parts of the organization that creates that value. The business model that drives cost cutting will drive for standardization and simplification when possible.

Embracing variation does not acquit us from the burden to manage the business processes. In fact, allowing more than one version of a business process to exist inherently increases the complexity in the business. More problematic though, it becomes important to identify and clearly define the similarities and principles that

记录这种差异[9]。官方的行动是"定义可能在您的公司发生变化的可能原因"。这种差异可能是以下情况之一[10]：

- 废弃的数量；
- 活动的过度消耗；
- 执行时间更长；
- 其他资源,等等。

遵循该方案的任何人最终都会有多种变化,这也体现在高水平的ERP系统定制中,而不是意识到由于业务功能、角色、服务等的重复而可能存在变化。另一个缺失的方面是缺乏识别组织独特而不是唯一的方面的能力。因此,与业务流程建模和管理的大多数方面一样,没有简单的方法来对业务流程差异建模和管理。实际上,这取决于建模和管理工作的目标。业务流程的开发、建模和管理有不同的原因,不同的利益相关者有不同的期望。例如,工程部经理可能完全了解机械部件和电子部件的故障分析之间的差异,但没有考虑到为足够重要的差异构建单独的流程。对于设计和开发机械和电子产品的工程师来说,这些差异具有最大的意义,甚至可能是特定的团队或团队对组织贡献的独特价值。故障分析过程的两种变化之间的差异不一定要取决于所执行的步骤,但可以取决于所使用的信息类型或如何捕获和结果呈现。然而,这两种情况都是从流程的角度来看的。N. Tsikriktsis和J. Heineke发现,流程中的不一致性对客户满意度有显著影响,特别是当平均服务质量较低时[11]。显然,进行变更是一个最谨慎的决定。

这个例子旨在展示不同的目标如何展示允许业务流程变更的决策以及允许的程度。同样,进行决策的基础将返回到组织的业务模型。实践已经证明,大多数组织,特别是非常成功的组织,都同时具有多个有效的业务战略以保持其竞争优势[12]。其中,与向客户交付独特价值有关的模型很可能包含创建该价值的组织部分中的业务流程差异。如果可能的话,驱动成本削减的业务模型将驱动标准化和简化。

接受变化并不能减轻我们管理业务流程的负担。事实上,允许一个业务流程的多个版本固有地存在会增加业务的复杂性。但问题更为严重的是,识别和明确定义流程所有版本中的相似性和原则变得非常重要,将思维扩展到业务流程差异

must be maintained in all versions of a process. It is also easy to extend the thinking toward a scenario in which the differences between two variations of a business process become so pronounced that it is difficult to explain why they are considered variations of the same process. Clearly then, the decision to allow variations of a business process to exist should be taken very carefully and should support business goals and objectives.

To make informed decisions regarding business process variation, the problem has to be understood. To summarize, four main challenges deal with business process variation:

- Deciding when and how much variation to allow;
- Defining and justifying business process variances;
- Capturing business process variation without introducing unnecessary complexity and ambiguity; and
- Managing business process variances.

SOLUTION DESCRIPTION

Business analysts and architects will naturally oppose allowing variation in business processes. It will seem like it makes the solution more complicated or even suboptimal. Instead, business process variance should be seen as an opportunity for the processes to more accurately reflect reality. More importantly, process variance is a way to capture techniques, knowledge, or other intricacies that are unique to a certain business competency or practitioner. Those details can be compared to the standardized process to identify that which creates the unique value to the business. Thus, business process variance is not only a viable option for an organization to identify its own unique value enablers, but also to exploit those enablers and hopefully build on them.

When Should Variation Be Allowed and How Much Is Enough?

We have established that variation in business processes introduces performance, modeling, and management challenges. Thus, we require guidance on when business process variance is desirable and how much should be allowed. What helps here is to find out when the organization should allow for uniqueness and thereby high variability, or when the level of standardization should be high and variation minimal. The challenge is that this information will not be found in the process itself, but rather in the business model showing the relevant business competencies being identified and calling upon their respective processes.[13] By exercising its business competencies, the business delivers value internally and externally, for example, value is delivered through business tasks, business functions, and services within a competency to those that benefit from the value created. Competencies may be essential to compete, in which case they are described as *core-competitive*; or they may differentiate the business from its customers, in which case they are

的两个变体之间也变得很容易和明显,以至于很难解释为什么它们被视为同一流程的变体。显然,允许业务流程存在变化的决策应该非常谨慎,并且应该支持业务目标。

要做出有关业务流程变化的明智决策,必须了解问题。总而言之,处理业务流程变化的四个主要挑战如下:

- 决定允许的变更时间和变化量;
- 定义和证明业务流程差异;
- 确定业务流程变化不会引入不必要的复杂性和模糊性;
- 管理业务流程差异。

4.3.4 解决方案描述

业务分析师和架构师自然会反对允许业务流程的变更,因为它看起来会使解决方案更复杂甚至不理想。相反,我们应把业务流程变更视为流程更准确地反映现实的机会。更重要的是,流程变更是一种捕获某些业务能力或从业者独有的技术、知识或其他复杂性的方法。我们可以将这些细节与标准化流程进行比较,以确定为业务创造有独特价值的流程。因此,业务流程差异不仅是一个可行的选择,一个组织可以确定自己独特的价值使能者,而且还可以利用该使能者,并有希望地在他们的基础上进行构建。

1. 什么时候应该允许变更,多少才足够?

我们已经确定,业务流程中的变更引起了性能、建模和管理方面的问题。因此,我们需要关于何时需要业务流程变更以及允许变更多少方面的指导。这里的指导是指找出什么时候组织应该保持唯一性前提下的高可变性,或者什么时候标准化水平应该高而变化最小。所面临的挑战是,这些信息将不会在流程本身中找到,而是在显示正在识别的相关业务能力并调用其各自流程的业务模型中找到[13]。企业通过行使其业务能力,可以在内部和外部提供价值,例如,通过业务任务、业务职能和服务内的服务将价值提供给那些从创造的价值中受益的人。竞争力可能是竞争的关键,在这种情况下,它们被称为核心竞争力,或者他们可能将业务与其客户区分开来,在这种情况下,他们是核心差异化的。

core-differentiating. The majority of competencies are simply necessary for the functioning of the business, and these are commonly referred to as *non-core* competencies. The ability to categorize competencies as either *core-differentiated*, *core-competitive*, or *non-core* is missing within contemporary process modeling and process architecture practice. The inability to identify competencies is the very reason why process experts and process architects have no insight as to which processes are a part of an organization's competitive aspects and which are not. This is also why they are not able to take into consideration the process variances in terms of where they should be, where they create value and where they should not be.

The link between the organization's competencies and process execution provides the means of identifying ways to appropriately standardize variances and thereby reduce cost, improve the effectiveness and efficiency of operations, or conversely to support value creation and thereby revenue growth. Without this context, there is no means to judge the "goodness" of a particular process or process variance design. For example, if it is not possible to detect that a process contributes value, it is best not to have any variances as it should be done in the cheapest way possible. Figure 1 shows a summary of the concepts for categorizing the three domains of business models, the competencies that enable the business models, and the type of practice standards that correspond to the different competencies.[14]

For noncore competencies it makes sense to adopt standard best practice, in an effort to optimize operations and minimize cost. Similarly, industry best practice may be adopted for core-competitive competencies, because the business only aims to compete effectively with its competitors and maximize its performance. However, to drive growth in revenue and value, new products and services have to be developed to give the business a competitive advantage. By its nature, an advantage requires something that is not offered elsewhere, thus the business strives toward developing and nurturing core-differentiating competencies. However, applying such differentiating competencies in a standardized business process will at best result in high performance, but not in differentiating value. As shown in Figure 1, true differentiating competencies

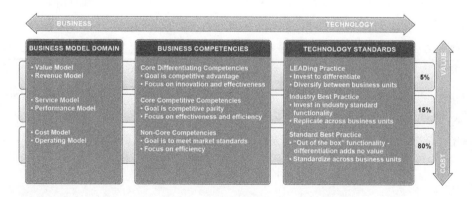

FIGURE 1

When to apply LEADing industry and best practice.[15]

大多数能力仅仅是业务运作所必需的,这些能力通常被称为非核心能力。在当代流程建模和流程架构实践中,流程专家和流程架构师如果不能将竞争力分为核心差异化、核心竞争力或非核心竞争力,则无法识别哪些流程是组织竞争方面的一部分,哪些不是。这也是他们不能考虑流程变化的原因,比如它们应该在哪里、它们在哪里创造价值、它们不应该在哪里。

　　组织能力和流程执行之间的联系提供了一种确定变更程度标准化的方法,从而降低成本,提高操作的有效性和效率,或者反过来支持价值创造,从而实现收入增长。没有这个背景关系,就无法判断特定流程或流程变更设计的"好坏"。例如,如果无法检测某个流程是否贡献了价值,那么最好不要使用任何变化,因为这是最适当的方法。图1显示了对业务模型的三个领域、支持业务模型的能力以及与不同能力相对应的实践标准类型分类的概念和介绍[14]。

　　对于非核心能力,采用努力优化运营并降低成本的方法是有意义的。同样,行业最佳实践也可用于核心竞争能力,因为企业的目标仅仅是与竞争对手有效竞争,并最大限度地提高其绩效。然而,为了推动收入和价值的增长,必须开发新的产品和服务,使企业具有竞争优势。从本质上讲,优势是指其他地方没有提供的东西,因此企业努力发展和培养核心差异化能力。然而,企业在标准化的业务流程中应用这样的差异化能力,这最多只会导致高性能,而不会导致价值差异化。如图1所示,真正的差异化能力通常只占业务的一小部分,尽管在真正的创新型企业中,这

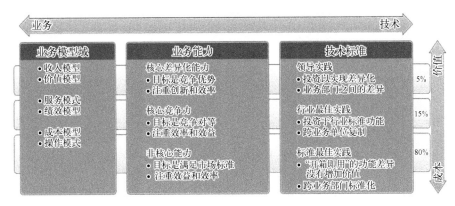

图1　何时应用领先行业和最佳实践[15]

typically compose a very small portion of the business, though it may be a much larger percentage in truly innovative enterprises. Therefore, in the relatively limited cases in which a business aims to offer unique or even market-leading products and services, it is crucial for the business to appreciate and embrace variation in the business processes that produce the characteristics that make the offering superior.

The challenge we see in most organizations is actually relating the business competencies to the processes. Table 1 illustrates how to link traditional process aspects with business competencies and Table 2 provides a step-by-step guide how to do it.

Table 1 The Possible Linkage between Process and Competencies

	Process #	What Specification:					Who/Whose Specification:		
		Business Process Area	Process Groups	Business Process	Process Steps	Process Activities	Stakeholder Involved	Process Owner	Roles/ Involved
Business competency 1	#								
Business competency 2	#								
Business competency N	#								

Table 2 A Table Showing that Process Objects Relate to Business Competency and the Tasks Associated with It

Business Competency: An Integrated and Holistic Set of Interconnected Knowledge, Skills, and Abilities, Related to a Specific Set of Resources (Including Persons and Organizations) that, Combined, Enable the Enterprise to Act in a Particular Situation	
Rules	(D) Process relates to Business Competency.
Tasks	• Identify in the business model or an operating model which business competencies are core-differentiated. • Identify in the business model or an operating model which business competencies are core-competitive. • Identify in the business model or an operating model which business competencies are non-core. • Associate and tie the business competencies to the business processes. • Associate and tie the business competencies to the process steps of the business process. • Associate and tie the business competencies to the process activities of the business process. • Associate and tie the business competencies to the stakeholders involved in the business process. • Associate and tie the business competencies to the process owner of the business process. • Associate and tie the business competencies to the managers involved in the business process. • Associate and tie the business competencies to the roles/resources involved in the business process.

一比例可能要大得多。因此,在相对有限的情况下,如果企业的目标是提供独特的甚至是市场领先的产品和服务,那么企业必须了解并接受业务流程中的变更,这些变更会产生使产品更优越的特征。

我们在大多数组织中看到的挑战实际上是将业务能力与流程联系起来。表1说明了如何将传统流程方面与业务能力联系起来,表2提供了一步一步的指导。

表1　过程和能力之间可能的联系

	流程#	什么是标准					谁/谁的标准		
		业务流程领域	流程组	业务流程	流程步骤	流程活动	利益相关者参与	过程所有者	角色/参与
业务能力1	#								
业务能力2	#								
业务能力3	#								

表2　显示流程对象与业务能力以及与之相关的任务相关的表

业务能力:一组相互关联的知识、技能和能力的集成和整体集合,这些知识、技能和能力与一组特定的资源(包括个人和组织)相关,这些资源的组合使企业能够在特定的情况下采取行动	
规则任务	(D)流程与业务能力有关 • 在业务模型或运营模型中确定哪些业务能力是核心差异化的 • 在业务模型或运营模型中确定哪些业务能力具有核心竞争力 • 在业务模型或运营模型中确定哪些业务能力是非核心的 • 将业务能力与业务流程联系起来 • 将业务能力与业务流程的流程步骤相关联并绑定 • 将业务能力与业务流程的流程活动相关联和联系 • 将业务能力与业务流程中涉及的利益相关者联系起来 • 将业务能力与业务流程的流程所有人相关联并绑定 • 将业务能力与业务流程中涉及的管理人员联系起来 • 将业务能力与业务流程中涉及的角色/资源相关联和绑定

Table 1 shows all the aspects that can be linked to a competency, including stakeholders, managers, process owners, and roles that are in a business competency area. The process–business competency matrix captures all aspects that can be linked between a business competency and business processes. See Table 2 for the semantic rules for this mapping and the tasks to establish the relationship.

It is important to note that the business competency type, that is, core-differentiated, core-competitive or non-core, needs to be derived directly from a business model, operating model, or a business competency matrix.[16] Figure 2 shows an example of an Oil and Gas business model showing the business competencies and the typical aspects of a business model, including the following[17]:

- The Business Competency Areas;
- The Business Competency Groups;
- The various Business Competencies.

This Oil and Gas Petroleum Engineering business model will be used to show the different options for modeling business process variance. The Technical Quality Control business competency will be used, because it was identified to be core-competitive with high-value potential, but poor performance. In addition to that, it needs to be standardized and needs evaluation and audit aspects. It is exactly such a case in which process variances and the people involved need to be identified to reduce unnecessary complexity and ensure high performance and value realization.

Defining and Justifying Business Process Variance

Once it has been determined when and how much variance will be allowed, these decisions should be documented to ensure that the rationale is captured. Furthermore, the variances that will be allowed should be defined to ensure that the business process modeling stays in line with the intention. Table 3 shows a matrix that may be used to document the decisions and define the variances to be developed.

As previously explained, process variances are justified by linking the process to a core business competency. Table 3 also allows business competency variances to be defined and justified and the number of variances that will be created. To ensure consistency in the way variances are identified, justified, and defined among different team members, Table 4 provides some rules and tasks to be followed by the business analysts, architects, and subject matter experts. These rules are similar to the reasons for process variance listed in Section 3 of this chapter.

Once a process or competency variant has been justified, the steps listed in Table 4 may be followed to complete Table 3. This will help that the necessary information for each variant is properly captured, resulting in consistency and repeatability in the business process modeling. Such a completed table will be an invaluable input into the modeling of business process variance.

表1显示了与能力相关的所有方面,包括利益相关者、经理、流程所有人和业务能力领域中的角色。"流程-业务能力矩阵"展示了业务能力和业务流程之间可以链接的所有方面。有关此映射的语义规则以及建立关系的任务可参阅表2。

值得注意的是,业务能力类型(即核心差异化、核心竞争力或非核心业务)需要直接从业务模型、运营模型或业务能力矩阵中获得[16]。图2的示例(石油和天然气业务模型)显示业务能力和商业模式的典型方面,包括[17]:

- 业务能力领域;
- 业务能力组;
- 各种业务能力。

这个油气石油工程业务模型将用于显示建模业务流程差异的不同选项。将使用被确定为核心竞争力具有高价值潜力但性能较差的技术质量控制业务能力。此外,它需要标准化,需要评价和审计方面工作。正是在这种情况下,需要识别流程差异和相关人员,以减少不必要的复杂性,进而确保高性能和价值实现。

2. 业务流程变更的定义和证明

一旦确定了允许变更的时间和数量,就应该记录这些决策以确保得到最优。此外,应该定义允许的变更方面以确保业务流程建模与意图保持一致。表3显示了一个矩阵,可以用来记录决策并定义要开发的变更。

正如前面所解释的,通过将流程与核心业务能力链接起来的流程变更是合理的。表3还允许定义和证明业务能力变更以及将要创建的变更的数量。为了确保在不同的团队成员之间识别、证明和定义变更的方式的一致性,表4提供了业务分析人员、架构师和主题专家需要遵循的一些规则和任务。这些规则类似于4.4.3节中列出的流程变更的原因。

一旦一个流程或能力变更被证明是合理的,表4中列出的步骤可能会被用来完成表3。这将有助于正确捕获每个变体的必要信息,从而在业务流程建模中实现一致性和可重复性。这样一个完整的表将是对业务流程变动建模的宝贵输入。

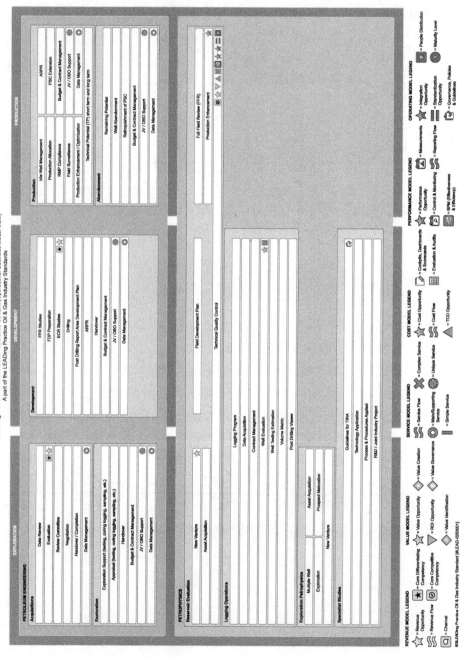

FIGURE 2

Example of Oil and Gas Petroleum Engineering business model.

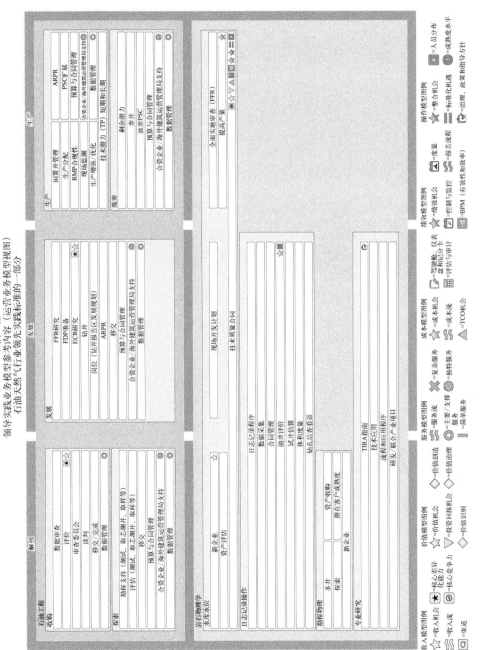

图 2　油气石油工程业务模型实例

Table 3 Business Process Variance Matrix

Process Variance Matrix	Variance #	Where Specification: (Business Model Relevant)			What Specification: (Process Model Relevant)		
		Business Competency Area	Business Competency Group	Business Competency	Business Process	Process Steps	Process Activities
Variance 1 List reason for variance	#						
Variance 2 List reason for variance	#						
Variance N List reason for variance	#						

Table 4 Example of Process Variance Rules and Tasks

	Process Variances: A Matrix Used by an Enterprise to Indicate Where They Have Variances and Indicate Why They Have Them
Rules	Process variance should be allowed based on the rules defined in the organization. The following are some examples of such rules: • The function of the resource varies (competency relevant). • Resource-based overhead efficiency variance (competency and process relevant). • Material-usage variance (competency and process relevant). • Move-based overhead efficiency variance (competency and process relevant). • Standard cost adjustment variance (competency and process relevant). • Processing-efficiency variance (process relevant). • Process policy, rule, and process compliance variance (process relevant). • Process-execution variance (process relevant).
Tasks	1. Identify and categorize the variance to the business competency areas. 2. Associate and tie the variance to the business competency groups. 3. Pinpoint the variances relevant for the business competencies. 4. Relate variances relevant for the business competencies to the business processes. 5. Identify and categorize the variance-only specific business processes. 6. Decompose business process variance to process steps and process activities. 7. Identify and categorize the variance only specific to process steps and process activities.

Modeling of Business Process Variance

Once the business process have been linked to the business competencies, classified as core-differentiated, core-competitive, or non-core, business process development relies on modeling the processes that need variances and standardize those that need more repetition. This modeling takes on different forms to achieve different critical success factors and business objectives. This is important, because the most appropriate way business process variances are captured and

表3　业务流程方差矩阵

流程变化矩阵	变化#	规范:(商业模式相关)			什么规格:(相关流程模型)		
		业务能力方面	业务能力组	业务能力	业务流程	流程步骤	流程活动
变化1:变化原因一览表	#						
变化2:变化原因一览表	#						
变化3:变化原因一览表	#						

表4　流程变更规则和任务的示例

过程变更:企业使用的一种矩阵,用于指示哪里有变更,以及为什么有变更	
规则	应根据组织中定义的规则允许流程变更。以下是这些规则的一些例子 • 资源的功能各不相同(与能力相关) • 基于资源的开销效率变更(与能力和流程相关) • 材料使用变更(与能力和流程相关) • 基于移动的开销效率变更(与能力和流程相关) • 标准成本调整变更(与能力和流程相关) • 处理效率变更(与过程相关) • 流程策略、规则和流程遵从性变更(与流程相关) • 过程执行变更(与流程相关)
任务	1. 识别并分类业务能力领域的变更 2. 将变更与业务能力组联系起来 3. 确定与业务能力相关的变更 4. 将与业务能力相关的变更与业务流程关联起来 5. 标识和分类仅针对变更的特定业务流程 6. 将业务流程变更分解为流程步骤和流程活动 7. 识别并分类仅特定于流程步骤和流程活动的变更

3. 业务流程变化的建模

一旦业务流程与业务能力(分类为"核心差异化"、核心竞争力或非核心)相关联,业务流程开发就依赖于对需要变化的流程进行建模,并对需要更多重复的流程进行标准化。此建模采用不同的形式来实现不同的关键成功因素和业务目标。这很重要,因为捕获和建模业务流程变化的最合适方式也取决于目标。可考虑以下

modelled also depends on the purpose. The following three options, or combinations thereof, can be considered:

1. Distinctive business process maps, diagrams, or models;
2. Separate processes with a master-and-variant type relationship; or
3. Single process with variances at the lower levels, such as activity or task variances.

These three options differ mostly at the level of detail at which the variation is captured. Variation in business processes can also entirely be removed from the process itself by capturing the differences in documents that accompany and support the process. Admittedly, this will lead to the desirable scenario in which only one business process model or document exists for each process, but simply delegates the burden to the domain of documentation management. This approach will also inevitably lead to unnecessarily complicated documents; therefore, this approach is not recommended here. These three listed options are briefly discussed to illustrate the differences.

Distinctive Processes

Creating new, distinctive business processes can be considered the most extreme measure for dealing with business process variations. Essentially, it eliminates the need for any special consideration of process variance, by rather increasing the number of business processes. It should also be noted that such distinct variances can be created at any level of decomposition. Business competencies or even competency groups may be variations of each other. In fact, it is common for organizations to establish new competencies as variations of current competencies, to develop new, specialized, or innovative products or services. It is very common for organizations to create new competency and process variants, then later consolidate to cut costs and standardize.

Figure 3 shows the Technical Quality Control business process of the Oil and Gas Petroleum Engineering business model. Four separate and specialized process steps are shown, with their own activities for failure analysis for different types of mechanical equipment.

These process steps may initially have come from a single failure analysis process, but now represent four distinct activity flows, with no dependencies or relationships actively maintained. The obvious result of such a separation is that eventually these processes may have nothing in common. Conversely, though, the possible benefit is that the unique details, such as process activities, skills, knowledge, and information can easily be captured.

Master-and-Variant Processes

If there is a desire to maintain commonality between different process variants, or at least traceability back to the master process, it is advisable not to create distinct process maps, diagrams, or models. Maintaining such relationships in documents is a laborious task and business process modeling methodologies and tools do not typically support it. Instead, the master process and its variants should be seen as

三种选择或其组合：

(1)独特的业务流程映射、图表或模型；

(2)使用主变量类型关系分离流程；

(3)具有较低程度变化的单个流程,如活动或任务。

这三个选项在捕获变化的细节级别上差别很大。通过捕获伴随和支持流程的文档中的差异,业务流程中的变化也可以完全从流程本身中删除。诚然,这将导致理想的场景,即每个流程只存在一个业务流程模型或文档,只是将负担转移到文档管理领域。这种做法也必然导致不必要的复杂文件。因此,这里我们不建议使用这种方法。综上,本节简要讨论列出的这三个选项,以说明它们之间的差异。

1)独特的流程

创建新的、独特的业务流程可以被认为是处理业务流程变化的最极端的方法。本质上,它通过增加业务流程的数量,消除对流程差异的任何特殊考虑。需要指出的是,可以在任何分解级别上创建这种不同的差异。业务能力甚至能力组可能是彼此的变体。事实上,组织通常会在现有能力的基础上建立新的能力,开发新的、专门的或创新的产品或服务。组织通常会创建新的能力和流程变体,然后进行整合以降低成本和标准化。

图3显示了油气石油工程业务模型的技术质量控制业务流程,显示了四个独立和专门的工艺步骤,以及针对不同类型机械设备的故障分析活动。

这些流程步骤最初可能来自一个单一的故障分析流程,但现在代表四个不同的活动流程,没有任何依赖关系或关系得到积极维护。这种分离的明显结果是：最终这些流程可能没有任何共同点。相反,可能的好处是：独特的细节(如流程活动、技能、知识和信息)可以很容易地被捕获。

2)主流程和分支流程

如果希望保持不同流程变体之间的共性,或者至少可以追溯回主流程,我们建议不要创建不同的流程图、图表或模型,因为在文档中维护这样的关系是一项艰巨的任务,业务流程建模方法和工具通常不支持它。相反,主流程及其变体因为具有多个利益相关者,所以应视为一个单元。在文档格式中,主文件及其变体通常占用

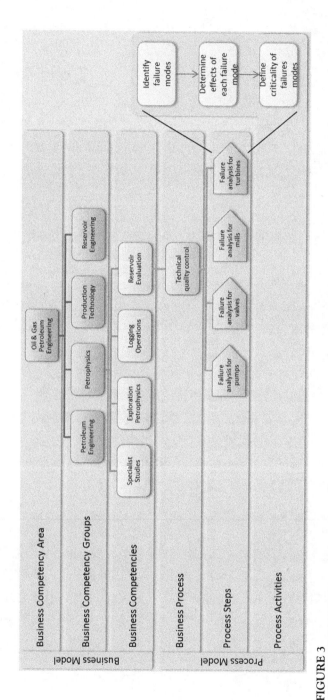

FIGURE 3

Breakdown of processes by failure analysis.

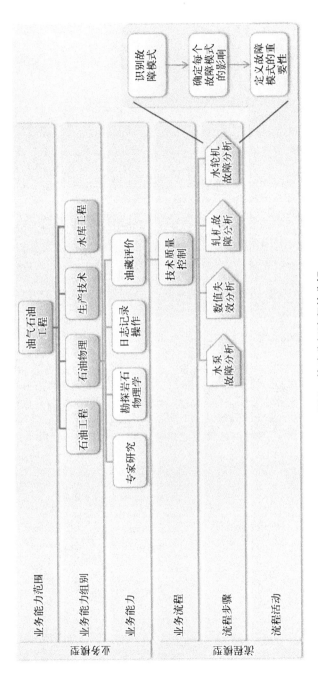

图3 故障分析流程的分解

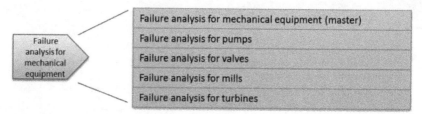

FIGURE 4

Master process with variants.

a single unit, with multiple stakeholders. In document format, the master and its variants will typically occupy a single document, with sections for the variants. Figure 4 shows a very simple illustration of how this configuration can be modeled for the same failure analysis process used previously.

When accessing the details of the failure analysis process, several options are presented. The user then has the option to access the details of the master process or any of its variants. Admittedly, this approach still does not force commonality between the process variances, but at least the traceability to the master process is very clearly maintained. Representing this type of business process with variants in a document format will ultimately result in a substantial document with sections for each of the variances.

Lower Level Variances

The final approach presented aims to maintain commonality of at least the process activities, or any lower level details of the process. Thus, traceability and alignment between the master and variance processes are maintained by essentially forcing the use of common process activities or tasks. This approach then only allows for differences in the inputs, outputs, controls, and mechanisms involved in the process. Figure 5 shows a very simple example of the lower level details of the same failure analysis process.

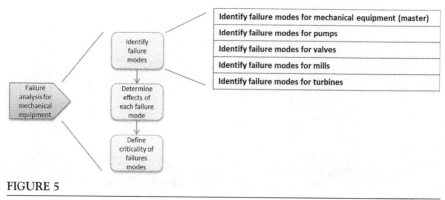

FIGURE 5

Common process with variation on activity level.

图4　带有分支的主流程

一个单独的文档,其中包含变体的部分。图4显示了一个非常简单的示例,说明如何为前面使用的相同故障分析流程建模此配置。

在查阅故障分析流程的详细信息时,提供了几个选项。然后,用户可以选择访问主流程或其任何分支的详细信息。诚然,这种方法仍然不强制流程变化之间的共性,但至少对主流程的可追溯性是非常清楚地保持的。用文档格式的变量表示这种类型的业务流程,最终将生成一个实质性的文档,其中包含每个变量的部分。

3)低水平变化

最后提出的方法至少旨在维护流程活动的共性或流程的任何低层细节。因此,通过本质上强制使用公共流程活动或任务来维护主流程和分支流程之间的可跟踪性和一致性。然后,这种方法只考虑流程中涉及的输入、输出、控制和机制的差异。图5显示了同一故障分析流程的底层细节的一个非常简单的示例。

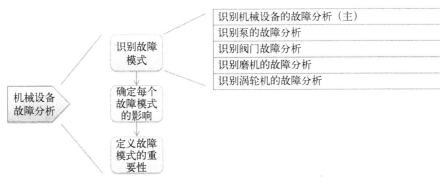

图5　行为水平变化的通用流程

Only one process is shown with a definitive set of activities. When attempting to access the details of the first activity, the user is presented with a menu of options, corresponding to the application of the failure analysis process to different types of equipment. When selecting one of the options, the user may be presented with tasks that are more detailed or perhaps with the inputs, outputs, controls, and mechanisms for that specific application of the activity. This approach suffers when presenting business processes in document format. Each process step or activity is accompanied by a matrix to show the different inputs, outputs, mechanisms, and controls of the variants.

Managing Business Process Variances

The three approaches to capturing and modeling business process variance unsurprisingly result in varying degrees of managerial burden. Simply put, the amount of administrative control necessary is proportional to the amount of content created and the need for accurate traceability. The first approach, in which distinctive processes are created, results in the most architectural content, but does not really lend itself to maintaining traceability between the processes. A generic master process may obviously be documented, but its relationship to the variant processes can at best be a text-based reference. The other two approaches are better suited to maintaining traceability to the master process. The second approach, in which variants of the master process are created, will result in significantly more business processes, but at least commonality is encouraged by keeping the processes together in the model. The third approach will result in the fewest business processes and least content to manage, but is very difficult to capture in document format.

When considering management of the complete process life cycle, the need for different approaches for different processes is further enforced. Figure 6 shows how the various business models drive the value life cycle, which in turn drives the process life cycle.

The value and revenue models target innovation and align to the analysis and design phases of the process lifecycle. Thus, in the context of business process variance, business value, and revenue creation will drive the identification of unique variations in process and how those variations deliver value to the business. Innovation will eventually make way for a focus on efficiency and effectiveness, once a product or service reaches its midlife. Thus, the performance and service models will drive business process improvement and standardization. Eventually then, cost and operating models will be introduced to drive optimization and simplification. Thus, the management of business process variance is not only dependent on the business model and strategies, but also on the life cycle of the specific process and its resulting product or service. Early in the life cycle, when innovation is encouraged and freedom is sought, variation should be allowed. When the innovation delivers value, the core-differentiating competencies should be captured and treated as recognized business process variances. This approach ensures that the

　　只有一个流程显示了一组确定的活动。在尝试访问第一个活动的细节时,用户会看到一个对应于故障分析流程在不同类型设备上应用的选项菜单,当选择其中一个选项时,可能会向用户显示更详细的任务,或者可能显示活动的特定应用程序的输入、输出、控制和机制。当以文档格式呈现业务流程时,这种方法会受到影响。每个过程步骤或活动都伴随着一个矩阵来显示变量的不同输入、输出、机制和控制。

4. 管理业务流程变动

　　捕获和建模业务流程变化的三种方法无疑会导致不同程度的管理负担。简单地说,所需的管理控制量与创建的内容量和准确可追溯性的需求成正比。第一种方法创建了单独的流程,它产生了最多的体系结构内容,但并不真正有助于维护流程之间的可跟踪性。　一个通用的主流程很明显是文档化的,但是它与不同流程的关系最多是一个基于文本的引用。另外两种方法更适合于维护主流程的可跟踪性。第二种方法(创建主流程的变体)将导致更多的业务流程,但是将流程放在模型中至少会鼓励通用性。第三种方法将导致最少的业务流程和最少的管理内容,但是很难以文档格式捕获。

　　在考虑整个流程生命周期的管理时,不同流程的不同方法的需求被进一步加强。图6显示了各种业务模型如何驱动价值生命周期,而价值生命周期又反过来驱动流程生命周期。

　　价值和收益模型以创新为目标,并与流程生命周期的分析和设计阶段相一致。因此,在业务流程变更的背景下,业务价值和收入创造将推动识别流程中的独特变更以及这些变更如何为业务带来价值。一旦产品或服务达到生命周期的中期,创新最终将成为关注效率和有效性的方式。因此,性能和服务模型将推动业务流程改进和标准化。最后,将引入成本和运营模型来推动优化和简化。因此,业务流程差异的管理不仅依赖于业务模型和策略,还依赖于特定流程及其产生的产品或服务的生命周期。在生命周期的早期,当鼓励创新和寻求自由时,应该允许变化。当创新带来价值时,核心差异化能力应被捕获并视为公认的业务流程差异。此方法

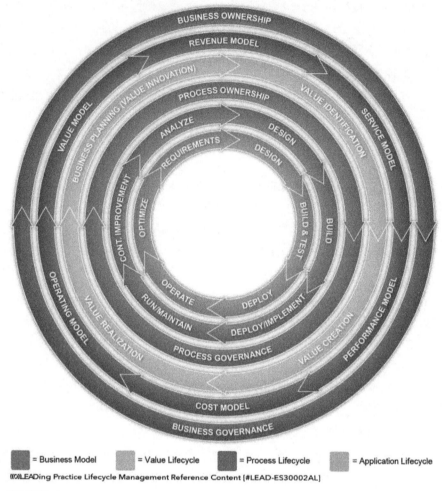

= Business Model   = Value Lifecycle   = Process Lifecycle   = Application Lifecycle

[(©)LEADing Practice Lifecycle Management Reference Content [#LEAD-ES30002AL]

FIGURE 6

Business models and life cycle alignment.[18]

justification for the variance is captured to enable informed management thereof later in the process life cycle.

Regardless of the life-cycle phase of a process, it should be measured and managed. The process performance indicators will typically be associated to individual variants, to allow for the comparison of the process variants. The more business process variance exists, the more management effort is necessary, because effectively the amount of architectural content is increased. As with all business processes the documentation, configuration, and interfaces of all variants must be managed. This task is significantly more difficult though; not only must alignment be maintained between the actual process, the documentation that describes it and what is

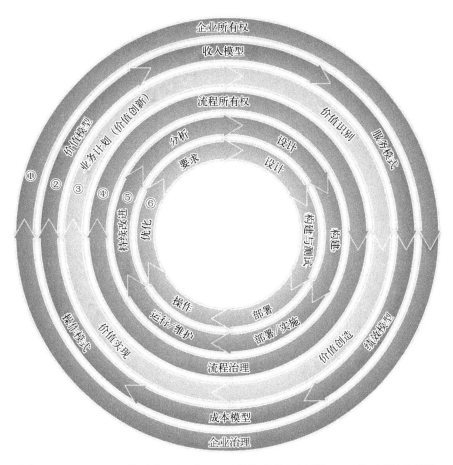

①②：业务模型　　　③：价值生命周期　　　④⑤：流程生命周期　　　⑥：应用生命周期

图6　业务模型和生命周期的一致性[18]

确保捕获差异的理由，以便在流程生命周期的后期实现信息化管理。

无论流程的生命周期处于何种阶段，都应该对其进行度量和管理。流程性能指标通常与单个变量相关联，以便实现对流程的比较。流程的变更增加了体系结构内容的数量，因此，业务流程存在的变动越多，就越需要进行管理。与所有业务流程一样，必须管理所有变体的文档、配置和接口。然而，这项任务要困难得多，不仅必须在实际流程、描述流程的文档和对流程的期望之间保持一致性，而且必须保持对主流程的可跟踪性和通用性（如果存在的话）。这需要在流程内容中建立和

expected of the process, but also traceability and commonality to the master process, if one exists. This requires establishment and maintenance of an additional relationship within the process content and appreciable attention from business process management.

COST CALCULATION OF PROCESS VARIANCES

The ability to calculate the cost of process variances is an important aspect for BPM, Information Management, and Operations Management. The typical information solution vendors like SAP[19] or Oracle have standards to calculate cost of process variances. Oracle lists the following ways to calculate cost of process variances[20]:

1. *Material Usage Variance*: The difference between the actual material issued and the standard material required to build a given assembly, calculated as follows:

   Standard material cost × (*quantified issued* − *quantified required*)
   Such a variance occurs when an organization over- or underissue components or use an alternate bill.

2. *Resource and Outside Processing Efficiency Variance*: The difference between the resources and outside processing charges incurred and the standard resource and outside processing charges required to build a given assembly, calculated as follows:

   (*applied resources units* × *standard or actual rate*) − (*standard resource units at standard resource rate*)
   This variance occurs when you use an alternate routing, add new operations to a standard routing during production, assign cost resources to No-direct charge operations skipped by shop floor moves, overcharge or undercharge a resource, or charge a resource at actual.

3. *Move-Based Overhead Efficiency Variance*: Move-based overhead efficiency variance is the difference between overhead charges incurred for move-based overheads (overhead basis of Item or Lot) and standard move-based overheads required to build a given assembly, calculated as follows:

   applied move-based overheads − *standard move-based overheads*
   This variance occurs when you use an alternate routing, add operations to a standard routing during production, or do not complete all the move transactions associated with the assembly quantity being built.

4. *Resource-Based Overhead Efficiency Variance*: Resource-based overhead efficiency variance is the difference between overhead charges incurred for resource based overheads (overhead basis of Resource units or Resource value) and standard resource-based overheads required to build a given assembly, calculated as follows:

   applied resource-based overheads − *standard resource-based overheads*
   This variance occurs when you use an alternate routing, add new operations to a standard routing during production, assign cost resources to No-direct charge

维护额外的关系,并且需要业务流程管理人员的显著关注。

4.3.5　流程变更的成本计算

计算流程变更成本的能力是BPM、信息管理和运营管理的一个重要方面。典型的信息解决方案供应商(如SAP[19]或Oracle)均有计算流程变动成本的标准。Oracle列出了以下计算过程变动成本的方法[20]。

(1)材料使用差异:实际发放的材料与制造给定组件所需的标准材料之间的差异,计算方法如下。

$$标准材料成本 \times (需要量化发行 - 量化)$$

当组织发行过多或过少的组件或使用替代票据时,就会出现这种差异。

(2)资源与外部处理效率差异:产生的资源和外部处理费用之间的差异以及构建给定程序集所需的标准资源和外部处理费用,计算方法如下。

$$(应用资源单位 \times 标准或实际利率) - 标准资源单位的标准资源率$$

当您使用备用工艺路线、在生产过程中将新操作添加到标准工艺路线、对资源进行多收费或少收费,或对资源进行实际收费时,就会出现这种差异。

(3)基于移动的费用效率差异:基于移动的费用效率差异是基于移动的费用(项目或批次的间接费用基础)和构建给定装配所需的标准的费用之间的差异,计算如下。

$$应用基于移动的费用 - 标准基于移动的费用$$

当您使用备用工艺路线、在生产过程中将操作添加到标准工艺路线或不完成与正在生成的装配数量相关联的所有移动事务时,就会发生这种差异。

(4)基于资源的管理费用效率差异:基于资源的管理费用效率差异是基于资源的间接费用(基于资源单位或资源价值的间接费用)与构建给定程序集所需的基于资源的标准管理费用之间的差异,计算如下。

$$应用的基于资源的日常开支 - 标准的基于资源的日常开支$$

这种差异出现在,当您使用备用工艺路线、在生产过程中将新操作添加到标准工艺路线、将成本资源分配给被车间移动而跳过的非直接收费操作、对资源进行多

operations skipped by shop floor moves, overcharge or undercharge a resource, or charge a resource at actual.

5. *Standard Cost Adjustment Variance*: Standard cost adjustment variance is the difference between costs at the previous standards and costs at the new standards created by cost update transactions.

cost of previous standards – cost of new standards

The following are some of the challenges doing cost calculations of process variances:

- The cost of the process variances can only be calculated if all the numbers exist before and after. Many organizations do not measure what they had, or they did not measure it the same way.
- The cost of calculating the cost of variances is very time and resource consuming. In other words, it is very costly to identify high-cost and inefficient process variances.
- To determine variances between production and planning involves not only the process but also the information flow. Although such an analysis has huge potential, it can be very time-consuming.
- The ability to show the causes of the variances and assign the variances to different variance categories depending on the cause.
- All of the above ways of evaluating cost of process variances do not really identify the specific process that is the root cause for the high cost and inefficient process variances. They only identify the high cost of doing it in a different way.

LESSONS LEARNED

For business process variance to deliver value to the enterprise, it is crucial to properly plan how it will be modeled and managed. As explained, variance introduces significant additional content and complexity to the business process landscape, resulting in increased management burden. Furthermore, it is important to consider whether the business process management function is up to the task of handling the increased burden. Ultimately, it is always a trade-off between more accurate representation of the core-differentiating competencies and increased complexity in the business processes.

The best approach is to introduce as little variation as possible and to make sure it is in the core business. If it is found that only the skills, knowledge, information, or tools differ between process variants, it should be entirely adequate to only have variances at process activity level. This will ensure that the desired commonality between the process variants is maintained and the amount of new content is minimized. Alternatively, if the business is targeting innovation and wants to allow its practitioners more freedom, it is probably more appropriate to create completely distinct and separate processes. Either way, the business processes should be formalized and documented, even if only at a low level of detail, to gain the wide-ranging benefits thereof.

收费或少收费,或对资源进行实际收费时。

(5)标准成本调整差异:标准成本调整差异是以前标准的成本与创建的新标准的成本之间的变动。

$$以前标准的成本 - 新标准的成本$$

以下是对流程变动进行成本计算的一些注意事项。

- 只有在所有数字都存在的情况下,才能计算流程变动的成本。许多组织不测量他们拥有的东西,或者他们不以同样的方式测量。
- 计算变动成本非常耗费时间和资源。换句话说,识别高成本和低效率的流程变动是非常昂贵的。
- 确定生产和计划之间的变动不仅涉及过程,还涉及信息流。虽然这样的分析有巨大的潜力,但可能非常耗时。
- 显示变动的原因,并将变动分配到不同的变动类别的能力需要视情况而定。
- 上述评估过程变动成本的所有方法都没有真正确定特定流程,这是造成高成本和低效率流程变动的根本原因。他们只发现用不同的方式做这件事的高成本。

4.3.6 经验教训

对于向企业交付价值的业务流程变更,正确规划如何建模和管理它至关重要。正如所解释的,变更给业务流程布局带来了显著的额外内容和复杂性,从而增加了管理负担。此外,重要的是要考虑BPM功能是否能够完成处理增加的负担的任务。归根结底,它总是在更准确地表示核心能力变动和增加业务流程复杂性之间进行权衡。

最好的方法是尽可能少地引入变更,并确保它在核心业务中。如果发现流程变量之间只有技能、知识、信息或工具不同,那么只在流程活动级别上存在变动就完全足够了。这将确保流程变体之间保持所需的共性,并最小化新内容的数量。或者,如果企业以创新为目标,并希望让其从业者有更多的自由,那么创建完全不同和独立的流程可能更合适。无论哪种方式,业务流程都应该正式化和文档化,即使只是在一个较低的细节级别上。

CONCLUSION AND SUMMARY

Business process variance should be seen as a viable way of allowing small differences in the way the core business functions are performed. It is advisable to only introduce variation in those business processes that represent the core-differentiating competencies of the organization. This will allow an enterprise to develop its own practice and deliver unique value to clients and other stakeholders. For non-core and core-competitive competencies, best practice and industry best practice should suffice.

Business process variance can be modeled three different ways, depending on what is expected thereof. If the aim is only to capture slight differences in the inputs, outputs, controls, and mechanisms of processes, it will be adequate to only create variances at the process activity or task levels. However, if the actual steps of the variant processes are different, true process variances can be used by presenting all the variances together in a single model or document or separate distinct processes may even be developed.

The modeling approach taken has a major impact on the management of the business processes and variances. When certain commonality between the master process and its variants is important, additional business process management techniques are necessary to maintain this traceability. This will require that a great deal of attention is given to establishing and maintaining the traceability links between the variants. Separate and distinct processes introduce more process content, but standard business process management is applied because traceability to the master process is unnecessary.

When introducing process variance, caution should be taken and the amount of variation should be minimized. If the development and modeling is not sufficiently controlled, the amount of additional and unnecessary content will very quickly become unmanageable. However, if it is done well, it is an excellent way for organizations to acknowledge and embrace their unique value enablers, without losing out on the many benefits of business process modeling and management.

End Notes

1. Dijkman R., La Rosa M., and Reijers H. A., "Managing large collections of business process models – current techniques and challenges," *Computers in Industry* 63, no. 2 (2012): 91–97, doi:10.1016/j.compind.2011.12.003.
2. *Quality Management Systems – Fundamentals and Vocabulary*, International Standard (Switzerland: The International Organization for Standardization, 2005).
3. Stephanie C. Payne, Satoris S. Youngcourt, and Watrous K. M., "Portrayals of F. W. Taylor across textbooks," *Journal of Management History* 12, no. 4 (2006): 385–407, doi:10.1108/17511340610692752.
4. Jason W. Osborne and Overbay A., "The power of outliers (and why researchers should ALWAYS check for them)," *Practical Assessment, Research & Evaluation* 9, no. 6 (March 2, 2004), http://pareonline.net/getvn.asp?v=9&n=6+.
5. Woodruff R. B., "Customer value: the next source for competitive advantage," *Journal of the Academy of Marketing Science* 25, no. 2 (March 1997): 139–153, doi:10.1007/BF02894350.

4.3.7 结论和总结

业务流程变更应被视为一种可行的方式,我们允许核心业务功能在执行方式上存在微小差异。建议只在那些表示组织的核心竞争力的业务流程中引入变更,这将允许企业开发符合自己特征的实践,并向客户和其他利益相关者交付独特的价值。对于非核心竞争力和核心竞争力方面,现有的最佳经验和行业最佳经验应该足够了。

根据对业务流程变更的期望,可以用三种不同的方式对其建模。如果目标只是捕获流程的输入、输出、控制和机制中的细微变化,那么仅在流程活动或任务级别上创建变更就足够了。然而,如果不同流程的实际步骤是不同的,那么可以通过在单个模型或文档中显示所有的流程变更来使用真正的流程差异,甚至可以开发单独的不同流程。

所采用的建模方法对业务流程和变更管理有重大影响。当主流程及其变体之间的某些共性很重要时,需要额外的BPM技术来保持这种可追溯性。这将需要非常注意建立和维护变体之间的可追溯性的链接。单独且不同的流程引入了更多流程内容,但是应用了标准BPM的主流程不需要可追溯性。

在引入流程变更时,应注意并尽量减少变化量。如果开发和建模没有得到足够的控制,那么额外的和不必要的内容的数量将很快变得无法管理。但是,如果做得很好,那么对于组织来说,这是一种很好认可和成为其独特的价值推动者,并且不会失去业务流程建模和管理许多好处的方式。

参考文献

[1] Dijkman R., La Rosa M., and Reijers H. A., "Managing large collections of business process models—current techniques and challenges," Computers in Industry 63, no. 2 (2012): 91–97, doi:10.1016/j.compind.2011.12.003.

[2] Quality Management Systems — Fundamentals and Vocabulary, International Standard (Switzerland: The International Organization for Standardization, 2005).

[3] Stephanie C. Payne, Satoris S. Youngcourt, and Watrous K. M., "Portrayals of F. W. Taylor across textbooks," Journal of Management History 12, no. 4 (2006): 385–407, doi:10.1108/17511340610692752.

[4] Jason W. Osborne and Overbay A., "The power of outliers (and why researchers should ALWAYS check for them)," Practical Assessment, Research & Evaluation 9, no. 6 (March 2, 2004), http://pareonline.net/getvn.asp?v=9&n=6+.

[5] Woodruff R. B., "Customer value: the next source for competitive advantage," Journal of the Academy of Marketing Science 25, no. 2 (March 1997): 139–153, doi:10.1007/BF02894350.

6. Rosenberg A., Business Processes Variants – SAP Modeling Handbook – Modeling Standards – SCN Wiki (January 4, 2014), http://wiki.scn.sap.com/wiki/display/ModHandbook/Business+Processes+Variants.

7. Rosenberg B., Processes Variants – SAP Modeling Handbook – Modeling Standards – SCN Wiki.

8. Work in Process Standard Cost Variances (Oracle Cost Management), accessed August 13, 2014, http://docs.oracle.com/cd/A60725_05/html/comnls/us/cst/stdvar01.htm.

9. Define Reasons for Variances – Process Order – SAP Library, accessed August 13, 2014, http://help.sap.com/saphelp_46c/helpdata/en/a9/e264b20437d1118b3f0060b03ca329/frameset.htm.

10. Ibid.

11. Tsikriktsis N. and Heineke J., "The impact of process variation on customer dissatisfaction: evidence from the U.S. Domestic Airline Industry," *Decision Sciences* 35, no. 1 (February 2004): 129–141, doi:10.1111/j.1540-5414.2004.02483.x.

12. David J. Teece, "Business models, business strategy and innovation," *Long Range Planning* 43, no. 2–3 (April 2010): 172–194, doi:10.1016/j.lrp.2009.07.003.

13. LEADing Practice Competency Modeling Reference Content (LEAD-ES20013BC).

14. Ibid.

15. Ibid.

16. Ibid.

17. LEADing Practice Business Model Reference Content (LEAD-ES20004BC).

18. LEADing Practice Lifecycle Management Reference Content, Standard (LEADing Practice, n.d.), accessed August 7, 2014.

19. http://help.sap.com/saphelp_45b/helpdata/en/90/ba667e446711d189420000e829fbbd/content.htm.

20. Work in Process Standard Cost Variances (Oracle Cost Management).

［ 6 ］ Rosenberg A., Business Processes Variants ─ SAP Modeling Handbook ─ Modeling Standards ─ SCN Wiki (January 4, 2014), http://wiki.scn.sap.com/wiki/display/ModHandbook/Business+Processes+Variants.

［ 7 ］ Rosenberg B., Processes Variants ─ SAP Modeling Handbook ─ Modeling Standards ─ SCN Wiki.

［ 8 ］ Work in Process Standard Cost Variances (Oracle Cost Management), accessed August 13, 2014, http://docs. oracle.com/cd/A60725_05/html/comnls/us/cst/stdvar01.htm.

［ 9 ］ Define Reasons for Variances ─ Process Order ─ SAP Library, accessed August 13, 2014, http://help.sap. com/saphelp_46c/helpdata/en/a9/e264b20437d1118b3f0060b03ca329/frameset.htm.

［ 10 ］ Ibid.

［ 11 ］ Tsikriktsis N. and Heineke J., "The impact of process variation on customer dissatisfaction: evidence from the U.S. Domestic Airline Industry," Decision Sciences 35, no. 1 (February 2004): 129−141, doi:10.1111/ j.1540-5414.2004.02483.x.

［ 12 ］ David J. Teece, "Business models, business strategy and innovation," Long Range Planning 43, no. 2−3 (April 2010): 172−194, doi:10.1016/j.lrp.2009.07.003.

［ 13 ］ LEADing Practice Competency Modeling Reference Content (LEAD-ES20013BC).

［ 14 ］ Ibid.

［ 15 ］ Ibid.

［ 16 ］ Ibid.

［ 17 ］ LEADing Practice Business Model Reference Content (LEAD-ES20004BC).

［ 18 ］ LEADing Practice Lifecycle Management Reference Content, Standard (LEADing Practice, n.d.), accessed August 7, 2014.

［ 19 ］ http://help.sap.com/saphelp_45b/helpdata/en/90/ba667e446711d189420000e829fbbd/content.htm.

［ 20 ］ Work in Process Standard Cost Variances (Oracle Cost Management).

Focusing Business Processes on Superior Value Creation: Value-oriented Process Modeling

Mark von Rosing, Mathias Kirchmer

INTRODUCTION

In this chapter, we will focus on Value-oriented Process Modeling, both in terms of what it is, how it is applied, as well as when it can be applied. This includes Value-oriented Process analysis, design, implementation, and governance considerations. Enabling organizations with the ability to interlink value engineering,[1] modeling, and architecture concepts with process aspects.

VALUE IS A DIFFERENT KIND OF CONCEPT FOR PROCESS TEAMS

Value planning, value identification, value creation, and value realization are not necessarily methods and approaches that are used by process teams today. However, the Global University Alliance[2] research around value modeling concepts has revealed that most organizations differentiate about 5% of their business competencies and have about 15% of their business competencies in the areas of core-competitive aspects. Competing head to head with the rest of the industry, the rest of the organization's business competencies are non-core and, thereby, commodity. This is very relevant for process modeling, as about 80% of the processes are commodity processes that do not add to the differentiation or competitiveness of the organization. A value-oriented process design and implementation considers this by focusing innovation and optimization initiatives, as well as company-specific software development, on the 20% high-impact processes, whereas commodity processes are designed based on industry reference models and implemented as far as possible through standard software. Design and implementation of processes target systematically on creating business value.

The chapter describes an approach to such a value-oriented segmentation, design, and implementation of businesses processes—transferring strategy into execution, at pace with certainty. The approach is explained using case examples.

The Complete Business Process Handbook. http://dx.doi.org/10.1016/B978-0-12-799959-3.00023-9

4.4 将业务流程聚焦于卓越的价值创造
——以价值为导向的流程建模

Mark von Rosing, Mathias Kirchmer

4.4.1 介绍

在本节中,我们将重点讨论面向价值的流程建模,包括它是什么、如何应用它以及何时可以应用它。这包括面向价值的流程分析、设计、实现和治理,使组织能够将价值工程[1]、建模和体系结构概念与流程方面联系起来。

4.4.2 对于流程团队来说,价值是一种不同的概念

价值规划、价值识别、价值创造和价值实现不一定是当今流程团队使用的方法和方式。然而,全球大学联盟[2]关于价值建模概念的研究表明,大多数组织在竞争力方面的差异约为5%,而在核心竞争力方面的业务能力差异约为15%。组织与其他行业竞争的业务能力属于非核心领域,因此也是商品。这与流程建模非常相关,因为大约80%的流程是与商品有关的流程,不会增加组织的差异性或竞争力。面向价值的流程设计和实现,可通过将创新和优化活动以及特定于公司的软件开发集中于20%的高影响流程来考虑。与之相比,商品流程由于是基于行业参考模型设计的,可尽可能通过标准软件实现。因此要系统地设计和实现以创造业务价值为目标的流程。

本节介绍一种以价值为导向的业务流程细分、设计和实现方法,以明确的速度将战略转化为行动。本文通过实例说明该方法。

Targeting Value

Organizations have to master the permanent change in our business environ-ment if they want to survive in the intermediate and long term. Dealing suc-cessfully with a volatile business environment, in general, means continuously "leveraging people to build a customer-centric performance-based culture."[3] Therefore, it is not only important to have a good strategy, hence to know *what* to do, but, in many organizations, the key challenge is about *how* to execute the strategy. To overcome this challenge, more and more organizations establish a value-oriented Business Process Management (BPM)[4] concept with a conse-quent process orientation across the company. This management discipline is about moving strategy into execution quickly with low risk. It enforces in par-ticular a customer and performance focus, because business processes deliver, by definition, a result of value for a client outside the process. A key component of such a BPM discipline is a structured value-oriented design of processes realizing the business strategy of an organization.[5,6]

This chapter presents an approach for business process design and implementa-tion that meets those requirements of targeting value. It is both focused on execut-ing the strategy of an organization and on being as resource efficient as possible. The result is a practical and effective approach to process design and implementation. The typical results of this approach embedded into BPM are transparency through-out an organization's processes, which enables achievement of quality and efficiency, agility and compliance, external integration and internal alignment, as well as inno-vation and conservation.

Research has shown that organizations compete with only about 5% of their processes, and another 15% are important processes supporting their competitive advantage. This means that 80% of the business processes are commodity processes, which can be carried out according to industry standards or common industry prac-tices. An industry average performance is sufficient. Sophisticated improvement approaches or even innovation initiatives targeting higher performance are not delivering real additional business value. Hence, process innovation and optimi-zation initiatives have to focus on the 20% high-impact processes, whereas other business processes can be designed and implemented using existing industry com-mon practices. Results are highly organization-specific business processes in which competitive advantages and processes are delivered following industry common practices when these are sufficient.

Targeting value systematically requires the appropriate segmentation of processes as basis for a differentiated design and implementation approach.[7] Process models developed during the process design need to reflect the requirements of those differ-ent process segments and the importance of the resulting business processes for the strategy of an organization. Different levels of sophistication regarding the improve-ment approaches are necessary.

The following process implementation, including the appropriate software sup-port, is executed accordingly to the process design based on the identified process

1. 目标价值

企业要想在中长期生存下去,就必须时刻掌握商业环境的持续变化。一般来说,成功地处理不稳定的业务环境意味着持续地"利用人们来构建以客户为中心的基于绩效的文化"[3]。因此,重要的不仅是要有一个好的战略,还应该知道该做什么,而且,在许多组织中,关键的挑战是如何执行战略。为了克服这一挑战,越来越多的组织建立了面向价值的BPM[4]概念,并在整个公司范围内建立了相应的面向流程的概念。这个管理原则是关于在低风险的情况下快速地将策略付诸实施。它特别关注客户和绩效,因为根据定义,业务流程为流程外的客户交付价值。此类BPM学科的一个关键组成部分是通过结构化价值导向的流程设计实现组织业务战略[5,6]。

本节介绍一种满足目标价值需求的业务流程设计和实现方法。它既关注于组织战略的执行,也关注于尽可能提高资源使用效率,得到的结果是一种实用有效的流程设计与实现方法。这种方法嵌入BPM的典型结果是透明度贯穿于整个组织流程,它能够实现质量和效率、灵活性和合规性、外部集成和内部协调以及创新和节约。

研究表明,组织只有大约5%的流程与竞争力有关,另外15%是支持其竞争优势的其他重要流程,这意味着80%的业务流程是可以根据行业标准或常见行业实践来执行的关于商品的流程。因此,只需满足行业平均水平就足够了,再复杂的改进方法,甚至以更高性能为目标的创新活动都不能带来真正的额外业务价值。因此,流程创新和优化活动必须关注20%的高影响流程,而其他业务流程可以使用现有的行业通用实践进行设计和实现。结果是得到高度特定于组织的业务流程,其中竞争优势和流程在满足条件的情况下按照行业通用实践交付。

系统地确定价值目标需要对流程进行适当的细分,并将其作为差异化设计和实现方法的基础[7]。在流程设计过程中开发的流程模型需要反映这些不同流程部门的需求以及由此产生的业务流程对组织战略的重要性。对于改进方法,不同的成熟度级别是必要的。

下面的流程实现(包括适当的软件支持),是根据基于所标识的各流程阶段的流程设计执行的。以价值为导向的设计常常导致需要使用不同的方法来开发软

segments. The value-oriented design leads often to different approaches to procure required enabling software. Highly organization-specific processes often require an individual development of software. Processes designed based on industry standards[8] lead in most cases to the use of standard software packages.

A value-oriented approach to design and implementation of processes enables organizations to use resources when they provide best value during design and implementation initiatives. People who are highly qualified in sophisticated process design and implementation methods, for example, focus on high-value areas. They can systematically target value as well as reduce the risk of project failure. They focus on moving the organization to the next level of performance, including the right degree of digitalization. This requires, in general, an "enlightened" Chief Information Officer (CIO),[9,10] who moves away from being a technical expert to becoming a driver of innovation and performance. The approach allows such a CIO to transition into a Chief Process Officer.[11]

The approach has been developed based on practical experience in large and mid-size organizations, mainly in the USA, South America, Japan, India, and Europe. It has been combined with academic research regarding value-oriented design and implementation methodologies, especially the LEADing practice value reference content.[12]

Segmentation of Business Processes

A business strategy needs to be operationalized to use it to drive process design and implementation. This is done by deriving strategic value drivers of an organization from its strategy. Those value drivers describe necessary achievements to make the strategy happen. The degree of realization of a value orientation is measured through key performance indicators (KPIs). A business process assessment based on the impact of a business process on strategic value drivers is the basis for the segmentation of processes into high-impact and commodity processes.[13,14] This process assessment is the key tool to align business strategy with process design and implementation. It enables the desired value-oriented approach and makes it part of a BPM discipline to transfer strategy into execution.

The value drivers are derived from the business strategy of the organization using value-driver-tree models (value-driver trees). This is a way of transferring the strategic intension of an organization into operational value-oriented business targets. An example for such a value-driver tree is shown in Figure 1. The value drivers themselves can again be weighted to focus the segmentation on the most important value drivers.

In practice, a three-step approach to developing a value-driver tree has proven successful. The strategy delivers the business priorities showing the overall direction toward which a company has to move. These priorities are decomposed into strategic objectives describing the key components of a business priority. Then one or several value drivers are identified for each objective, hence the operational achievements that make this objective happen.

件。高度贴合于组织的流程通常需要进行单独的个性化开发软件。在大多数情况下,可以使用标准软件模块构建行业标准[8]的流程。

面向价值的流程设计和实现方法使组织能够在设计和实现活动期间提供最佳价值时使用资源。例如,在复杂的流程设计和实现方法方面具有很高资质的人专注于高价值领域,他们可以系统地定位价值,同时降低项目失败的风险。他们专注于将组织提升到下一个绩效水平,包括正确的数字化程度。一般来说,这需要一位"开明的"从技术专家转变为创新和绩效的推动者CIO[9,10],这种方法允许这样的CIO转变为CPO[11]。

该方法是根据主要在美国、南美、日本、印度和欧洲的大中型组织的实践经验制定的,它与面向价值设计和实现方法的学术研究,特别是领先的实践价值参考内容相结合[12]。

2. 业务流程细分

需要对业务策略进行操作,以使用它来驱动流程设计和实现。这是通过从战略中获得组织的战略价值驱动来实现的。这些价值驱动因素描述了使战略发生的必要成就。价值导向的实现程度是通过KPI来衡量的。基于业务流程对战略价值驱动程序的影响的业务流程评估是将流程划分为高影响和商品流程的基础[13,14]。此流程评估是将业务策略与流程设计和实现相结合的关键工具。它使面向价值的方法成为可能,并使其成为BPM规程的一部分,实现将策略转化为行动。

价值驱动来自使用价值驱动树(value-driver trees)模型的组织的业务策略。这是一种将组织的战略意图转化为以价值为导向的业务目标的方法。图1中显示了这种价值驱动程序树的示例。可以再次对价值驱动因素本身进行加权,以将重点集中在最重要的价值驱动因素上。

实际上,开发价值驱动程序树的三步法已被证明是成功的。该方法提供了业务优先级,显示了公司必须采取的总体方向。这些优先级被分解为描述业务优先级关键组件的战略目标,然后为每个目标确定一个或多个价值驱动因素,从而实现此目标的运营成果。

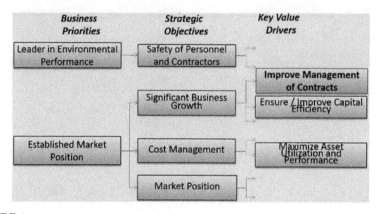

FIGURE 1

Value-driver tree (excerpt).

For a full-value-driver tree and reference content that can be used by organizations, please see the Value Tree.[15]

The business processes of an organization are then evaluated based on their total impact on the specific value drivers. The result are two segments of business processes: high-impact and commodity processes. "High-impact" processes are the ones that are key to make the business strategy of the organization happen: the "competitive" processes and supporting core processes. They are the most important link of business strategy to execution. This approach is visualized in Figure 2.

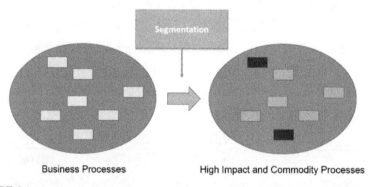

FIGURE 2

High impact and commodity processes.

The value drivers can be weighted by their importance. Minor changes and adjustments in strategy can then be reflected through adjustments of those weights. Larger strategy changes result in different or additional value drivers. This update of value drivers and their weights enables an agile adjustment of process priorities to updated strategies reflecting changing business environments.

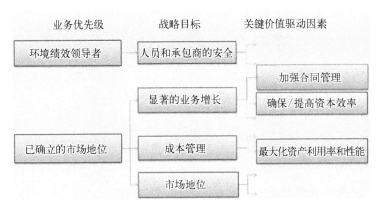

图1   价值驱动树(摘录)

有关组织可以使用的完整价值驱动树和参考内容,请参见"价值树"[15]。

根据组织的业务流程对特定价值驱动程序的总体影响进行评估,评估结果可把业务流程分成两个部分:高影响力流程和商品流程。高影响力流程是实现组织业务战略的关键,它分为竞争流程和支持核心流程。它们是业务战略与执行之间最重要的联系。这种方法的可视化见图2。

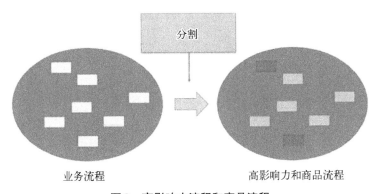

图2   高影响力流程和商品流程

价值驱动因素可以根据其重要性进行加权,然后,通过调整这些权重,可以反映出战略中的微小变化和调整。较大的战略变化会导致不同或额外的价值驱动因素。价值驱动因素及其权重的更新使流程优先级能够灵活调整,以反映不断变化的业务环境的更新策略。

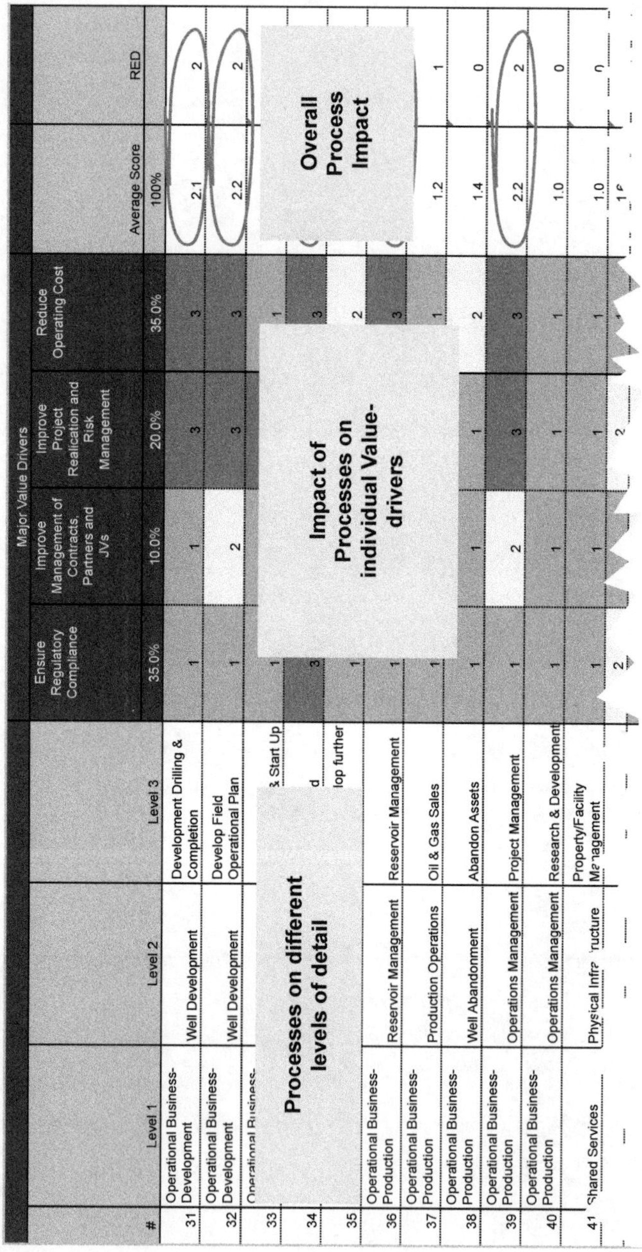

#	Level 1	Level 2	Level 3	Major Value Drivers				Average Score	RED
				Ensure Regulatory Compliance	Improve Management of Contracts, Partners and JVs	Improve Project Realication and Risk Management	Reduce Operating Cost		
				35.0%	10.0%	20.0%	35.0%	100%	
31	Operational Business-Development	Well Development	Development Drilling & Completion	1	1	3	3	2.1	2
32	Operational Business-Development	Well Development	Develop Field Operational Plan	1	2	3	3	2.2	2
33	Operational Business-		& Start Up	1			1		
34			d	3			3		
35			lop further	1			2		
36	Operational Business-Production	Reservoir Management	Reservoir Management	1	1	1	1	1.2	1
37	Operational Business-Production	Production Operations	Oil & Gas Sales	1				1.4	0
38	Operational Business-Production	Well Abandonment	Abandon Assets	1	1	2	2	2.2	2
39	Operational Business-Production	Operations Management	Project Management	1	2	3	3	1.0	0
40	Operational Business-Production	Operations Management	Research & Development	1	1	1	1	1.0	0
41	Shared Services	Physical Infrastructure	Property/Facility Management	1		2		1.?	0
				2	2		2		

Processes on different levels of detail

Impact of Processes on individual Value-drivers

Overall Process Impact

FIGURE 3

Process assessment matrix (excerpt).

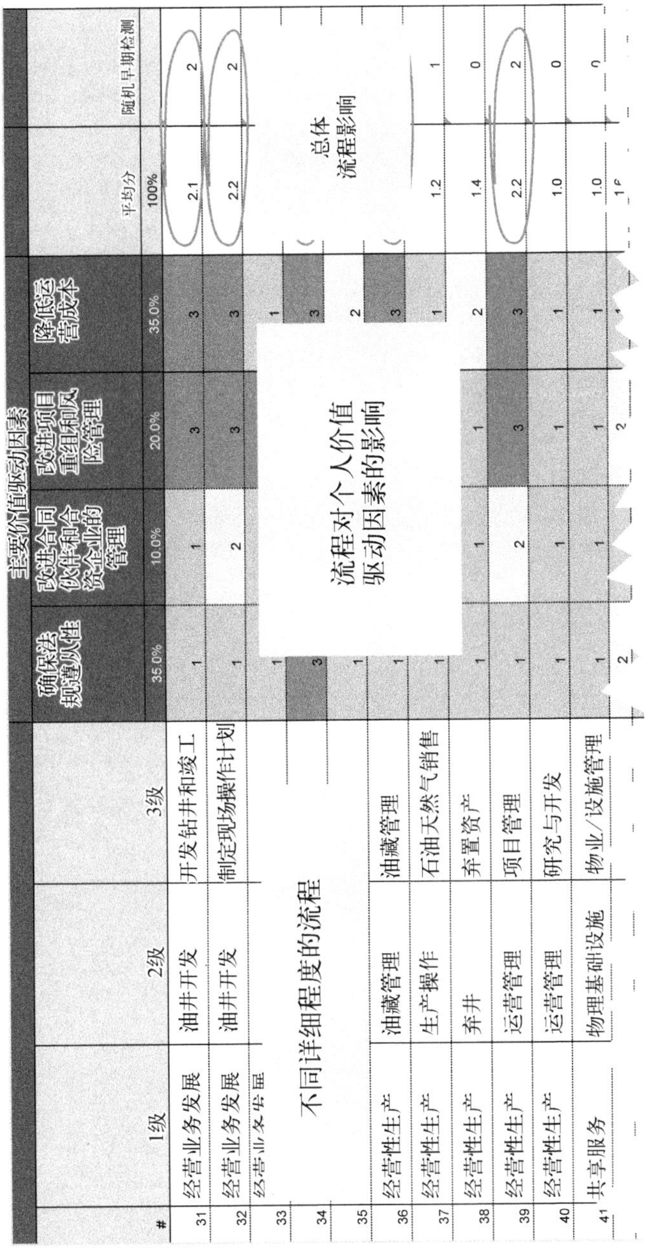

主要价值驱动因素

#	1级	2级	3级	确保长期储量/储量 35.0%	改进合同获得和资金的管理 10.0%	改进项目重组和风险管理 20.0%	降低经营资本 35.0%	平均分 100%	随机早期检测
31	经营业务发展	油井开发	开发钻井和竣工	1	1	3	3	2.1	2
32	经营业务发展	油井开发	制定现场操作计划	1	2	3	3	2.2	2
33	经营业务发展			1			1		
34				3	1		3		
35				1			2		
36	经营性生产	油藏管理	油藏管理				3		
37	经营性生产	生产操作	石油天然气销售	1	1	1	1	1.2	1
38	经营性生产	弃井	弃置资产				2	1.4	0
39	经营性生产	运营管理	项目管理	3	2	3	3	2.2	2
40	经营性生产	运营管理	研究与开发	1			1	1.0	0
41	共享服务	物理基础设施	物业/设施管理	2		2	1	1.8	1

不同详细程度的流程

流程对个人价值驱动因素的影响

总体流程影响

图 3　流程评估矩阵（摘录）

For each process, it has to be defined if it has no (0), low (1), medium (2), or high (3) impact on each of the value drivers. Then the overall impact is calculated in a process assessment matrix by multiplying impact with the weight of the appropriate value driver and calculating the total of all impacts of a process. An example of a process assessment matrix is shown in Figure 3.

The high-impact processes have then to be evaluated based on general industry practices, for example, through benchmarks or purely qualitative evaluations. In those ways, you identify the high-impact "high-opportunity" business processes. Improvements have the biggest value potential, because the processes have a high impact on the strategy, but they currently perform only in, or even below, the industry average of these processes.

Practice experience with different companies has shown that the processes should be identified on a level of detail so that 150–200 process definitions describe the entire organization. This is often referred to as "level 3" (L3). This level is detailed enough to obtain differentiated results, but high level enough to avoid high work efforts. Using the results of the process assessment matrix, the 20% of the processes that are classified as high impact can be identified. The others are considered the commodity processes.

In practice, a "gray" area of processes could be in either group. Hence, approximately 20% of the processes are in the high-impact segment. This issue has to be resolved in a case-by-case basis reflecting the specific situation of an organization, its business strategy, and the overall business environment in which it works (see Figure 4).

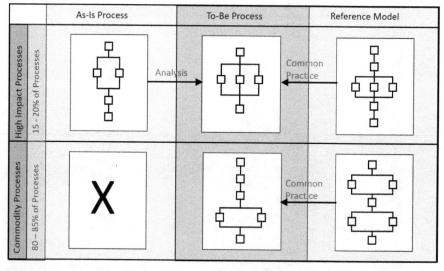

FIGURE 4

Conceptual view of the value-oriented process approach.

对于每个流程,必须定义它是否对每个价值驱动因素都"没有(0)、低(1)、中(2)或高(3)"影响。然后在流程评估矩阵中,通过将影响乘以适当价值驱动因素的权重,并计算一个流程的所有影响的总和来计算总体影响。流程评估矩阵的示例如图3所示。

然后,必须根据一般行业惯例,例如,通过标准或纯粹的定性评估,对高影响过程进行评估。通过这些方式,您可以确定高影响的高机会业务流程。改进具有最大的价值潜力的流程,因为这些流程对策略的执行有很大的影响,但是它们目前有可能仅在这些流程的行业平均值甚至低于行业平均值中运行。

不同公司的实践经验表明,为了能够使用150 ～ 200个流程定义描述整个组织,流程应该根据详细级别进行标识,这通常被称为三级分类(L3)。这一层次的详细程度足以获得差异化的结果,但高层次的详细程度足以避免高强度的工作。利用流程评估矩阵的结果,可以识别出20%的高影响的流程。其他流程被认为是商品流程。

在实践中,流程的"灰色"区域可以在任何一组中。因此,大约20%的流程属于高影响力细分部分。此问题必须根据具体情况逐一解决,以反映组织的具体情况、业务战略以及组织的整体业务环境(图4)。

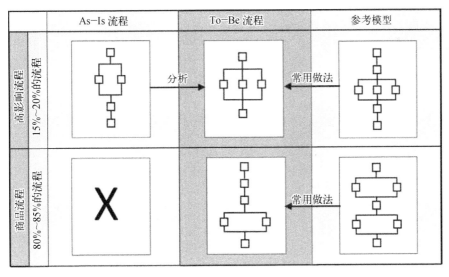

图4 以价值为导向的流程方法的概念视图

Value-oriented Design

The high-impact processes (or at least high–impact, high-opportunity processes, if further prioritization is necessary, say, due to budgets) are subject to detailed process innovation and optimization activities focusing on the previously identified value drivers. The degree of achievement is measured through KPIs that relate to the identified value drivers. The check of the quality of a process design through KPIs can be used in agile or top-down waterfall design approaches. Depending on the specific process and the culture of the organization, both approaches or a combination of both can be relevant.[16] The design approach uses formal modeling methods like Event-driven Process Chains (EPC), the Business Process Modeling Notation (BPMN) or VDML—Value Delivery Modeling Language to model the value and the activities and thereby facilitate the integration of process design and implementation.

Product and market-oriented design approaches have been proven effective, because they link processes with their value drivers to the offerings a client is looking for. The product and market-oriented design supports an integrated product (offering) and process innovation. Such an approach is especially important for the processes that are highly relevant for the strategic positioning of an organization, hence the top 5%. To identify these business processes, another segmentation of the high-impact processes is required distinguishing between strategic and nonstrategic high-impact processes. The focus is on high-impact strategic processes. These are perfect targets for innovation initiatives. As an example, a compressor company may deliver "compressed air as a service" instead of just selling compressors. Offering as well as related sales, delivery, and maintenance processes change simultaneously, reflected in the integrated value-oriented design.

Also new technologies, especially information technologies (IT) relevant for specific processes, have to be evaluated in a business-driven way. You can, for example, model different process scenarios representing various degrees of automation. The best scenario is chosen based on the expected value of the relevant KPIs.

For all high-impact processes, techniques like process model-based simulations and animations are helpful to come up with best-suited design solutions based on KPIs. Often even the transparency created through those information models is sufficient by itself to discover relevant improvement or even innovation opportunities.

Traditional improvement methods like Lean or Six Sigma[17] can be applied in selected cases. However, these approaches do not generally support focused innovation or a full-blown optimization of processes, including automation opportunities. Hence, they are more targeted to bring less-strategic, people-intensive processes to better efficiency, in most cases resulting in cost or time reductions.

The starting points for the design of the 80% commodity processes are industry or functional reference models. These models are available, for example, through industry organizations or consulting and software companies. In many cases, they are already developed using standard modeling methods. The industry common practices reflected in those models are only adjusted to the specific organization when this is absolutely necessary, for example due to legal requirements in country subsidiaries or specific logistics requirements through the product.

3. 价值导向设计

高影响流程（或者至少是高影响、高机会流程，如果需要进一步的选择优先顺序，如预算流程）将受到详细的流程创新和优化活动的制约，这些活动的重点是预先确定的价值驱动因素。通过与确定的价值驱动因素相关的KPI衡量实现程度。通过KPI检查流程设计的质量，以及可以用于敏捷或自上而下的"瀑布式"设计方法。根据具体的流程和组织文化，这两种方法或两者的组合都是相关的[16]。设计方法使用EPC、业务流程建模标记法（BPMN）或价值交付建模语言（VDML）等正式的建模方法对价值和活动进行建模，从而促进流程设计和实现的集成。

产品和面向市场的设计方法已被证明是有效的，因为它们将流程与其价值驱动因素联系起来，以满足客户所寻求的产品。产品和面向市场的设计支持产品集成和流程创新。这种方法对于与组织的战略定位高度相关的流程尤其重要，因此排在前5%。为了识别这些业务流程，需要对高影响流程进行另一个细分以区分战略性和非战略性高影响流程，重点是高影响力的战略流程。这些都是创新举措的完美目标。例如，压缩机公司可以提供"压缩空气作为服务"而不仅仅是销售压缩机，如果这样的话，提供以及相关的销售、交付和维护流程都会同时发生变化，因此要在集成的价值导向设计中进行体现。

此外，新技术，特别是与特定流程相关的IT，必须以业务驱动的方式进行评估。例如，您可以对表示不同自动化程度的不同流程场景进行建模。根据相关KPI的预期值选择最佳方案。

对于所有具有高影响力的流程，即使通过这些信息模型创建的透明度本身也足以发现相关的改进甚至创新机会，但是基于流程模型的模拟和动画等技术有助于提出基于KPI的最佳设计解决方案。

传统的改进方法（如精益或六西格玛[17]）可以应用在选定的情况下。然而，这些方法通常不支持包括自动化机会在内的重点创新或流程的全面优化。因此，它们更有针对性地提高战略意义较低、人员密集的流程的效率，这在大多数情况下会导致成本或时间的减少。

80%商品流程设计都可以参考行业或功能模型，这些模型可以通过行业组织、咨询公司或软件公司获得。在许多情况下，它们已经使用标准建模方法开发出来了。这些模型中反映的行业通用做法仅在绝对必要时针对特定组织进行调整，例如，要遵循国家法律要求或特定物流要求。

The process design work focuses on "making the industry standard happen." If process areas are identified when the industry standard cannot be applied, for example, due to product specifics, only those areas will be designed in a company-specific way, keeping the adjustments as close to the industry standard as possible. Process solutions can here often be found through a simple application of the mentioned traditional improvement methods like Lean and Six Sigma, because a pure efficiency focus is in most cases justified here. However, it is important to keep in mind that it is, in general, not worth improving the above industry average performance.

This value-oriented process design approach is visualized in Figure 5. It shows that, also for the design of high-impact processes, reference models can be used as an input. Nevertheless, this is only one component of getting all information together to come up with real innovative and optimized solutions regarding the KPIs and the value drivers to which they relate.

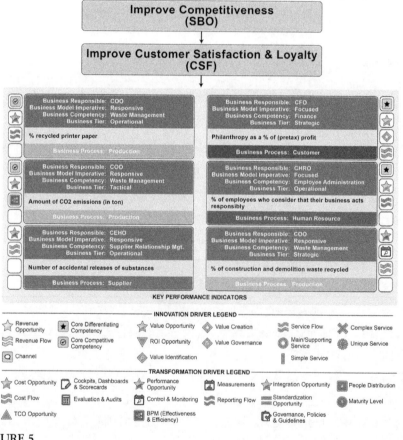

FIGURE 5

Value oriented process modeling.[18]

流程设计工作的重点是"实现行业标准"。如果由于产品的特殊性等无法应用行业标准确定流程领域,需要仅以公司特定的方式设计这些领域,使调整尽可能接近行业标准。在这里,通常可以通过简单应用上述传统的改进方法(如精益和六西格玛)找到流程解决方案,因为在大多数情况下,纯效率焦点在这里是合理的。然而,重要的是要记住,总体而言不值得将其水平提高到高于行业平均水平。

这种面向价值的流程设计方法如图5所示。这表明,对于高影响流程的设计,参考模型也可以作为输入。尽管如此,这仅仅是将所有信息汇集在一起,是KPI及其相关的价值驱动因素提出真正的创新和优化解决方案的一个组成部分。

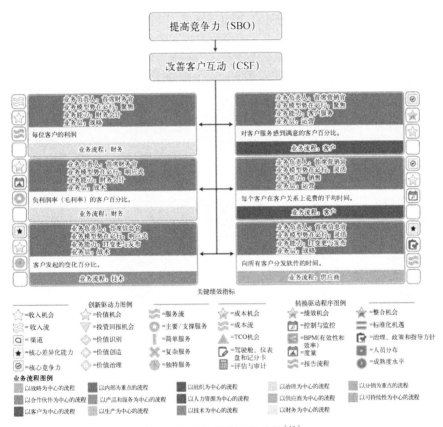

图5 面向价值的流程建模[18]

In both cases, process models are developed until the level of detail still provides relevant business information through the design. The decomposition of the function "Enter Customer Order" into "Enter First Name", "Enter Last Name", and so on, would from a business point of view not add any additional relevant content (but may be necessary later for the development of software). When reference models are used, this can mean that in areas in which the design deviates from the initial industry model a higher level of modeling detail is required than in other "standard" areas.

Both, high-impact and commodity processes are part of overlying end-to-end business processes. Process interfaces in the underlying detailed processes reflect this overall context and make sure that the various process components or subprocesses fit together. Hence, during the process improvement work cause-and-effect considerations have to take place to avoid fixing issues in one area while creating new ones in other processes.

Value-oriented Process Modeling

The value-oriented process modeling concepts require more consideration to the design and modeling aspects than traditional process design and modeling. For the most part, because Value-oriented Process Modeling needs a formalized breakdown of strategic business objectives (SBOs) into critical success factors (CSFs), with their associated KPIs and process performance indicators (PPIs), only then, the right measurements can be put in place in a manner that ensures that they are integrated and strategically aligned. They then can be linked to the proper responsible decision-making bodies, in a way that they allow performance improvement to occur. This brings support to this complex task by providing the discussed value tree, as shown in Figure 6, a taxonomy of the previously mentioned value indicators and

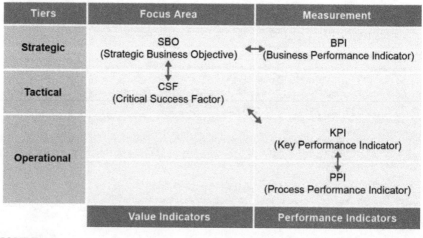

FIGURE 6

Value indicators and performance indicators.[19]

　　在这两种情况下,流程模型都要开发到通过设计提供相关业务信息的详细级别,例如,将"输入客户订单"功能分解为"输入姓名""输入姓氏"等功能。从业务的角度来看,并不会增加任何额外的相关内容(但在以后的软件开发中可能是必要的)。当使用参考模型时,这可能意味着在设计偏离初始行业模型的领域中,比在其他标准领域中需要更高层次的建模细节。

　　高影响流程和商品流程都是覆盖端到端业务流程的一部分。底层详细流程中的流程接口反映了这一总体前后关系,并确保各种流程组件或子流程适合匹配在一起。因此,在流程改进工作中必须考虑因果关系,以避免在一个领域中修复问题的同时,无意中在其他流程中产生新的问题。

4. 价值导向的流程建模

　　与传统的流程设计和建模相比,面向价值的流程建模需要更多考虑的方面是设计和建模。在大多数情况下,由于价值导向的流程建模需要将SBO正式细分为CSF及其相关的KPI和PPI,只有这样,正确的衡量标准才能确保它们与战略方向的一致。然后,他们可以与适当的负责任的决策机构联系在一起,通过这种方式,他们可以实现绩效改进,这就为这项复杂的任务提供了支持,如图6所示,其方法是提供我们讨论过的价值树(对前面提到的价值指标和绩效指标进行分类,以及

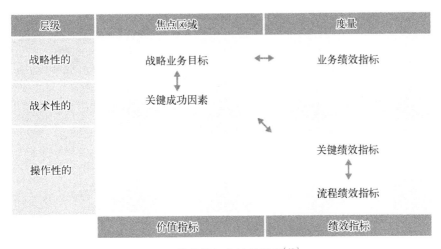

图6　价值指标和绩效指标[19]

performance indicators and how they relate to each other.[20] Enabling organizations to categorize and classify their value indicators and performance indicators according to the enterprise tiers, focus areas, and existing measures.

Many organizations realize that traditional process design does not consider the value-oriented aspects of one's organization. Executives who ask themselves what it takes to move from traditional process design to value-oriented process design have to consider the strategic role that value-oriented aspects play in their organization, but also how and when to apply the concepts. The ability to succeed with one's value-oriented initiatives is directly related to the ability to connect the defined value drivers (SBO's and CSF's) and the performance drivers (KPI's and PPI's), as well as how the organization applies them to their competencies, processes, and services.

As illustrated in Figure 5, the core aspects of Value-oriented process modeling are therefore about linking the various aspects together; this includes:

1. Value drivers (SBO's and CSF's)
2. Performance drivers (KPI's and PPI's)
3. Organizational components (relevant Business Competencies)
4. The responsible person
5. The relevant business tier, that is, strategic, tactical, or operational
6. The appropriate and related process that links to all above points
7. Specification of the innovation and transformation aspects

Once the process has been sorted according to the value-oriented aspects, the organization now fully understands the value of their process investments, the relationship to their organizational components (relevant business competencies), the responsible persons or owners involved, and thereby also a link to evidence-based decision making.

Value-oriented process modeling, in addition, enables a whole new way of interlinking to the enterprise innovation and transformation aspects. This thereby enables not only Value-oriented process analysis, design, building, and implementation, it ensures that the business innovation and transformation happens alongside the progression. The link to innovation and transformation, however, prerequisites that all the processes involved need to be mapped to the value and performance indicators. The reason this is so vital is that, as illustrated in Figure 7, different strategies will have different critical success factors, all supporting the same strategy. To ensure consistency of value-oriented process modeling and to make sure that the strategies are executed, all relevant processes must be included. If not, it will be a siloed view of strategy execution. That is good enough for value-oriented process *design*, but not good enough for full value-oriented process *modeling*, which must include aspects of innovation and transformation.

Value-oriented Implementation

The organization-specific process models for high-impact business processes are, in general, implemented using people and highly flexible next-generation process

表示它们如何相互关联[20]），使组织能够根据企业层次、重点领域和现有措施，对其价值指标和绩效指标进行分类。

许多组织意识到传统的流程设计没有考虑组织面向价值的方面。那些问自己从传统的流程设计到面向价值的流程设计需要什么东西的管理人员必须考虑面向价值的方面在他们的组织中扮演的战略角色，以及如何、何时应用这些概念。成功实施价值导向计划的能力直接与连接已定义的价值驱动程序（SBO和CSF）和性能驱动程序（KPI和PPI）的能力，以及组织如何将它们应用于其能力、流程和服务相关。

如图5所示，面向价值的流程建模的核心方面是将各个方面联系在一起，这包括：

（1）价值驱动因素（SBO和CSF）；

（2）绩效驱动因素（KPI和PPI）；

（3）组织组成部分（相关业务能力）；

（4）负责人；

（5）相关业务层，即战略、战术或运营层面；

（6）与上述所有要点相关的适当和相关的过程；

（7）创新和转型方面的规范。

一旦按照价值导向方面对流程进行了分类，组织现在就完全了解了：其投资流程的价值、与组织组成部分（相关业务能力）的关系、涉及的负责人或所有者，从而也与基于证据的决策建立了联系。

此外，以价值为导向的流程建模实现了与企业创新和转型方面相互联系的全新方式。因此，这不仅可以实现以价值为导向的流程分析、设计、构建和实施，还可以确保业务创新和转型与发展同步发生。然而，创新和转型的关联是所有相关过程需要映射到价值和绩效指标的先决条件。如此重要的原因是（如图7所示）：不同的策略将有不同的关键成功因素，所有这些都支持相同的策略。为确保以价值为导向的流程建模的一致性并确保策略得以执行，所有相关流程都必须包含在内。如果没有，它将是战略执行的孤岛，这对于以价值为导向的流程设计来说已经足够了，但对于完全以价值为导向的流程建模来说还不够，后者必须包括创新和转型的各个方面。

5. 价值导向的实施

通常，我们可以使用人员和高灵活度的下一代流程自动化引擎来实现组织中高影响力的个性化流程模型。在大多数情况下，实现这一点需要开发特定的应用

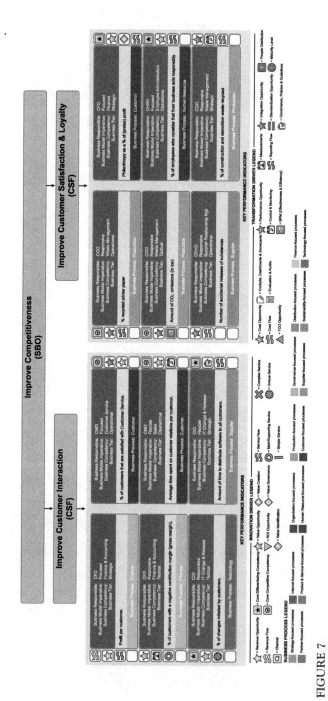

FIGURE 7

Value-oriented process modeling and the link to common strategy but different value and performance indicators.[21]

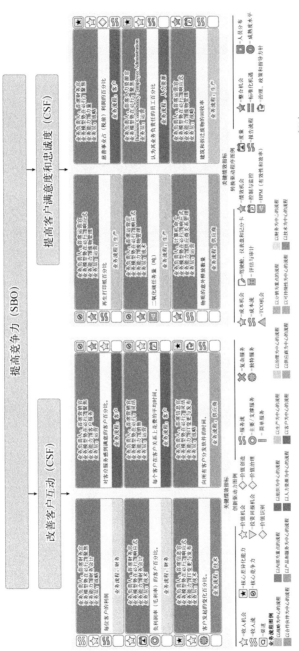

图 7 以价值为导向的流程建模和与共同战略的联系，但不同的价值和绩效指标[21]

automation engines. The implementation requires in most cases the development of specific application software components. The process models reflecting the optimized KPIs regarding the relevant value drivers are the entrance points for the more detailed modeling of the underlying software. They enable a consistently value-oriented process implementation and automation. At this point, the modeling method can change, for example, to the Unified Modeling Language (UML), reflecting the desired software structure to support the high-impact processes. In addition, the work-flow engine of next-generation process automation engines can be configured based on those models, depending on the underlying modeling repository and execution technology even automatically or semi-automatically. The integration between process modeling and execution tools can be extremely beneficial in this situation, especially because it enables the flexible value-oriented adjustment of processes.

The overall architecture of such next-generation process automation environments is often referred to as Service-oriented Architecture (SOA). In such architecture, the "execution software" and the "process logic" (work flow) are separated.[22,23] Hence, the developed process models can on one hand be used to configure the work flow and on the other hand to develop the software services that are not available in existing libraries. Existing software services may include detailed process reference models that can be re-used in the process design. This architecture of next-generation process automation environments is visualized in Figure 8.

The key advantage of such architecture is the high degree of flexibility in adjusting process flows and functionality. This can be crucial for a company looking for agility and adaptability. The main disadvantage is the effort for providing the appropriate governance while running such an environment, as well as information modeling efforts in the building phase.

The process models of the commodity processes are used to select or at least evaluate preselected "traditional" software packages like Enterprise Resource Planning (ERP) systems, Supply Chain Management (SCM), or Customer Relationship Management (CRM) systems. These can become part of the overall next-generation architecture, representing one software component. Then those models from the process design are used to drive a process-oriented implementation of the software packages across the various organizational units involved in the business processes in scope.[24,25] Ideally, one uses already industry-specific software-reference models during the process design. This means one procures the reference models to be used from the software vendor. If this is possible, one benefits from the "business content" of the software and minimizes design and modeling efforts. Using other industry reference models (different from the software-based model) may lead to design adjustments and with that to rework, once the software is selected.

Figure 9 shows the architecture of such traditional software. Here, process definition and software functionality are linked in a static way. This means the software more or less dictates how a process has to be executed (allowing only predefined variants through the software configuration). This is fine for commodity processes, but often causes issues in strategic high-impact processes that need to be company specific. Consequently, we have used another implementation approach for those strategic processes. However, in some cases it is also possible to develop add-on

程序软件组件。反映有关相关价值驱动因素的优化 KPI 的流程模型是基础软件更详细建模的入口点,它们实现了价值导向的流程实施和自动化。此时,建模方法可以更改为 UML,以反映支持高影响流程所需的软件结构。此外,下一代流程自动化引擎的工作流引擎可以基于这些模型进行配置,具体取决于底层建模存储库和执行技术,甚至可以是自动或半自动的。在这种情况下,流程建模和执行工具之间的集成可能非常有益,特别是因为它能够灵活地对流程进行面向价值的调整。

　　这种下一代流程自动化环境的总体体系结构通常称为 SOA。在这种体系结构中,"软件执行"和"流程逻辑"(工作流)是分开的[22,23],因此,开发的流程模型一方面可以用来配置工作流,另一方面可以用来开发现有库中不可用的软件服务。现有的软件服务可能包括详细的流程参考模型,这些模型可以在流程设计中重用。下一代流程自动化环境的体系结构如图 8 所示。

　　这种体系结构的主要优点是在调整流程流和功能方面具有高度的灵活性。这对于寻找敏捷性和适应性的公司来说是至关重要的。主要的缺点是在运行这样的环境时需要提供适当的治理,以及在构建阶段需要进行信息建模。

　　商品流程的流程模型用于选择或评估"传统"软件系统,如 ERP 系统,供应链管理(SCM)或 CRM 系统。这些可以成为整个下一代架构的一部分,代表一个软件组件。然后,流程设计中的这些模型被用来驱动面向流程的软件包实现,这些软件包跨越业务流程所涉及的各个组织单元的范围[24,25]。理想情况下,在流程设计过程中使用行业特定的软件参考模型,这意味着您可以从软件供应商处获得要使用的参考模型。如果可行,可以从软件的"业务内容"中获益,并最大限度地减少设计和建模工作。使用其他行业参考模型(基于软件模型的区别)可能会导致设计调整,并在选择软件后进行返工。

　　图 9 显示了这种传统软件的体系结构。在这里,流程定义和软件功能以静态方式链接。这意味着软件或多或少地规定了流程的执行方式(只允许通过软件配置进行预先定义的变体)。这适用于商品流程,但通常会导致公司内的战略性高影响流程出现问题。因此,我们对这些战略流程使用了另一种实施方法。但是在某些情况下,还可以通过开发附加软件以支持高影响力流程,并将它们集成到更大的

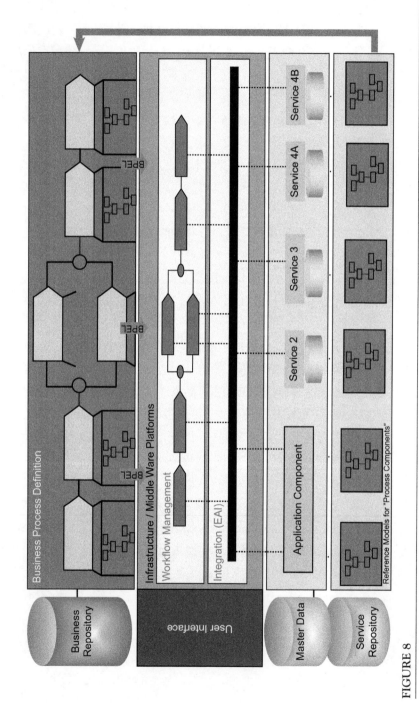

FIGURE 8

Next generation process automation.

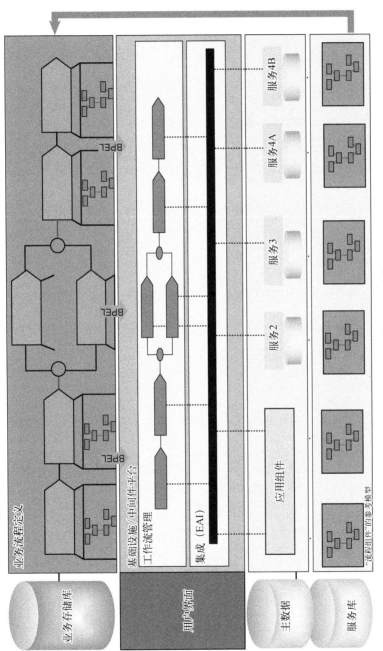

图8　下一代流程自动化

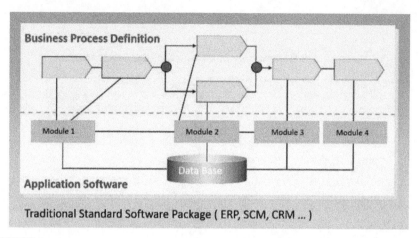

FIGURE 9

Traditional software architecture.

software to support high-impact processes and integrate them into the larger software package, for example, the ERP system.

Advantages and disadvantages are just the opposite as explained for next-generation process automation approaches. Hence, in practice, a combination of both implementation technologies and approaches is in most cases the solution that delivers best value.

The process interfaces in the different process models guide the software integration. This can be supported from a technology point of view through appropriate enterprise application integration environments—in general, included in SOA environments. Such software or middle-ware tools reduce the efforts for interface development to a necessary minimum. Their efficient use is again driven through the appropriate process models, specifically, the integration of the various process components.

The implementation of processes includes as a main component the preparation of the people involved for the new work environment. They have to learn new manual processes and how to use the automation technologies in the specific process context. The necessary change management is carried out using the same process design as a basis that was used to drive the development and configuration of the IT components. Information, communication, and training are supported through the information models of the process design.[26,27] The integrated implementation of people- and IT-based processes leads to a "digital organization" that really delivers additional business value.

The implementation of the business processes can again be based on an agile approach, developing several "intermediate" prototypes or a top-down waterfall approach. In most cases a combination of both is best suited, because this avoids a possibly "endless" number of development cycles created by agile development

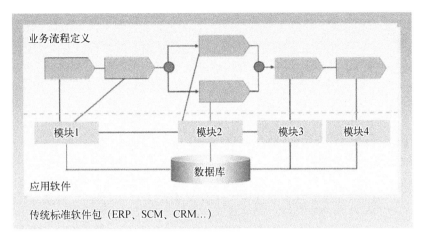

图9　传统软件架构

软件包中,如ERP系统。

正如下一代流程自动化方法所解释的那样:优点和缺点正好相反。因此,在实践中,实现技术和方法的结合在大多数情况下都是提供最佳价值的解决方案。

不同流程模型中的流程接口可以指导软件集成,这可以从技术的角度通过SOA环境中包含的企业应用程序集成环境得到支持。这种软件或中间软件工具将接口开发的工作减少到了最低限度。它们的有效使用再次通过适当的流程模型(特别是各种过程组件的集成)来驱动。

流程实施内容的一个主要组成部分是:为新的工作环境准备相关人员。他们必须学习新的手动流程,以及如何在特定的流程环境中使用自动化技术。必要的变更管理是以相同的流程设计为基础,用于驱动IT组件的开发和配置。通过流程设计的信息模型支持信息、通信和培训[26,27]。基于人员和IT的流程的集成实现了一个真正提供额外业务价值的"数字化组织"。

业务流程的实现也可以基于敏捷方法,开发一些"中间"原型或自顶向下的"瀑布"方法。在大多数情况下,两者的结合是最合适的,因为这样可以避免敏捷开发所创建的"无限期"的开发周期,或者避免在自上向下的"瀑布式"开发过程

or developments getting stuck on their way top–down of waterfall development models.[28]

The results are end-to-end business processes based on a value-oriented process design and an appropriate integrated automation. The approaches provide the necessary flexibility by which they deliver real business value and the required efficiency when possible.

Value-oriented Process Governance to Sustain Value

Once business processes have been designed and implemented targeting business value, the results need to be sustained and governed. To control one's business processes, especially the high-impact processes, if the KPIs remain in an acceptable range, adjust design or implementation as necessary. In addition, changes in business strategy need to be reflected. Therefore, the value-oriented design and implementation approach needs to be part of the larger BPM concept and discipline, the management discipline focused on moving strategy to execution, quickly, and at low risk. This BPM discipline is established through an appropriate "process of process management" in the BPM Center of Excellence (see the BPM CoE chapter) that manages the work, monitoring, and continuous improvement through the life cycle (see the BPM life cycle chapter).

Providing appropriate process governance is especially important to make BPM Governance a reality and keep processes focused on creating value (see the BPM governance chapter). This means that the process ownership, accountability, and responsibility, as well as a mechanism to take decisions and execute resulting actions across organizational boundaries, are defined.[29,30] In many successful organizations, the "process of process management" is owned and focused on value by a chief processes officer (CPO) and operationally managed by a BPM Center of Excellence (CoE) with various operational BPM roles (see the BPM roles chapter). Business processes require roles like process owners and supporting operative roles to be kept on target over time. These roles can be decentralized in business units or centralized, project-based, or permanent, in-house or out-sourced.

The approach of value-oriented process modeling allows an organization to move its strategy systematically into execution. It aligns the modeling and implementation efforts with the strategic direction of the organization after which the value-oriented process governance starts.

First experiences with real live companies showed that this approach helps on one hand to dramatically reduce process design and implementation times due to the efficient handling of commodity processes. Companies estimated more than 50% savings in time and effort. On the other hand, it enables real strategic advantage though the innovation and optimization of high-impact process areas based on the KPIs and the related strategic value drivers. To sustain the value creation and realization in the process execution, the following aspects illustrated in Figure 10 need to governed.

中停滞不前[28]。

结果是基于面向价值的流程设计和适当的集成自动化的端到端业务流程。这些方法提供了必要的灵活性,使它们能够提供真正的业务价值,并在可能的情况下提供所需的效率。

6. 以价值为导向的流程治理以维持价值

一旦针对业务价值设计和实施业务流程,就需要进行维护和管理。如果KPI保持在可接受的范围内,则要控制业务流程,特别是影响较大的流程,要根据需要调整设计或实现,此外,还需要反映业务策略的变化。因此,管理规程关注于快速、低风险地将策略转移到执行,面向价值的设计和实现方法需要成为更大的BPM概念和规程的一部分。这个BPM规程是通过BPM CoE中适当的"流程管理流程"(参见BPM CoE章节)建立的,该流程管理通过生命周期管理、监视和持续改进(参见BPM生命周期章节)。

提供适当的流程治理对于实现BPM治理并使流程专注于创建价值尤为重要(请参阅BPM治理一节)。这意味着定义了流程所有权、标准和责任,以及跨组织边界进行决策和执行结果操作的机制[29,30]。在许多成功的组织中,"流程管理的流程"由CPO拥有并关注,并由BPM CoE通过各种可操作的BPM角色进行操作管理(请参阅BPM角色一节)。业务流程要求流程所有人和支持操作角色等角色随着时间的推移保持在目标上。这些角色可以分散在业务单位,也可以集中、基于项目,也可以是永久性的、内部的或外包的。

面向价值的流程建模方法允许组织系统地将其策略转移到执行中,它使建模和实现工作与组织的战略方向保持一致,然后开始面向价值的流程治理。

公司的经验表明,基于对商品流程的有效处理,面向价值的流程建模方法一方面有助于显著减少流程设计和实现时间(估计节省了50%以上的时间和精力);另一方面,通过基于KPI和相关战略价值驱动因素的高影响过程域的创新和优化,实现真正的战略优势。为了维持流程执行中的价值创造和实现,图10中所示的以下方面需要进行管理。

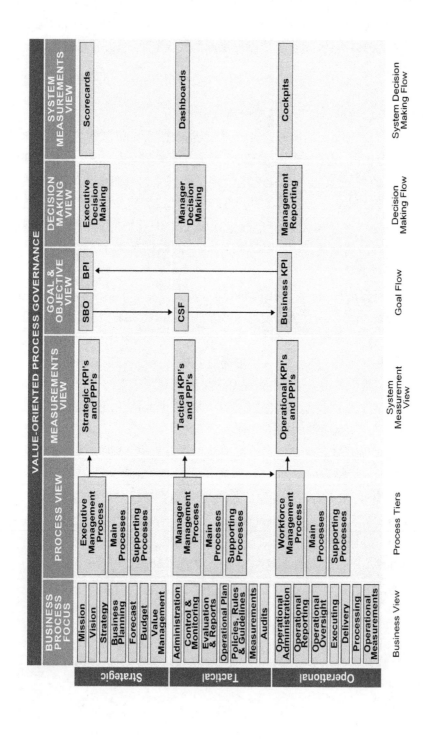

FIGURE 10

Example of a value oriented process governance.[31]

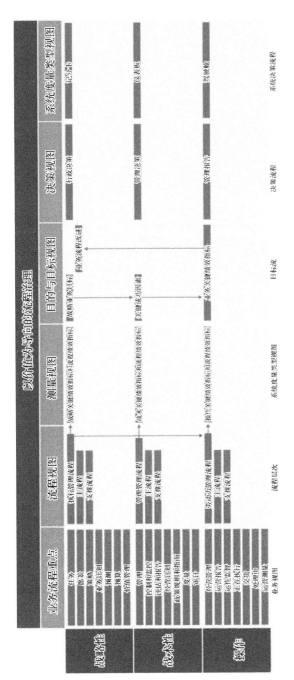

图 10 面向价值的流程治理示例[31]

CONCLUSION

In this chapter, we have elaborated on what value-oriented process modeling is and how it differentiates to the traditional process analysis, process design, process implementation, and process governance considerations. We furthermore illustrated practical examples on who applies value-oriented process modeling and how it enables the link to innovation and transformation. It enables organizations with the ability to address their processes as core enterprise assets in a whole new way.

End Notes

1. Elzina G., Lee, *Business Engineering* (Norwell: Springer, 1999).
2. http://www.globaluniversityalliance.net/.
3. Mitchel R., van Ark, *The Conference Board – CEO Challenge 2014: People and Performance, Reconnecting with the Customer and Reshaping the Culture of Work. The Conference Board Whitepaer, New York, e.a. 2014* (2014).
4. Stary, *S-BPM One – Scientific Research. 4th International Conference, S-BPM ONE 2012, Vienna, Austria, April 2012, Proceedings* (2012).
5. Rummler R., Rummler, *White Space Revisited – Creating Value through Processes* (San Francisco: Wiley, 2010).
6. Burlton, "Delivering Business Strategy through Process Management," in *Handbook on Business Process Management 2 – Strategic Alignment, Governance, People and Culture* ed. J. Vom Brocke, M. Roemann (Berlin, New York, e.a.: Springer, 2013).
7. Hendrickx H. H. M., Daley S. K., Mahakena M., and von Rosing M, *Defining the Business Architecture Profession* (IEEE Commerce and Enterprise Computing, September 2011), 325–332.
8. Scheer, ARIS – Business Process Frameworks, 2nd ed. (Berlin, e.g.: Springer, 1998).
9. LEADing, *The Leading Practice Value Reference Content #Lead-ES20007BCPG – Avalue Ontology and Value Semantic Description – Views, Stakeholders and Concerns. Version: Lead 3.0* (Leading Practice publication, 2014).
10. Scheer, "Tipps fuer den CIO: Vom Tekki zum Treiber neuer Business Modelle," in *IM+IO – Das Magazin fuer Innovation, Organisation und Management* (Sonderausgabe, December 2013).
11. Franz K., *Value oriented business Process Management – The Value-Switch for Lasting Competitive Advantage* (New York, e.a.: McGraw-Hill, 2012).
12. http://www.leadingpractice.com/enterprise-standards/enterprise-modelling/value-model/.
13. Kirchmer F., *The BPM-Discipline – Enabling the Next Generation Enterprise* (London, Philadelphia: BPM-D Executive Training Documentation, 2014).
14. Kirchmer, "Value oriented Design and Implementation of Business Processes – From Strategy to Execution at Pace with Certainty," *Accepted for publication in: BMSD'14 Proceedings* (Luxembourg, June 2014).
15. http://www.leadingpractice.com/wp-content/uploads/2013/05/LEAD-Value-Reference-Framework-Enterprise-Value-Tree.pdf.
16. Morris, *Architect, Design, Deploy, Improve (ADDI) – A BPMS Development Methodology* (Chicago: Wendan Whitepaper, 2014).
17. George, *The Lean Six Sigma Guide to Doing More with Less – Cut Costs, Reduce Waste, and Lower your Overhead.* (New York, e.a.: McGraw-Hill, 2010).

4.4.3　结论

在本节中，我们阐述了什么是面向价值的流程建模，以及它与传统的流程分析、流程设计、流程实现和流程治理考虑事项的区别。此外，我们进一步说明了谁应用价值导向过程建模的实际例子，以及它如何实现与创新和转型的联系。它使组织能够以全新的方式将其流程作为企业核心资产来解决。

--- 参考文献 ---

［ 1 ］ Elzina G., Lee, Business Engineering (Norwell: Springer, 1999).
［ 2 ］ http://www.globaluniversityalliance.net/.
［ 3 ］ Mitchel R., van Ark, The Conference Board — CEO Challenge 2014: People and Performance, Reconnecting with the Customer and Reshaping the Culture of Work. The Conference Board Whitepaer, New York, e.a. 2014 (2014).
［ 4 ］ Stary, S-BPM One — Scientific Research. 4th International Conference, S-BPM ONE 2012, Vienna, Austria, April 2012, Proceedings (2012).
［ 5 ］ Rummler R., Rummler, White Space Revisited — Creating Value through Processes (San Francisco: Wiley, 2010).
［ 6 ］ Burlton, "Delivering Business Strategy through Process Management," in Handbook on Business Process Management 2 — Strategic Alignment, Governance, People and Culture ed. J. Vom Brocke, M. Roemann (Berlin, New York, e.a.: Springer, 2013).
［ 7 ］ Hendrickx H. H. M., Daley S. K., Mahakena M., and von Rosing M, Defining the Business Architecture Profession (IEEE Commerce and Enterprise Computing, September 2011), 325−332.
［ 8 ］ Scheer, ARIS — Business Process Frameworks, 2nd ed. (Berlin, e.g.: Springer, 1998).
［ 9 ］ LEADing, The Leading Practice Value Reference Content #Lead-ES20007BCPG — A Value Ontology and Value Semantic Description — Views, Stakeholders and Concerns. Version: Lead 3.0 (Leading Practice publication, 2014).
［10］ Scheer, "Tipps fuer den CIO: Vom Tekki zum Treiber neuer Business Modelle," in IM+IO — Das Magazin fuer Innovation, Organisation und Management (Sonderausgabe, December 2013).
［11］ Franz K., Value oriented business Process Management — The Value-Switch for Lasting Competitive Advantage (New York, e.a.: McGraw-Hill, 2012).
［12］ http://www.leadingpractice.com/enterprise-standards/enterprise-modelling/value-model/.
［13］ Kirchmer F., The BPM-Discipline — Enabling the Next Generation Enterprise (London, Philadelphia: BPM-D Executive Training Documentation, 2014).
［14］ Kirchmer, "Value oriented Design and Implementation of Business Processes — From Strategy to Execution at Pace with Certainty," Accepted for publication in: BMSD'14 Proceedings (Luxembourg, June 2014).
［15］ http://www.leadingpractice.com/wp-content/uploads/2013/05/LEAD-Value-Reference-Framework-Enterprise-Value-Tree.pdf.
［16］ Morris, Architect, Design, Deploy, Improve (ADDI) — A BPMS Development Methodology (Chicago: Wendan Whitepaper, 2014).
［17］ George, The Lean Six Sigma Guide to Doing More with Less — Cut Costs, Reduce Waste, and Lower your Overhead. (New York, e.a.: McGraw-Hill, 2010).

18. Value & Performance Management Model, LEADing Practice Value Model Reference Content #LEAD-ES20007BCPG.

19. Ibid.

20. Taken from the Value Model Reference Content (Enterprise Standard ID# LEAD-ES20007BCPG), http://www.leadingpractice.com/enterprise-standards/enterprise-modelling/value-model/.

21. See note 18 above.

22. Kirchmer, *High Performance through Process Excellence – From Strategy to Execution with Business Process Management*, 2nd ed. (Berlin, e.a.: Springer, 2011).

23. Slama N., Enterprise BPM – Erfolgsrezepte fuer unternehmensweites Prozessmanagement. (Heidelberg: dpunkt.verlag, 2011).

24. Kirchmer, 1999a. *Business Process Oriented Implementation of Standard Software – How to achieve Competitive Advantage Efficiently and Effectively*, 2nd ed. (Berline, e.a.: Springer).

25. Kirchmer, "Market- and Product-oriented Definition of Business Processes," in *Business Engineering* ed. D. J. Elzina, T. R. Gulledge, C-Y. Lee (Norwell: Springer, 1999b).

26. Kirchmer, *High Performance through Process Excellence – From Strategy to Execution with Business Process Management*, 2nd ed. (Berlin, e.a.: Springer 2011).

27. Franz K., *Value oriented business Process Management – The Value-Switch for Lasting Competitive Advantage* (New York, e.a.: McGraw-Hill, 2012).

28. Morris, *Architect, Design, Deploy, Improve (ADDI) – A BPMS Development Methodology* (Chicago: Wendan Whitepaper, 2014).

29. Kirchmer H., "Value oriented process Governance – Wettbewerbsvorteile durch die richtige Processorganisation," in *IM+io Fachzeitschrift fuer Innovation, Organisation und Management* (Germany, March, 2013).

30. Kirchmer, "How to create successful IT Projects with Value oriented BPM," in *CIO Magazine Online*, (February 27, 2013).

31. See note 18 above.

［18］Value & Performance Management Model, LEADing Practice Value Model Reference Content #LEAD-ES20007BCPG.

［19］Ibid.

［20］Taken from the Value Model Reference Content (Enterprise Standard ID# LEAD-ES20007BCPG),http://www.leadingpractice.com/enterprise-standards/enterprise-modelling/value-model/.

［21］See note 18 above.

［22］Kirchmer, High Performance through Process Excellence — From Strategy to Execution with Business Process Management, 2nd ed. (Berlin, e.a.: Springer, 2011).

［23］Slama N., Enterprise BPM — Erfolgsrezepte fuer unternehmensweites Prozessmanagement. (Heidelberg: dpunkt.verlag, 2011).

［24］Kirchmer, 1999a. Business Process Oriented Implementation of Standard Software — How to achieve Competitive Advantage Efficiently and Effectively, 2nd ed. (Berline, e.a.: Springer).

［25］Kirchmer, "Market- and Product-oriented Definition of Business Processes," in Business Engineering ed. D. J. Elzina, T. R. Gulledge, C-Y. Lee (Norwell: Springer, 1999b).

［26］Kirchmer, High Performance through Process Excellence — From Strategy to Execution with Business Process Management, 2nd ed. (Berlin, e.a.: Springer 2011).

［27］Franz K., Value oriented business Process Management — The Value-Switch for Lasting Competitive Advantage (New York, e.a.: McGraw-Hill, 2012).

［28］Morris, Architect, Design, Deploy, Improve (ADDI) — A BPMS Development Methodology (Chicago: Wendan Whitepaper, 2014).

［29］Kirchmer H., "Value oriented process Governance — Wettbewerbsvorteile durch die richtige Processorganisation," in IM+io Fachzeitschrift fuer Innovation, Organisation und Management (Germany, March, 2013).

［30］Kirchmer, "How to create successful IT Projects with Value oriented BPM," in CIO Magazine Online, (February 27, 2013).

［31］See note 18 above.

Sustainability Oriented Process Modeling

Gabriella von Rosing, David Coloma, Henrik von Scheel

INTRODUCTION

Sustainability is becoming a part of organizations and thus strategy, branding, and customer orientation and all of their processes. Executives, managers, business analysts, process experts, process architects, and process owners are a few types of profession that have an interest in sustainability and how it relates to business process management (BPM). Sustainability, achieving endurance of systems and processes, is one of the most relevant challenges that organizations and societies need to address. The ability to meet this challenge is hindered by the complexity of integrating sustainability into the strategy, the business model, and the different business functions that will execute the strategy. The 5 years of research work by the Global University Alliance on Enterprise Sustainability, which resulted in the development of an enterprise standard in terms of an enterprise sustainability reference content[1] and how it relates to BPM, is presented in this chapter. Areas touched by and of concern in sustainability oriented process modeling, including process design and operations, the link to strategy, and flows, roles involved, relevant rules, and compliance aspects as well as process automation, measurements, and reporting are discussed.

SITUATION, COMPLICATIONS, AND THE MAIN QUESTIONS

Since the publication of *The Limits of Growth*, sustainability has been a growing issue in the global agenda. Other significant landmarks are the publication of *Our Common Future* by the World Commission for Environment and Development, the Rio Earth Summits of 1992 and 2012, and the Kyoto Protocol (1997).

During this time, the world has seen some formidable environmental challenges, such as

- Ozone layer depletion, avoided thanks to the ban on chlorofluorocarbons
- The climate change challenge, still to be resolved
- The Fukushima, Three Mile Island, and Chernobyl nuclear disasters.
- Oil spills such as the Deepwater horizon rig and the *Exxon Valdez* tanker
- Waste and pollution threats as the Great Pacific garbage patch

The Complete Business Process Handbook. http://dx.doi.org/10.1016/B978-0-12-799959-3.00024-0

4.5　面向可持续发展的流程建模

Gabriella von Rosing, David Coloma, Henrik von Scheel

4.5.1　介绍

可持续发展正在成为组织的一部分,进而成为战略、品牌和以客户为导向及其所有流程的一部分。可持续性及其与BPM的关系的相关职务包括:企业高管、经理、业务分析师、流程专家、流程架构师和流程责任人。

可持续性、实现系统和流程的持久性,是组织和社会需要解决的最相关挑战之一。现实中,不同业务职能中的复杂性阻碍了将可持续性整合到战略、业务模式和执行战略。本节介绍全球大学企业可持续性联盟(Global University Alliance on Enterprise Sustability)5年的研究成果,该研究促使企业标准的制定,包括企业可持续性参考内容[1]及其与BPM的关系,讨论面向可持续性的流程建模中涉及和关注的领域,包括:流程设计和操作、运维和流程的链接、涉及的角色、相关规则和合规方面以及流程自动化、评估和报告。

4.5.2　情况、并发症和主要问题

自《增长的极限》(*The Limits of Growth*)出版以来,可持续性一直是全球议程中一个日益重要的问题,其他重要的里程碑包括世界环境与发展委员会(World Commission for Environment and Development)出版的《我们共同的未来》(*Our Common Future*)、1992年和2012年里约热内卢地球首脑会议和《京都议定书》(1997年)。

在这段时间里,世界面临着一些严峻的环境挑战,例如:

- 禁止使用氯氟烃,避免了臭氧层的消耗;
- 气候变化挑战仍有待解决;
- 福岛、三里岛和切尔诺贝利核灾难;
- 石油泄漏,如深水地平线钻井平台和埃克森瓦尔迪兹油轮;
- 垃圾和污染威胁,如太平洋大垃圾带;

- Deforestation and desertification
- The depletion of fisheries and the destruction of biodiversity

This range of problems presented poses tremendous challenges to humanity as the environmental services on which we rely on are severely compromised. Simultaneously, nongovernmental organizations (NGOs) have arisen as prominent players in the global arena by reacting against some of the most questionable facets of the pursuit of indefinite economic growth in a closed system. This in turn has given rise to promotion of the idea that economic development has to come about in ways that respect the rights and needs of workers and communities as well as alleviating negative side effects to other third parties. This mindset has resulted in the surge of the idea of corporate social responsibility.

In view of this, two relevant concepts have emerged. The first is the idea of sustainable development, popularized in the *Our Common Future* report by the World Commission for Environment and Development, which describes it as the "development that meets the needs of present generations without compromising the ability of future generations to meet their needs."

As an extension of this concept, the second one is the Elkington's *Triple Bottom Line*, which asserts that business performance has to deliver value in three key areas: economic, environmental, and social. Any outcome that fails to accomplish positive impacts in all three criteria will be clearly unsustainable in the long term.

This situation makes paramount the role of businesses in finding solutions, because they are the main agents that supply products and services to our society.

In the face of these demands, the data show that the time is ripe for solutions and the value of achieving more sustainable performance. Data from the International Organization for Standardization show[2] that in 2009 223,149 organizations were certified with ISO 14,001 from 159 countries, far from the 4433 organizations in 1997. Also, the need for sustainable management solutions was clearly shown by a Forrester Research study[3] that estimated that the global sustainability consulting market was worth $2.7 billion US in 2010 and that it would grow to $9.6 billion by 2015.

From the executive viewpoint, a United Nations Global Compact study[4] revealed that 93% of 766 global chief executive officers (CEOs) surveyed believed that sustainability was critical to the future success of their organizations. However, in this same study 49% of CEOs cited complexity of implementation across functions as the most significant barrier to implementing an integrated, organization-wide approach to sustainability, whereas competing strategic priorities was second, with 48% of respondents.

CONDITIONS, CIRCUMSTANCES, AND COMPLEXITY

Strong social and business demand for wide sustainability creates a scenario in which organizations, public and private, feel more pressure to improve in sustainability performance to achieve:

- 森林砍伐和荒漠化；
- 渔业的枯竭和生物多样性的破坏。

由于我们所依赖的环境服务受到严重损害，这一系列问题对人类构成了巨大的挑战。与此同时，非政府组织已逐步成为全球舞台上的重要参与者，它们对在封闭体系中追求无限经济增长的一些最令人质疑的方面做出了反应。这反过来又促进了这样一种观点，即经济发展必须以尊重人和社区的权利和需要以及减轻对其他第三方的不利副作用的方式来实现。这种心态导致了企业社会责任理念兴起。

鉴于此，出现了两个相关概念：第一个是可持续发展的理念，在世界环境与发展委员会发表的《我们共同的未来》中得到普及，该报告将其描述为"在不损害后代满足其需要的能力的前提下，满足当代人需要的发展"。

作为这一概念的延伸，第二个是Elkington（埃尔金顿）的《三重底线》(Triple Bottom Line)，它断言业务绩效必须在三个关键领域提供价值：经济、环境和社会。从长远来看，任何未能在所有三个标准中取得积极影响的结果显然都是不可持续的。

这种情况至关重要的是企业在寻找解决方案中的作用，因为它们是为我们的社会提供产品和服务的主要供应商。

面对这些需求，数据显示解决问题的时机已经成熟，实现可持续绩效的价值也已经成熟。ISO的数据显示[2]，2009年有223 149个组织获得了来自159个国家的ISO14001认证，远远高于1997年的4 433个组织。此外，Forrester Research的研究[3]清楚地表明了对可持续管理解决方案的需求：2010年全球可持续性咨询市场价值27亿美元，到2015年将增长到96亿美元。

从行政角度来看，联合国全球契约研究[4]显示，接受调查的766位全球CEO中有93%认为可持续性对其组织未来的成功至关重要。然而，在同一项研究中，49%的CEO表示，跨职能部门的实施复杂性是实施整体组织范围可持续性方法的最大障碍，而48%的受访者表示，竞争性战略优先级则排在第二位。

4.5.3 条件、环境和复杂性

对广泛可持续性的强大社会和企业的需求创造了这样一种情景：公共和私人组织感到更大的压力，需要提高可持续性绩效，以实现以下方面。

- Cost efficiencies, in which eco-efficiency is taking an increasing role as it delivers a reduction in raw materials and energy consumption, in greenhouse and other taxed emissions and waste
- Easier accommodation to stricter environmental standards and regulations, whether governmental or voluntary
- Exploitation of business opportunities in the form of new products and services or new revenue streams brought about by the monetization of emission rights, sale of by-products, or license of know-how on sustainability. Here, we can find some new products and services such as electric and hybrid cars, renewable energy projects, and smart grids and cities
- Radical transformation of those business models most affected by sustainability concerns, such as logistics, oil, power, mining, and forest products, among others
- Improved reputation as well as proactive response to customer and stakeholder requirements of environmental- and community-friendly products, services, and processes

Currently, there are several approaches to environmental and social responsibility (ISO 14,001, EMAS (Eco-Management and Audit Scheme), GRI (Global Reporting Initiative), ISO 26,000, etc.). Unfortunately, their focus is narrow because they seek only specific aspects of sustainability in organizations, addressing areas such as reporting, communication of impacts to stakeholders, process controls, compliance, or performance improvement, but typically only an environmental or social context. When any of these approaches takes on the larger view, it is at best addressed in a partial and restricted way. Then, wider opportunities for sustainable value creation remain mostly untouched because of a lack of integration of those considerations in the organization's (or indeed, the enterprise's) overall operations.

THE MAIN QUESTIONS COVERED

To deal with these primordial challenges, we propose a sustainability oriented process modeling framework as a response to the need for innovative and transformed strategies, business models, processes, and various end-to-end process flows to achieve drastic improvement not only in the economic value area but also in the environmental and social areas.

Here, BPM is an especially powerful approach to sustainability problems, because it implies the achievement of an organization's objectives through the improvement, management, and control of essential processes.

The sustainability oriented process modeling concept brings a unified way of thinking, working, modeling, governing, implementing, and training to organizations willing to implement sustainability.

- A *way of thinking* is provided because it articulates the definitions and guiding principles to capture, design, plan, and structure process-relevant objects and

- 成本效率,其中生态效率发挥着越来越大的作用,因为它减少了原材料和能源消耗,减少了温室气体和其他征税排放和废物。
- 更容易适应更严格的环境标准和法规,无论是政府的还是自愿的。
- 以新产品和服务的形式开发商业机会,或通过排放权货币化、副产品销售或可持续性技术许可带来的新收入流。在这里,我们可以找到一些新的产品和服务,如电力和混合动力汽车、可再生能源项目、智能电网和城市。
- 对那些最受可持续性问题影响的商业模式进行彻底变革,如物流、石油、电力、采矿和林业产品等。
- 提高声誉、积极响应客户和利益相关者对环境和社区友好型产品、服务和流程的要求。

目前,环境和社会责任有多种方法[ISO14001、生态管理和审计方案(eco-management and audit scheme, EMAS)、全球报告倡议(global reporting initiative, GRI)、ISO26000等]。不幸的是,他们的关注点是狭隘的,因为他们只寻求组织中可持续性的特定方面,解决诸如报告、对利益相关者的影响的沟通、过程控制、遵从性或性能改进等领域,但通常只是环境或社会背景。当这些方法中的任何一种具有更大的视图时,最好以一种局部的和受限制的方式来处理它。然后,可持续价值创造的更广泛的机会仍然基本没有受到影响,因为在组织(或者实际上,企业)的整体操作中缺乏对这些考虑的集成。

4.5.4 涵盖的主要问题

为了应对这些基本挑战,我们提出一个面向可持续发展的流程建模框架,以满足对创新和转型战略、业务模型、流程和各种端到端流程流的需求,从而实现激烈的工作。不仅在经济价值领域,而且在环境和社会领域都有改善。

在这里,BPM是解决可持续性问题的一种特别强大的方法,因为它意味着通过改进、管理和控制基本流程来实现组织的目标。

面向可持续性的流程建模概念为愿意实施可持续性的组织带来了统一的思维、工作、建模、管理、实施和培训方式。

- 提供一种思维方式,因为它阐明了捕获、设计、计划和构建与流程相关的对象和制品的定义和指导原则,以便了解问题并为不同的流程域带来解决

artifacts to understand problems and bring about solutions to different process domains.

- A *way of working* is supported, as the sustainability oriented process modeling principles, structuring the tasks to be performed in the development and implementation of process initiatives.
- A *way of modeling* is supplied as the sustainability oriented process modeling principles allow the description of the relevant process entities and relationships in the business sustainability domain so that solutions can be communicated to diverse stakeholders with different sustainability views and business concerns, and insights can be elicited.
- A *way of implementing* is provided so that appropriate paths of process transformation are made available.
- A *way of governing* is enabled by defining the expectations, intent, authority and responsibilities, and performance and sustainable process architecture to ensure value identification and creation.
- A *way of training* employees involved in sustainability management is brought about in the BPM certificate of excellence (CoE) and process owners and the rest of the organization.

THE ANSWER

The solution proposed to face these challenges is an integrated approach in the form of sustainability oriented process modeling principles that allows organizations to incorporate that triple-faceted concept of sustainability value into the business model, link it to the strategy, and develop the right performance measures as well as integrate it with its competencies and processes.

These sustainability oriented process modeling principles emerge from the Global University Alliance article entitled "Initial thoughts on a sustainability framework," written in 2009 by Mark von Rosing, Maria Hove, and Henrik von Scheel. It tried to tackle the complexity cited by executives as the main barrier to transitioning to more sustainable business models. By easing the transformation of the business processes from a sustainability point of view, it can be a powerful enabler.

THE WAY OF THINKING AROUND SUSTAINABILITY ORIENTED PROCESS MODELING

The way of thinking articulates the sustainability oriented process modeling principles by defining its scope and underlying principles. It deals with a set of objects and elements so that they can be properly identified, described, and further developed as needed:

方案。

- 支持一种工作方式,作为面向可持续性的流程建模原则,构建在流程计划的开发和实施中要执行的任务。
- 提供一种建模方式,因为可持续发展导向的流程建模原则允许描述业务可持续性领域中相关流程的实体和关系,以便可以将解决方案传达给具有不同可持续性观点和业务问题的不同利益相关者,并且可以获得洞察力。
- 提供一种实施方式,以便提供适当的流程转换路径。
- 通过定义期望、意图、权力和责任、绩效和可持续流程架构来确保价值识别和创造,从而实现管理方式。
- 培训参与可持续发展管理的员工的方式是在流程所有人以及组织的其他成员中实现BPM卓越证书。

4.5.5　答案

针对这些挑战提出的解决方案是一种以可持续性为导向的流程建模原则形式的综合方法,允许组织将可持续性价值的三方面概念纳入业务模型,将其与战略联系起来,制定正确的绩效衡量标准,并将其能力和流程整合。

这些面向可持续发展的流程建模原则来自全球大学联盟题为《可持续发展框架的初步想法》的文章,这篇文章由Mark von Rosing、Maria Hove和Henrik von Scheel于2009年撰写。它试图解决高管们所说的向更可持续的商业模式过渡的主要障碍——复杂性,从可持续性的角度来看,简化业务流程的转换可以成为一个强大的推动者。

4.5.6　围绕面向可持续性的过程建模的思考方式

这种思维方式通过定义面向可持续性的流程建模原则的范围和基本原则来阐明这些原则。它处理一组对象和元素,以便根据需要正确地标识、描述和进一步开发:

- External sustainability drivers and forces, as well as stakeholders and their requirements
- Mission, vision, goals, and value drivers
- Strategy, cost, operating, service, revenue, value, and performance models
- Sustainability processes and workflows, as well as their rules
- Overall governance and compliance
- Competencies and other relevant business and sustainability objects
- Organization chart, sustainability roles, and ownerships
- Performance measures and reporting, balanced scorecard, and governance
- Sustainability media and channels
- Objects needed in the infrastructure, application, and data layers below the sustainable business and process layers

We define, as many others in the market, sustainability in reference to three key focus areas (economic, environmental, and social), that have to be addressed in the business model through processes and resources that strive to reach eco-efficiency, social responsibility, and well-balanced growth goals (Figure 1).

The economic sustainability focus area has to attend to economic value creation, so that an effective and efficient use of economic resources brings economic success to the organization and prosperity to the society.

Environmental sustainability is about the wise use of natural resources, in the sense that they are preserved to meet the future needs of the organization and society. Here, cardinal notions and tools such as the ecological footprint, life cycle assessment, and ecological cost accounting, among others, are used.

The environmental area would be incomplete if no room were allowed for respect for other living forms besides the carrying capacity considerations. Here, biodiversity preservation, pollution prevention, and animal rights are to be considered.

Finally, the social sustainability focus area deals with conducting the business operations in a form that respects people inside and outside the organization, and tries to avert avoidable negative impact to third parties.

SUSTAINABILITY ORIENTED PROCESS MODELING: THE WAY OF WORKING

Steps to sustainability principles within BPM include:

1. Understanding the organization's sustainability personality
2. Building and transforming the BPM CoE and organization's culture toward sustainability
3. Developing the organization's business model
4. Developing the sustainability's life cycle
5. Building the organization's sustainability maturity model

- 外部可持续发展的驱动因素和力量,以及利益相关者及其要求;
- 使命、愿景、目标和价值驱动因素;
- 战略、成本、运营、服务、收入、价值和性能模型;
- 可持续发展流程和工作流程及其规则;
- 整体治理和合规性;
- 能力和其他相关业务及可持续发展目标;
- 组织结构图、可持续发展角色和所有权;
- 业绩计量和报告、平衡计分卡和治理;
- 可持续发展媒体和渠道;
- 可持续业务和流程层以下的基础结构、应用程序和数据层所需的对象。

与市场上的许多其他市场一样,我们参照三个关键重点领域(经济、环境和社会)来定义可持续性,这些领域必须在商业模式中通过努力实现生态效率、社会效率的流程和资源来解决责任以及均衡的增长目标(图1)。

经济可持续性重点领域必须关注经济价值创造,以便有效和高效地利用经济资源,为组织和社会繁荣带来经济上的成功。

环境可持续性是关于自然资源的明智使用,因为它们被保留以满足组织和社会的未来需求。在这里,使用了诸如生态足迹、生命周期评估和生态成本核算等基本概念和工具。

如果除了承载能力考虑因素之外不允许其他生命形式的空间,则环境区域将是不完整的。在这里,应考虑生物多样性保护、污染预防和动物权利。

最后,社会可持续性重点领域以尊重组织内外人员的形式开展业务运营,并努力避免对第三方造成负面影响。

4.5.7　面向可持续发展的流程建模：工作方式

BPM中可持续发展原则的步骤包括:

(1)了解组织的可持续性个性;

(2)建立并转变BPM CoE和组织的文化,实现可持续发展;

(3)发展组织的商业模式;

(4)发展可持续发展的生命周期;

(5)建立组织的可持续发展成熟度模型;

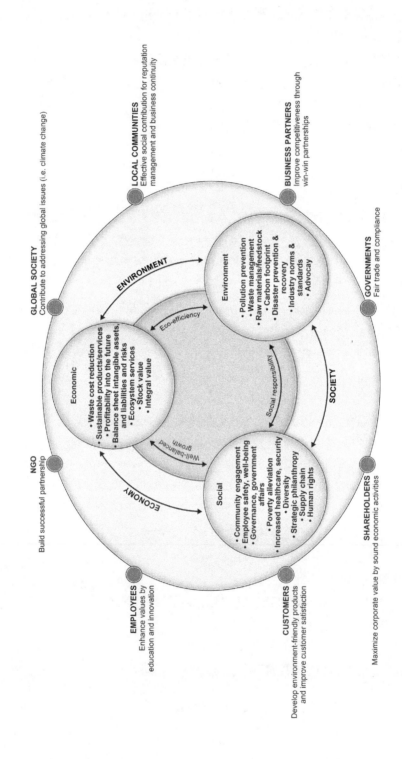

FIGURE 1

Business model for sustainability.[5]

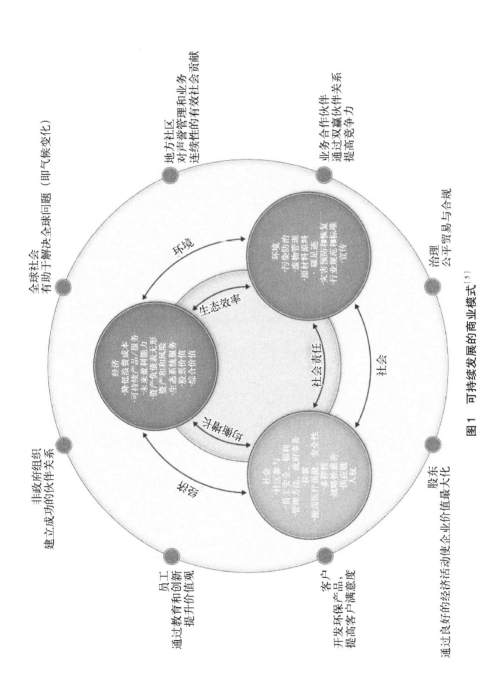

图 1　可持续发展的商业模式[5]

6. Developing the sustainability value model
7. Developing the sustainability revenue model

Details of these steps are provided below.

Understanding Your Organization's Sustainability Personality

The first step in the path to sustainability is to understand what the organization's personality is with regard to sustainability. That personality is derived from two dimensions: the attitudes and intentions toward sustainability and the activities and initiatives.

If the organization has initiated some sort of evolution toward sustainability, four sustainable personalities can be identified:

- Sustainability awareness, with low levels of both initiatives and intentions. Here, the organization probably has put in place some initial, unstructured measures
- Sustainability understanding, in which the initiatives level remains low but intentions have grown to higher levels. In this personality the organization has initiated sustainable strategies, policies, and initiatives
- Sustainability acceptance, with a high level of initiatives but a low level of intentions. Here, the organization would have implemented sustainable portfolio and performance management
- Sustainable commitment, in which the organization has both high intentions and initiatives levels. The sustainability is integrated in the strategy, sustainable innovation is developed, and there are aggressive sustainable goals and impacts

Building and Transforming the BPM CoE and Organization's Culture Toward Sustainability

To climb up the attitude ladder from compliance to leadership, the organization needs to provide motivation to create engagement. Therefore, the next step is to build within the BPM CoE a business case for sustainability that articulates an understanding of the business value at stake for the organization as well as the external forces and drivers of change that can affect the business process operating context.

Developing the Organization's Business Model

This phase in the road to sustainability provides the definition of the business model. Here, it is described in a hierarchical top-down way cascading into more detail. Thus, decomposition of the business model is broken down into the deeper and more concrete layers of competency groups, sustainable competencies, competency elements, process categories, process groups, processes, and activities.

（6）制定可持续发展价值模型；

（7）制定可持续性收入模型。

下面提供这些步骤的详细信息。

1. 了解组织的可持续性个性

实现可持续性的第一步是了解组织在可持续性方面的个性。这种个性来自两个维度：对可持续性的态度和意图以及活动和主动性。

如果组织已经开始向可持续性发展，那么可以确定以下四种可持续性人格。

- 可持续性意识、主动性和目的性较低。在这里，组织可能已经实施了一些初始的、非结构化的度量。
- 可持续性理解，即主动性水平仍然很低，但意图已经上升到更高的水平。在这种个性下，组织发起了可持续的战略、政策和行动。
- 可持续性接受，具有高水平的主动性，但意图较低。在这里，组织将实现可持续的投资组合和绩效管理。
- 可持续承诺，即组织具有高度的意愿和主动性。可持续发展被纳入战略，可持续创新被发展，有积极的可持续目标和影响。

2. 构建和改造BPM CoE和面向可持续性的组织文化

为了从合规上升到领导态度的阶梯，组织需要提供参与创造的动力。因此，下一步是在BPM CoE内构建一个可持续发展的商业案例，阐明对组织所涉及的业务价值的理解，以及可能影响业务流程运营环境的外部力量和变革驱动因素。

3. 发展本组织的商业模式

可持续发展道路的这一阶段提供了商业模式的定义，在这里，它被描述为以层次自上而下的方式分解到更详细的内容。因此，业务模型的分解被分解为能力组、可持续能力、能力要素、流程类别、流程组、流程和活动等更深更具体的层次。

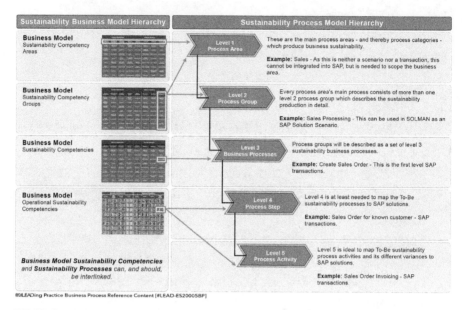

FIGURE 2

The sustainability business model to process model hierarchy.

In addition to the cascading detail of the business model, scenarios are elaborated that describe the processes needed to achieve a sustainability outcome (Figure 2).

Developing the Sustainability's Life Cycle

The subsequent point in the path to sustainability is to establish the organization's life cycle through which the transformation is to be carried out.

The sustainability lifecycle is based on the LEADing[1] practice life cycle model whose graphical representation is provided in Figure 3. It ensures that the different business model domains (service, revenue, cost, operating, value, and performance models) are addressed through a set of life cycles.

Following and exploiting these life cycles allows innovation and transformation of the sustainable business with a series of phases at the business process level and the applications level that support them, along with their required governance and value management.

The process life cycle corresponds to the sustainability personality. The analysis phase corresponds with the awareness personality, the design phase with understanding, implementation with acceptance, and continuous improvement with commitment.

This life cycle delivers alignment at the strategic, organizational, technological, and process levels.

---

[1] LEAD stands for Layered Enterprise Architecture Development.

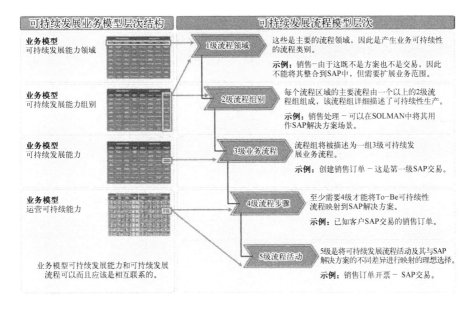

图2 用于处理模型层次结构的可持续性业务模型

除业务模型的级联细节之外,该分解模型还详细描述了实现可持续性结果所需的流程的场景(图2)。

4.发展可持续发展的生命周期

可持续发展道路上的后续要点是建立组织的生命周期,通过该生命周期进行转型。

可持续性生命周期基于领先的实践[1]生命周期模型,其图形表示如图3所示。它确保通过一组生命周期来处理不同的业务模型域(服务、收入、成本、运营、价值和性能模型)。

遵循和利用这些生命周期可以在业务流程级别和支持它们的应用程序级别,以及它们所需的治理和价值管理的一系列阶段,实现可持续业务的创新和转型。

流程生命周期对应于可持续性人格。分析阶段与意识人格、理解设计阶段、接受实施阶段、承诺持续改进阶段相对应。

此生命周期在战略、组织、技术和流程级别实现了一致性。

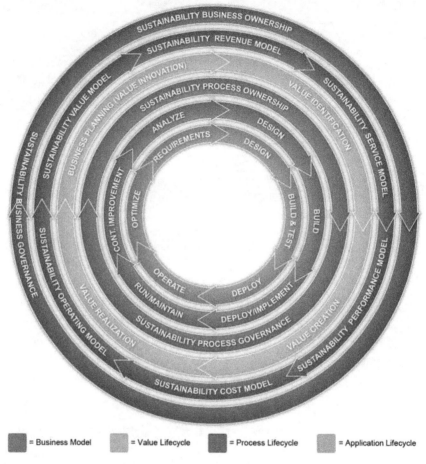

FIGURE 3

Business process management life cycle.[6]

Building the Organization's Sustainability Maturity Model

A tool is needed to guide evolution toward higher sustainability. Here, a sustainability maturity model (Figure 4) provides a benchmark tool against which the organization's status on the path to sustainability can be understood and from which guidelines and a roadmap to prioritize actions to evolve can be derived, as well as critical success factors (CSFs) and measurements produced.

The sustainability maturity model establishes five levels:

1. Recognize. First green processes come up, but in an ad hoc, poorly formalized, and uncontrolled manner.

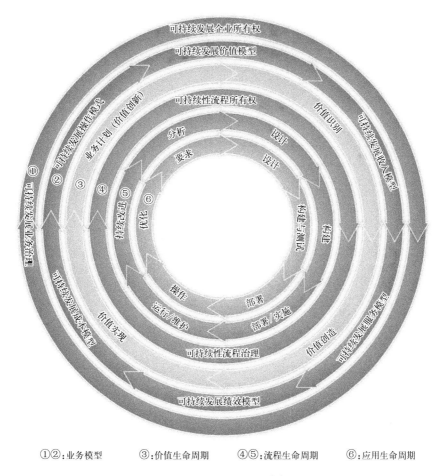

①②:业务模型　　　③:价值生命周期　　　④⑤:流程生命周期　　　⑥:应用生命周期

图3　BPM生命周期[6]

5.构建组织的可持续性成熟度模型

需要一种工具来引导进化到更高的可持续性。在这里,可持续性成熟度模型(图4)提供了一个基准工具,可以根据该工具了解组织在可持续性道路上的状态,并从中获得指导方针和路线图,以确定要演进的行动的优先级,以及产生的CSF和度量。

可持续性成熟度模型建立了以下五个层次。

(1)识别。首先出现的是绿色过程,但是是以一种特别的、不规范的和不受控制的方式出现的。

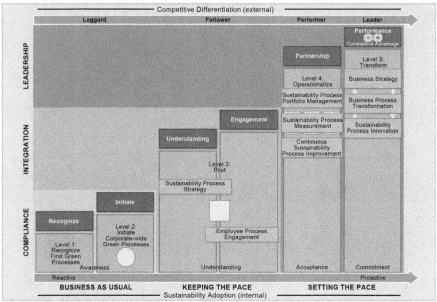

((0)LEADing Practice of the Sustainability Reference Framework

FIGURE 4

Maturity levels and personality profiles.[7]

2. Initiate. Sustainability processes are defined so that they become repeatable and consistent, although with room for improvement in rigor. Here, most initiatives seek compliance to rules or are "me too" respectable but unsound motivations.

3. Pilot. A sustainability strategy is established as well as standard sustainability processes and consistency in sustainability is achieved. The organization personality here evolves from awareness to understanding and starts to acknowledge the competitive importance of sustainability.

4. Operationalize. At this point, organization-wide sustainable portfolio management is set up while controlling the green sustainability initiatives and establishing sustainability capabilities.

5. Transform. In this stage, the organization focuses on improving sustainable innovation as sustainability is fully integrated into the business strategy.

Developing the Sustainability Value Model

The next activity in the sustainability transformation is elaboration of the sustainability value model, in which the different value viewpoints are considered from the triple bottom line perspectives.

The value model describes the external and internal value drivers, the value proposition, and expectations while translating the business strategy into strategic business objectives (SBOs) and CSFs.

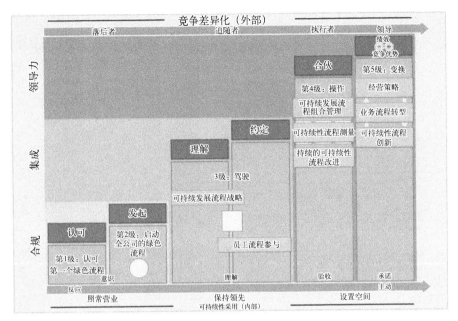

图4　成熟度级别和个性化简介[7]

（2）启动。可持续性流程被定义成可重复的和一致的，尽管在严格性方面还有改进的空间。在这里，大多数举措寻求遵守规则，或者"我也是"受人尊敬但动机不健全。

（3）试验。制定了可持续发展战略，建立了标准的可持续发展流程，实现了可持续发展的一致性。这里的组织人格从意识发展到理解，并开始认识到可持续性的竞争重要性。

（4）实施。在这一点上，在控制绿色可持续性计划和建立可持续性能力的同时，建立了组织范围内的可持续性项目组合管理。

（5）改进。在这一阶段，组织的重点是改进可持续创新，因为可持续性已经完全融入业务战略中。

6.发展可持续价值模型

可持续性转型的下一个活动是阐述可持续性价值模型，其中从三重底线的角度考虑不同的价值观点。

价值模型描述了外部和内部价值驱动因素、价值主张和期望，同时将业务战略转化为SBO和CSF。

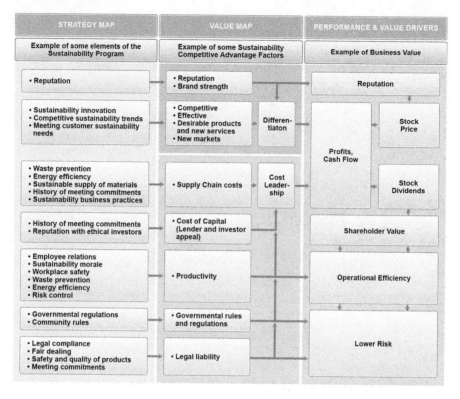

FIGURE 5

Relation among elements in sustainability, competitive advantage factors, and value drivers.[8]

One tool also used is selection of the competitive strategies (cost leadership, differentiation, or focus) that result in uniqueness and competitive advantages. Here, some considerations have to be made because current studies show that in our times, instead of choosing among exclusive strategies, some organizations are using blended strategies with successful results.

In creating these competitive advantages, construction of a sustainable balanced scorecard and a strategy map is essential because it helps to understand which pivotal competencies participate in the causal relationships among key sustainability value drivers.

In Figure 5 a map of the relations among sustainability strategy elements, value levers, and performance is provided.

The sustainability oriented process modeling concepts require more consideration of the value management discipline, because the two other sustainability value areas (environmental and social) have to be addressed in addition to the economic one.

Here, as described in the chapter on Value-Oriented Process Modeling, value management requires a formalized breakdown of (SBOs) into (CSFs), their

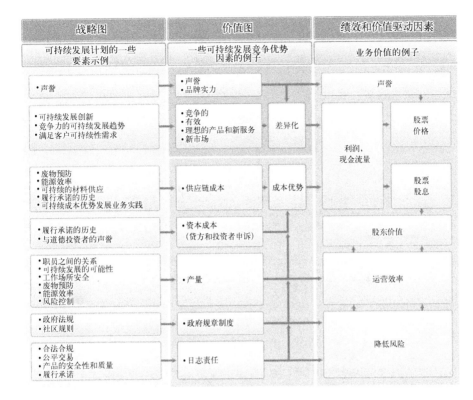

图5　可持续发展要素、竞争优势要素、价值驱动要素之间的关系[8]

我们还需要使用的一个工具是选择竞争战略（成本领先、差异化或集中化），从而产生独特性和竞争优势。在这里，必须考虑一些因素，因为目前的研究表明，在我们这个时代，一些组织正在使用具有成功结果的混合策略，而不是在独家策略中进行选择。

在创造这些竞争优势时，构建可持续的平衡计分卡和战略地图至关重要，因为它有助于了解哪些关键能力参与关键可持续性价值驱动因素之间的因果关系。

图5提供了可持续发展战略要素、价值杠杆和绩效之间关系的图。

面向可持续性的过程建模概念需要更多地考虑价值管理原则，因为除经济领域之外，还必须解决其他两个可持续性价值领域（环境和社会）。

这里，如价值导向流程建模章节所述，价值管理需要将SBO正式分解为CSF、它们的相关KPI和PPI。只有这样，才能以一种方式实施正确的衡量标准，

associated key performance indicators (KPIs), and process performance indicators (PPIs). Only then can the right measurements be put into place in a manner that ensures that they are integrated and strategically aligned as well as linked to the proper sustainability owners and decision-making bodies so that they allow performance improvement to occur. This brings support to this complex task by providing a green BPM value tree, a taxonomy of the previously mentioned value indicators.

Many organizations realize that traditional process design does not consider the sustainability aspects of one's organization. Executives who ask themselves what it takes to move from traditional process design to sustainability oriented process design have to consider the strategic role that sustainability has in their organization, but also how and where to apply the sustainability concepts. The ability to succeed with one's sustainability initiatives is directly related to the ability to connect the defined sustainability value drivers (SBOs and CSFs) and the sustainability performance drivers (KPIs and PPIs), as well as how the organization applies them to their competencies, processes, and services.

As illustrated in Figure 6, the core aspects of sustainability oriented process modeling are about linking the various aspects together:

1. Sustainability value drivers (SBOs and CSFs)
2. Sustainability performance drivers (KPIs and PPIs)
3. Organizational components (relevant business competencies)
4. The responsible person
5. The relevant business tier, i.e., strategic, tactical, or operational
6. The appropriate and related process that links to all five points
7. Specification of aspects of innovation and transformation

Once the right metrics are available, the value management life cycle can be operated—with the value planning, identification, creation, and realization phases—so that organizations can understand the value of their sustainable investments and impacts.

Then decisions can be properly made because their impact on sustainability is understood. These decisions encompass areas such as operational efficiency, risk management, market positioning, innovation, human capital development, community acceptance, supply chain management, access to resources, access to capital, and corporate governance.

Developing the Sustainability Revenue Model

Once value implications are understood, it is the time to describe revenue generation through the revenue model. This model defines how the organization makes money thanks to its value proposition and pricing model.

A sustainable business can generate income from sources other than the usual sales of goods and services, from other revenue opportunities such as the sale of by-products, emission rights, sustainable technologies, and process management know-how, among many others.

以确保这些衡量标准的整合和战略一致性，并与适当的可持续性所有者和决策机构相联系，从而实现绩效改进。这通过提供一个绿色的BPM价值树（前面提到的价值指标的分类）来支持这个复杂的任务。

许多组织意识到传统的流程设计没有考虑组织的可持续性方面。那些问自己如何从传统的流程设计转向面向可持续发展的流程设计的高管，必须考虑可持续性在其组织中的战略作用，以及如何和在何处应用可持续性概念。在可持续发展举措中取得成功的能力直接关系到将确定的可持续性价值驱动因素（SBO和CSF）与可持续性绩效驱动因素（KPI和PPI）联系起来的能力，以及组织如何应用他们的能力、流程和服务。

如图6所示，面向可持续性流程建模的核心方面是将各个方面联系在一起，这些方面包括：

（1）可持续性价值驱动因素（SBO和CSF）；

（2）可持续性绩效驱动因素（KPI及PPI）；

（3）组织构成部分（相关业务能力）；

（4）负责任的人；

（5）相关业务层，即战略的、战术的或操作的；

（6）连接到所有这五个点相关的过程；

（7）创新和转型方面的规范。

一旦有了正确的度量标准，价值的生命周期管理就可以与价值规划、识别、创建和实现阶段一起运行，以便组织能够了解其可持续投资的价值和影响。

我们可以根据对可持续性的影响的理解作出正确的决定，这些决定的领域包括：业务效率、风险管理、市场定位、创新、人力资源开发、社区接受（community acceptance）、供应链管理，以及资源、资本的存取和公司治理等。

7. 发展可持续发展的收入模式

一旦理解了价值的含义，就可以通过收入模型描述收入模式。这个模型定义了组织如何通过其价值主张和定价模型来获取利润。

一个可持续的企业还可以从商品和服务的正常销售之外的其他收入机会中获得收入，例如：从副产品的销售、排放权、可持续技术和流程管理知识等方面。

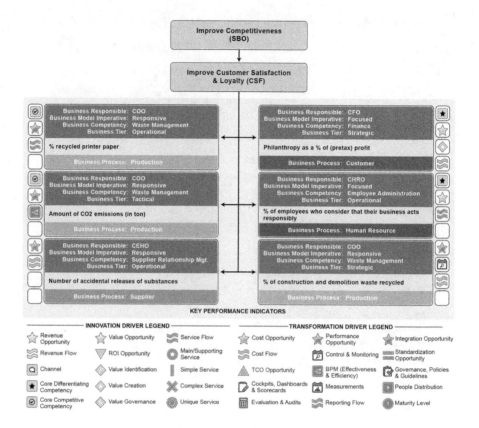

FIGURE 6

Example of sustainability oriented process modeling.[9]

In addition, a sustainable organization will increase in sales because of the increased appeal of sustainable products and services and the reputation effects (over price premiums and volume) of a sustainable brand.

A sustainable revenue model has to be linked to the development of a sustainable brand as well as sustainable marketing and communications to make customers and stakeholders aware of its sustainable performance.

SUSTAINABILITY ORIENTED PROCESS MODELING: WAY OF MODELING

As illustrated in Figure 6, sustainability oriented process modeling enables modeling of innovation and the transformation components within the business. Once the way of working is defined, the objects and their relationships can be modeled with

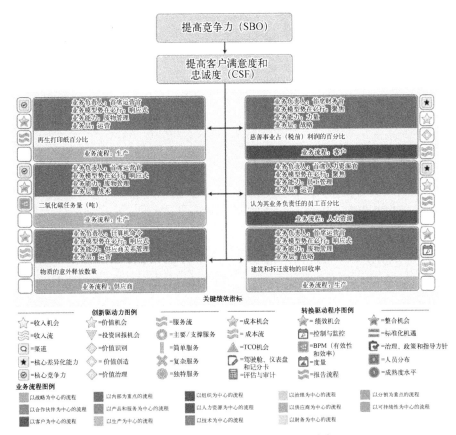

图6   面向可持续性的流程建模示例[9]

此外,由于可持续产品和服务的吸引力以及可持续品牌的声誉效应(价格溢价和销量过大)的增加,可持续组织的销售额也将随之增加。

可持续收益模式必须与可持续品牌的发展以及可持续营销和沟通联系起来,以使客户和利益相关者了解其可持续的绩效。

4.5.8   面向可持续性的过程建模:建模方法

如图6所示,面向可持续性的流程建模支持对业务中的创新和转型组件进行建模。一旦定义了对象的工作方式,就可以使用必要的细节和技术来构建模型及

the detail and techniques necessary to facilitate their deployment in the different layers. This can be modeled with the following aspects:

Innovation drivers
- Revenue model
- Value model
- Service model

Transformation drivers
- Operating model
- Cost model
- Performance model

Because of space limitations, we have chosen to focus on developing a sustainable operating model and within it, interlinking cost model concepts.

The sustainable operating model is developed, describing how the organization uses its resources and delivers performance and value across business competencies, functions, process, organization and technology.

Efficient sustainable execution has become highly reliant on information technology (IT) because the alignment of strategy, organization, and processes depends on technology. Nevertheless, IT has an ambivalent role regarding the global carbon footprint. Information technology is the cause of 2% of global CO_2 emissions with estimations of rapid growth,[10] but it has a huge potential because of its role as an enabler of efficiencies that can reduce the carbon footprints of the other 98% of emissions.

There are seven sustainable competency areas in which IT-enabled solutions can produce efficiencies (Table 1):

1. Energy and carbon. Here, the consumption of energy and the carbon emissions can be tracked and reported so that they can be reduced by a combination of initiatives in areas such as energy savings, energy mix optimization, carbon abatement initiatives, and carbon trading
2. Product safety and stewardship. These solutions empower products and services that are compliant and safe, and are designed with sustainable criteria that facilitate recycling and reuse and a low life cycle environmental footprint
3. Sustainable supply chain. Visibility throughout the supply chain can be warranted so that procurement, logistics, operations, assets, and product life cycle management activities can be optimized, and suppliers that most support the sustainability objectives can be selected
4. Environment, health, and safety. Here, the environmental performance and issues regarding the occupational health of workers, industrial hygiene and safety, and emergency management issues are addressed
5. Sustainable workforce. With these solutions, the organization can enforce and ensure compliance with labor rights and respect for diversity—both inside and outside the environment—and properly manage talent
6. Information technology infrastructure, further developed below
7. Sustainability performance. This covers the capabilities to manage performance on sustainability

其关系,以便于在不同层中部署它们。这可以用以下方面进行建模。

创新驱动因素:

- 收入模式;
- 价值模型;
- 服务模式。

转型驱动因素:

- 操作模式;
- 成本模型;
- 性能模型。

由于空间有限,我们选择专注于开发可持续运营模式,并在其中建立相互关联的成本模型概念。

开发可持续运营模式,描述组织如何使用其资源,并在业务能力、职能、流程、组织和技术方面提供绩效和价值。

因为战略、组织和流程的一致性取决于技术,可持续的高效执行已经变得高度依赖于IT。尽管如此,IT在全球碳足迹方面仍具有矛盾的作用,快速增长的IT占据全球二氧化碳排放量2%[10],但其由于作为效率的推动因素,可以减少其他98%排放的碳足迹,具有巨大的潜力。

在七个可持续的能力领域,IT支持的解决方案可以提高效率(表1)。

(1)能源和碳。在这方面,可以跟踪和报告能源消耗和碳排放,以便通过节能、能源组合优化、碳减排举措和碳交易等领域的综合举措来减少能源和碳排放。

(2)产品安全和管理。这些解决方案支持符合要求和安全的产品和服务,并采用可持续的标准设计,便于回收和再利用,并在环境中具有较低的生命周期。

(3)可持续供应链。可以保证整个供应链的可见性,以便优化采购、物流、运营、资产和产品生命周期管理活动,并选择最支持可持续发展目标的供应商。

(4)环境、健康和安全。在此,解决了有关工人职业健康、工业卫生和安全以及应急管理问题的环境绩效和问题。

(5)可持续的劳动力。有了这些解决方案,组织可以在内部和外部环境中执行并确保遵守劳工权利和尊重多样性,并妥善管理人才。

(6)IT基础设施,下文进一步阐述。

(7)可持续的绩效。这包括管理可持续发展绩效的能力。

Table 1 *Sustainable Competency Areas Supported by Information technology (IT) Solutions (SAP Example)*

Energy & Carbon	Energy-efficient Assets	Energy Management		Carbon Management	Smart Grids	
Product Safety & Stewardship	Product Compliance	Material & Product Safety	Recycling & Re-use	Recall Management	Environmental Footprint	Sustainable Design
Sustainable Supply Chain	Procurement	Traceability	Commodity Trade & Risk Management	Resource Optimization		Supply Chain Optimization
Environment, Health & Safety	Environmental Performance	Occupational Health		Industrial Hygiene & Safety	Emergency Management	
Sustainable Workforce	Labor Compliance & Rights		Diversity	Talent Management		
IT Infrastructure	Availability, Security, Accessibility & Privacy			Sustainability Oriented IT		
Sustainability Performance Management	Assured Reporting/Compliance	Benchmarks & Analytics		Strategy & Risk	Financial Performance	

In mixing the operating model, one can develop a sustainability cost model. It describes the costs incurred in operating the organization so that the service and revenue models are supported. Also, at this time the route for the transformation is developed, depending on the objectives and constraints of the organization. Three different transformation paths can be selected depending on the intent of the organization.

- Revolutionary approach, a big bang strategy that wants to achieve maximum improvement in the shortest time span. It is a high-cost bet that can bring fast and radical improvement but at a high risk and high disruption over daily operations
- Step-by-step approach, in which improvement is produced project by project. It delivers cost savings with lower risks but in a slower fashion. Here, dependencies between projects may not be properly addressed
- Evolutionary approach, in which improvement is tackled in a process-by-process mode. This approach mitigates risks but realization of benefits is slow

SUSTAINABILITY ORIENTED PROCESS MODELING: WAY OF IMPLEMENTING

This realm of the sustainability oriented process modeling framework is tackled mainly through a sustainability performance model that allows translation of the vision and strategy into performance measures and balanced scorecards as well as

表1　IT解决方案支持的可持续能力领域（SAP示例）

能源与碳	节能资产		能源管理		碳管理	智能网格
产品安全与管理	产品合规性	材料与产品安全	回收和再利用	召回管理层	环境足迹	可持续设计
可持续供应链	采购	可追溯性	商品贸易与风险管理		资源优化	供应链优化
环境，健康与安全	环境绩效		职业健康	工业卫生与安全		应急管理
可持续劳动力	劳工合规与权利		多样		人才管理	
IT基础设施	可用性，安全性，可访问性和隐私			面向可持续发展的IT		
可持续发展绩效管理	保证报告/合规		基准与分析	战略与风险		财务绩效

在混合运营模式的情况下，可以开发可持续成本模型。它描述了在组织的运营过程中产生的成本，以便支持服务和收入模型。另外，此时根据组织的目标和制约因素，制定了转型的路线。根据组织的意图，可以选择以下三种不同的转换路径。

- 革命性的方法。一个大爆炸式的战略，希望在最短的时间内实现最大的改进。这是一个高成本的赌注，但可以带来快速和彻底的改善，但会对日常运营产生高风险和高干扰。
- 分步改进的方法。改进是由项目产生的。它以更低的风险，但以更慢的方式节约成本。在这里，项目之间的依赖关系可能无法正确解决。
- 循序进化方法。在这种方法中，改进是在一个接一个的过程模式下进行的。这种方法可以降低风险，但实现效益的速度很慢。

4.5.9　面向可持续性的流程建模：实现方法

面向可持续性的流程建模框架主要通过可持续性绩效模型来解决，该模型允许将愿景和战略转化为绩效度量标准和平衡记分卡，以及政策、行为准则、报告和责任体系等其他结构。

other constructs as policies, codes of conduct, and reporting and responsibility lines.

Specific reporting guidelines need to be developed so that auditability and transparency to third parties is ensured. For more information, we reference the process implementation and deployment chapter.

SUSTAINABILITY ORIENTED PROCESS MODELING: WAY OF GOVERNING

In this part, an approach to sustainability governance and continuous improvement is developed. A sustainability board is established and sustainability owners are defined and mapped.

Stakeholders and their concerns and requirements are also described and mapped. Special attention to them should be given in addressing sustainability, because the nature of the subject implies more complexity than in other domains. Therefore, the needs of the supply chain actors (suppliers, customers, channels, and partners), employees, investors and lenders, governments, communities, and NGOs have to be carefully defined and considered. For more information, we reference the BPM governance chapter.

BENEFITS OF COMBINING BPM AND SUSTAINABILITY ORIENTED PROCESS MODELING

The benefits of combining BPM and sustainability oriented process modeling are multiple. Some of the concepts evolve the organization's sustainability maturity in that it enables the organization to:

- Understand the sustainability personality and transform its culture
- Develop the business model concepts around sustainability
- Establish the sustainability life cycle that aligns the different elements
- Progress toward sustainability with the help of a maturity tool
- Define the sustainability value, revenue, operating, performance, and cost models, as well as ensure green IT processes
- Build a sustainability governance model, including a board

The proposed approach to sustainability and BPM results in technical and business advantages to the applying organizations. Because executives cite complexity as the main barrier in transitioning to more sustainable business models, the sustainability oriented process modeling principles covered can enable benefits:

- Business value aspects:
  - Improved responsiveness and agility, including the ability to change business processes and business model in response to a changing environment
  - Improved innovation in products, services, processes, and business model

我们还需要制定具体的报告准则,以确保第三方的可审计性和透明度。有关更多信息,我们参考3.5.5小节流程实现和部署。

4.5.10　面向可持续性的过程建模: 治理方式

在这一部分中,我们建立了发展可持续性治理和持续改进的方法。包括: 建立可持续发展委员会、定义和确定可持续发展责任人。

我们还描述和映射了利益相关者的关注点和需求。在解决可持续性问题时应特别注意这些问题,因为主体的性质比其他领域更复杂。因此,必须仔细定义和考虑供应链参与者(供应商、客户、渠道和合作伙伴)、员工、投资者和贷款人、政府、社区和非政府组织的需求。有关更多信息,请参阅5.4节BPM治理。

4.5.11　结合BPM和面向可持续性的流程建模的优势

结合BPM和面向可持续性的流程建模的优势是多方面的。一些概念发展了组织可持续性的成熟度,因为它使组织能够:

- 了解可持续发展的特征并改变其文化;
- 围绕可持续发展商业模式概念;
- 建立可持续性生命周期,使不同的元素保持一致;
- 在成熟工具的帮助下向可持续性发展;
- 定义可持续价值、收益、运营、绩效和成本模型,并确保绿色IT流程;
- 建立可持续性治理模型,包括董事会。

提出的可持续性和BPM方法为应用组织带来了技术和业务优势。由于高管将复杂性视为向更可持续的业务模式过渡的主要障碍,因此涵盖可持续性的流程建模原则可以带来以下好处。

(1)商业价值方面:

- 提高业务响应能力和敏捷性,包括根据不断变化的环境更改业务流程和业务模型的能力;
- 改进产品、服务、流程和业务模式的创新;

- Integrated stakeholder management requirements in the business processes
- Reduced costs and increased employee and resource productivity
- Reduced risk and improved compliance to regulations and internal policies
- Improved quality
- Reduced waste and nonvalue-added activities
- Improved performance and control over processes
- Improved coordination
- Improved reputation
- Technical value aspects that result in diminished costs of implementation and operation of business processes:
  - Improved efficiency of processes
  - Component and process reuse, with lowered implementation time-to-market
  - Improved coordination among functions
- Better application integration
- Environmental value aspects:
  - Lower ecological footprint, with less energy and materials consumption and greenhouse gases emission
  - Lowered pollution levels
  - Enhanced product safety and traceability
  - Biodiversity preservation
- Social aspects:
  - Social rights of labor and communities respected
  - Improved diversity in workforce
  - Improved occupational health and safety

CONCLUSIONS

In this chapter we have deliberated on the various aspects of sustainability and BPM. Because of space constraints, not all aspects could be covered that are needed for BPM and sustainability oriented process modeling and all of the tasks and artifacts to achieve the right way of thinking, working, modeling, implementation, and governance.

We focused on the definition of sustainability that encompasses economic, environmental, and social value, why it is needed, the relevance to the business model and the process model, and a guided step-by-step approach to sustainability oriented process modeling, including strategy, competencies, value, performance, measurement, and reporting, all linked to processes.

End Notes

1. The LEADing Practice Enterprise Sustainability Reference Content (LEAD-ES20018AL).
2. International Organization for Standardization (ISO), "The ISO Survey 2009,".

- 业务流程中的综合利益相关者管理要求；
- 降低成本，提高员工和资源生产率；
- 降低风险并改善对法规和内部政策的遵守情况；
- 提高质量；
- 减少浪费和非增值活动；
- 改进的性能和对流程的控制；
- 改善协调；
- 提高声誉。

（2）导致业务流程实施和运营成本降低的技术价值方面：

- 提高流程效率；
- 组件和流程重用，降低上市时间；
- 改善职能之间的协调。

（3）更好的应用程序集成。

（4）环境价值方面：

- 更低的生态足迹，更低的能源和材料消耗以及温室气体排放；
- 降低污染水平；
- 增强产品安全性和可追溯性；
- 生物多样性保护。

（5）社会方面：

- 劳动和社区的社会权利得到尊重；
- 改善劳动力的多样性；
- 改善职业健康和安全。

4.5.12 结论

在本节中，我们讨论了可持续性和BPM的各个方面。由于空间的限制，并未包含BPM和面向可持续性的流程建模所需的所有方面，也不是实现正确的思维、工作、建模、实现和治理方式所需的所有任务和制品。

我们重点讨论了可持续性的定义，包括经济、环境和社会价值，为什么需要它，与业务模型和流程模型的相关性，以及可持续性导向流程建模的逐步指导方法，包括策略、能力、价值、性能、评估和报告，所有这些都与流程相关。

3. "Capitalizing on the Sustainability Consulting Services Opportunity," *Forrester Research,* (October 2010).

4. Accenture CEO Study (2010, 2011, 2012), "A New Era of Sustainability," (UN Global Compact).

5. Source: www.LEADingPractice.com

6. LEADing Practice Business Process Reference Content #LEAD-ES20005BP.

7. LEADing Practice Maturity Reference Content #LEAD-ES60003AL.

8. LEADing Practice Value Architecture Reference Content #LEAD-ES40003PG.

9. LEADing Practice Business Process Reference Content #LEAD-ES20005BP.

10. McKinsey & Company, "Pathway to a Low-Carbon Economy: Version 3 of the Global Greenhouse Gas Abatement Cost Curve," (2009): 7.

参考文献

［1］The LEADing Practice Enterprise Sustainability Reference Content (LEAD-ES20018AL).

［2］International Organization for Standardization (ISO), "The ISO Survey 2009,".

［3］"Capitalizing on the Sustainability Consulting Services Opportunity," Forrester Research, (October 2010).

［4］Accenture CEO Study (2010, 2011, 2012), "A New Era of Sustainability," (UN Global Compact).

［5］Source: www.LEADingPractice.com

［6］LEADing Practice Business Process Reference Content #LEAD-ES20005BP.

［7］LEADing Practice Maturity Reference Content #LEAD-ES60003AL.

［8］LEADing Practice Value Architecture Reference Content #LEAD-ES40003PG.

［9］LEADing Practice Business Process Reference Content #LEAD-ES20005BP.

［10］McKinsey & Company, "Pathway to a Low-Carbon Economy: Version 3 of the Global Greenhouse Gas Abatement Cost Curve," (2009): 7.1 LEAD stands for Layered Enterprise Architecture Development.

Information Modeling and Process Modeling

Hans-Jürgen Scheruhn, Mark von Rosing, Richard L. Fallon

INTRODUCTION

Business process modeling, the activity of recording and representing the processes of an enterprise, is an important part of information modeling, which is the recording and depiction of the persistent and future arrangement of information assets of an organization in a structured or formal manner. Information modeling is often incorrectly understood to be concerned only with data modeling. In reality, information modeling is composed of not only data modeling but also other aspects such as process modeling as well as value- or service-oriented modeling. The resulting information models, covering the strategic, tactical, and operational tier, can ultimately form a single integrated enterprise information model (see Figure 1). The message of this figure is that there must be integration of the strategic, tactical, and operating information models as well as integration into all phases of the business process life cycle. The information models and the record of their content fulfill the purpose of mapping not only the dynamic aspects of the business processes and data flows within an organization, but also the static characteristics of the information space on which the dynamic (time-dependent) aspects build.[1] The purpose of these models is varied; among other things, they provide a record of the information assets of the enterprise, the idea of creating a shared understanding of the business, and thus are important in problem solving and executing change.

Business process modeling tools should be used to depict current business processes ("as-is" modeling) as well as to develop the design of the new business process blueprint[2] ("to-be" modeling). Interlinking the business and application layers and their information meta objects can be organized and their content represented using a range of current information modeling techniques. These models apply concepts already discussed in the Extended Business Process Model and Notation (xBPMN) in Chapter[3], event-driven process chain (EPC), Unified Modeling Language (UML), information engineering (IE), and entity-relationship (ER) modeling, among others. We will then provide evidence via a case study of how these different modeling techniques complement each other through practical examples of their use.

INTENDED AUDIENCE

This topic is interesting to individuals who use only one of these information modeling techniques in their daily work, or professionals seeking to gain insight into how these modeling techniques can be put into practice in a real-world situation.

The Complete Business Process Handbook. http://dx.doi.org/10.1016/B978-0-12-799959-3.00025-2

4.6　信息建模和流程建模

Hans-Jürgen Scheruhn, Mark von Rosing, Richard L. Fallon

4.6.1　介绍

业务流程建模是记录和表示企业流程的活动,是信息建模的重要组成部分。它以结构化或正式的方式记录和描述组织信息资产的持续性和未来安排。信息建模通常被错误地理解为只关心数据建模,实际上,信息建模不仅由数据建模组成,还包括如流程建模以及价值或面向服务的建模等其他方面,由此产生的覆盖战略、战术和操作层的信息模型最终可以形成单一的综合企业信息模式(图1)。图1展现的信息是必须将战略、战术和操作模型进行整合,并集成到业务流程生命周期的所有阶段。信息模型及其内容记录不仅实现了映射组织内业务流程和数据流的动态方面,而且实现了动态(时间依赖)方面构建所依据的空间信息的静态特性[1]。这些模型的目的各不相同,除其他事项外,它们提供了企业信息资产的记录、创建对业务的共享理解的想法,因此在解决问题和执行变更中非常重要。

业务流程建模工具应该用于描述当前业务流程(“as-is”状态)以及新开发的业务流程规划(“to-be”状态)[2]。我们可以使用一系列当前的信息建模技术来组织业务层和应用层及其信息元对象,并展示它们的内容。这些模型应用了《扩展业务流程模型和符号(xBPMN)》一文中已经讨论论过的概念[3],包括:EPC、UML、信息工程(information engineering, IE)和实体关系(ER)建模等。我们将通过案例研究提供证据,说明这些不同的建模技术如何在其使用的实际例子中相互补充、协作。

4.6.2　目标受众

对于在日常工作中仅使用这些信息建模技术之一的个人,或者寻求深入了解这些建模技术如何在实际情况中付诸实践的专业人士而言,此主题很有趣。“答

((O)LEADing Practice Business Process Reference Content [#LEAD-ES30002AL]

FIGURE 1

Process life cycle and enterprise tiers.[4]

The models in the section "The Answer" show how each of the different modeling techniques can in fact complement each other and thus provide a set of integrated enterprise information models that contain all aspects of the business process that has been modeled.

PROCESS LIFE CYCLE

The view of a process life cycle is not new. Several authors[4,5–8] have looked at the problem of defining these cycles and proposed a number of different approaches; for example, Verner proposed a process life cycle containing seven individual stages to an iteration:

Analyze → Design → Build/Develop → Deploy → Operate → Maintain/Continuous Improvement.

However, for our working examples we will use the definition of the process life cycle (see Figure 2) as defined by the LEADing Practice framework[9] because the LEAD standards offer a paradigm shift in the goal of producing a truly open all-encompassing standard (LEAD standards include interfaces to other frameworks, methods, and approaches such as TOGAF, Zachman, FEAF, ITIL, Prince2, COBIT,

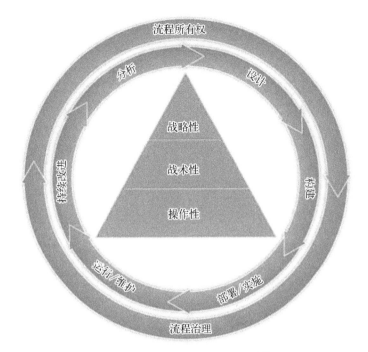

图1　流程生命周期和企业层[4]

案"部分中的模型显示了每种不同的建模技术如何能够相互补充,从而提供一组集成的企业信息模型,其中包含已建模的业务流程的所有方面。

4.6.3　流程生命周期

流程生命周期的观点并不新鲜。几位作者[4-8]已经研究了定义这些周期的问题,并提出了许多不同的方法。例如,L. Verner提出了一个流程生命周期,其中包含七个单独的阶段:

分析→设计→构建/开发→部署→操作→维护/持续改进。

但是,对于我们的工作示例,我们将使用由主要实践框架定义的流程生命周期定义(图2)[9]。因为LEAD标准为制定真正开放的无所不包的标准提供了范式转换(LEAD标准包括与其他框架、方法和方式的接口,如TOGAF、Zachman、FEAF、ITIL、Prince2、COBIT和DNEAF)[10]。对于我们的工作示例,我们将使用BPM生

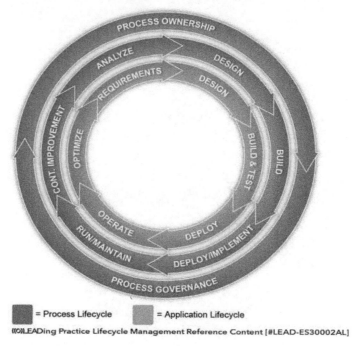

= Process Lifecycle   = Application Lifecycle

((O)LEADing Practice Lifecycle Management Reference Content [#LEAD-ES30002AL]

FIGURE 2

Process life cycle with the application life cycle.

and DNEAF).[10] For our working examples, we will use the definition of the process life cycle discussed in the chapter "BPM Life Cycle."

The diagram above (Figure 2) illustrates the cyclical nature of the process and application life cycle.

Analyze (and Discover)

The goal of process analysis is to detect implicit knowledge that exists in the organization about existing or as-is processes and make this knowledge available in an as-is model so as to organize and represent this knowledge.[11] Thus, the analysis phase and documentation are the first steps in providing a complete discovery of existing (as-is) business processes, closely followed by the capture, decomposition,[12] and documentation of all relevant related information objects, properties, and relationships. This procedure is commonly known as business process analysis (BPA).[13]

Above all, the processes, together with the related dynamic and static business structures, should ultimately support and execute the strategic business objectives and critical success factors of the organization. Thus, these strategic aspects are a part of the business direction and therefore value expectations and business requirements

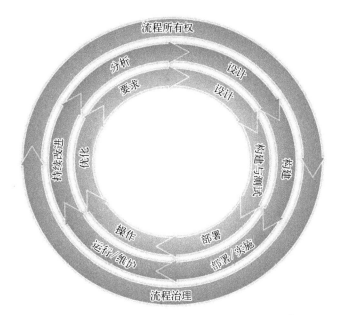

外圈、中圈：流程生命周期　　　　内圈：应用生命周期

图2　使用应用程序生命周期处理生命周期

命周期一节中讨论的过程生命周期的定义。

上面的图（图2）说明了流程和应用程序生命周期的周期性。

1. 分析（和发现）

流程分析的目标是检测组织中存在的关于现有或现状的隐性知识，并在现状模型中提供这些知识，以便组织和表示这些知识[11]。因此，提供分析阶段和文档是对现有（现状）业务流程的完整发现的第一步。紧随其后的是捕获、分解[12]和记录所有相关的相关信息对象、属性和关系，此过程通常称为业务流程分析（BPA）[13]。

最重要的是，这些流程以及相关的动态和静态业务结构应最终支持和执行组织的战略业务目标和关键成功因素。因此，这些战略方面是业务方向的一部分。

that have to be considered in the analysis and organizational design of the associated strategic information objects and their relationships to the processes. The connection between these objects and the processes must be must be both identifiable and verifiable. This connection occurs only through each of the members of the array of integrated and holistic sets of related knowledge, skills, and abilities that, combined, enable the enterprise to act in its environment—the enterprise's competencies. These competencies are important in executing the structured analysis of a process because they provide the context in which to judge the optimization criteria to be used when designing a process, whether centered on value maximization or cost minimization. It is therefore critical to distinguish at an early stage between core competitive, core differentiating, and non-core competencies and thereby the related processes.[14,15] Core competitive competencies and all related processes are essential for an enterprise to compete and core differentiating competencies and all related processes are those that differentiate the business to its customers. In both cases, the processes involved are the tasks that create value, whereas anything that is non-core but that must be done should be done for as little cost as possible.

For a correct and complete analysis of a business process (as-is model), all relevant information objects and their relationships to each other must be identified and documented. This includes consideration of value and business process flow, business competency, service, and data flow. In addition to the dynamic flows, the enclosing static (hierarchical) structures (value, competency, service, process, application, and data) should also be considered. In the case of any of these, the process expert or process engineer must make a thorough decomposition and analysis of the business process. Decomposition is the procedure by which the objects are broken down into their simpler forms. For example, a business process is decomposed into one or more process step(s), whereas a process step is decomposed into one or more process activities(s) and a process activity is decomposed into one or more transaction(s). The result of several successful iterations of the discover/analyze cycle is the completed as-is model.

The manual process of analyzing, decomposing, and documenting business processes from a previous successful run of a process life cycle (assuming process maturity greater than "3" or "standardized") can also be assisted using tools such as SAP Reverse Business Engineering and SAP Solution Manager or ARIS Process Performance Management. However, some tools provide only part of the information required about the processes' state and the relationships between the relevant objects. These tools are even less successful when determining process flows with business rules as well as static structures and hierarchies needed to obtain a full understanding of their design and properties, and lead to incomplete designs that often do not work as needed.

Design

In this phase the new business process flow and business process structures (to-be status) are designed.

所以,在分析和组织设计相关战略信息对象及与流程的关系时必须考虑价值预期、业务需求。这些对象与流程之间的连接必须是可识别和可验证的。这种联系只能通过一系列相关知识、技能和能力的综合及整体集合中的每一个成员来实现,这些知识、技能和能力的组合使企业能够在其环境中发挥企业的能力。这些能力对于执行流程的结构化分析非常重要,因为它们提供了在设计流程时要使用的判断优化标准的语境,无论是以"价值最大化"还是"成本最小化"为中心。因此,必须在核心竞争力、核心差异化和非核心竞争力之间进行区分,从而区分相关流程[14,15]。核心竞争能力和所有相关流程对于企业的竞争至关重要,核心差异化能力和所有相关流程都是将业务与客户区分开来的流程。在这两种情况下,所涉及的过程都是创造价值的任务,而任何非核心但必须完成的任务都应该以尽可能低的成本完成。

为了对业务流程(as-is模型)进行正确和完整的分析,必须标识和记录所有相关的信息对象及其相互之间的关系。这包括对价值和业务流程流、业务能力、服务和数据流的考虑。除了动态流之外,还应该考虑封闭的静态(分层)结构(值、能力、服务、流程、应用程序和数据)。在这些情况下,流程专家或流程工程师必须对业务流程进行彻底的分解和分析。分解是将对象分解成更简单形式的过程,例如,业务流程分解为一个或多个流程步骤,而流程步骤分解为一个或多个流程活动,流程活动分解为一个或多个事务。发现/分析周期的几个成功迭代的结果是完成的as-is模型。

使用SAP逆向业务工程(reverse business engineering)和SAP解决方案管理器(solution manager)或ARIS流程绩效管理(process performance management)等工具,还可以帮助分析、分解和记录流程生命周期中以前成功运行的业务流程(假设流程成熟度大于3或标准值)。然而,一些工具只提供了流程状态和相关对象之间关系所需的部分信息。当使用业务规则以及静态结构和层次结构来确定流程流时,这些工具的成功率甚至更低,这些静态结构和层次结构需要充分了解其设计和属性,并导致不完整的设计,而这些设计往往无法按需工作。

2. 设计

在此阶段,将设计新的业务流程流和业务流程结构(to-be status,未来状态)。

Depending on the scope of the project, the design work can involve anything from altering the complete process flow to adding and/or deleting business processes, or just to small changes in basic behavior. A similar range of the scope of change can occur with the information objects and the related dynamic and static structures contained within each of the business processes. This is relevant because the information model must be created throughout the end-to-end process flows. Therefore, output/product of this phase is the successful composition of the new to-be design, captured in a model.[16]

Build

The process build phase is concerned with applying the to-be models defined within the process design phase, including all related dynamic and static structures, to create the operating system (manual or automated).

In a purely manual situation, the build phases are addressed through work design, training, and the preparation of documentation. In an automated environment build, the activity may include programming, configuration, or other work within the software that performs or enables the work. Obviously, in many cases both types of work will be required and must be coordinated to complete the build to achieve the results that are required from the new operations.

Depending on the size and scope of the software-oriented build and the quality of the process models produced in advance, a so-called model-driven design can be used.[4,17] However, more comprehensive process models and methods are required when deploying enterprise and Web/restful services than are the case for implementing or customizing corporate standard software. In the latter case, for example an ERP system, partial automation can also be obtained through such tools as SAP Solution Manager and SAP Business Workflow or BPM systems such as SAP BPM or Software AG webMethods support.

The needs of the business analysts who have produced the specification of the to-be business and the technical application developers who implement the system are not always the same.[18] The challenge and problems associated with producing a successful combined system of work that fully implements the to-be models must therefore lie in collaboration between the members of these two groups. Part of the problem is finding balance between the parts of the work that should be done by machine and those best done by humans, and how best to establish the interface between the two; often the problems are related to matters of precision, which, with BPM and automated business processes, can lead to implementation that does not accurately fulfill the business requirements. When considered in total, the result is the description of a system of work in which human work is efficiently and effectively enabled by the roles and capabilities of the applications. Often attempts at a solution to this problem try to use UML diagrams. However, these are more suited to technical designers and less to business analysts, and they suffer from the fact they do not capture the information needed to provide a complete solution to this problem. In the section on the UML model, we detail a to-be example based on UML class diagrams.

根据项目的范围,设计工作可以涉及任何内容,从更改完整的流程流到添加、删除业务流程,或者只是对基本行为进行小的更改。每个业务流程中包含的信息对象和相关的动态及静态结构都可能发生类似的变化。这是相关的,因为必须在整个端到端流程中创建信息模型。因此,该阶段的产出/产品是在模型中捕获的新设计的成功组合[16]。

3. 构建

为创建手动或自动的操作系统,流程构建阶段涉及应用在流程设计阶段定义的未来模型(to-be models),包括所有相关的动态和静态结构。

在纯粹的手动情况下,构建阶段通过工作设计、培训和准备文档来处理。在自动化环境生成中,活动可能包括执行或启用该工作的软件中的编程、配置或其他工作。显然,在许多情况下,这两种类型工作只有相互协调才能完成构建以实现新操作所需的结果。

根据面向软件的构建的程度和范围以及预先生成的流程模型的质量,可以使用所谓的"模型驱动设计"[4-17]。但是,部署企业和Web/restful服务时需要比实现或定制企业标准软件更全面的流程模型和方法。在后一种情况下(如ERP系统),也可以通过SAP的解决方案管理器和SAP的业务流程系统(Business Workflow)或BPM系统(如SAP BPM或Software AG webMethods支持)等工具实现部分自动化。

制定面向未来业务规范的业务分析师和实施该系统的技术应用程序开发人员的需求并不总是相同[18]。因此,与建立一个充分实施未来模式的成功的综合工作制度有关的挑战和问题在于这两个群体的成员之间必须紧密合作。部分问题在于:最好在机器应该完成的部分和人类完成的部分之间找到平衡,以及如何最好地在两者之间建立对接。这些问题通常与对问题的认识有关,使用BPM和自动化业务流程可能会导致无法准确满足业务需求的实现。总的来说,结果是对工作系统的描述,其中通过应用程序的角色和功能有效地实现人类工作。解决此问题通常都尝试使用UML图,但是,这些更适合技术设计人员而不适合业务分析人员,并且他们不会得到为此问题提供完整解决方案所需的信息。在UML模型的部分中,我们详细介绍一个基于UML类图的示例。

Deploy/Implement

This is the phase where processes based on the to-be models are put into effect to be used by the business. The process models and the information models within them can be a basis for testing and can be used to offer a high level of support during the implementation phase.

Run/Maintain (Monitoring)

This phase is concerned with the successful operation of business processes and their enablers in a production environment. During this phase, efforts must be made to guard the process to ensure its operations remain consistent with the design objectives. Without oversight, the process may be sub-optimized or otherwise modified in ways that needlessly increase cost or reduce value. This is the main task of the process-monitoring phase, which is the final phase and ultimately is the input to the analysis phase in the next iteration of the cycle. Whereas the analysis phase is concerned with determining possible weaknesses of the dynamic and static structures of the business processes and their interrelation, the monitoring phase is concerned solely with one aspect: measurement of process performance indicators (PPIs) together with time, cost, and quality to verify the status of the process. The Gartner group quoted by Verner[19] coined the term "business activity monitoring" to describe the ability to produce real-time performance indicators to assess speed and effectiveness of business operations.

Continuous Improvement

Once the new business processes are operational, ongoing work is necessary to verify whether the intended goals have been met through a continual effort to learn from and improve on the design of the process to achieve its design goals. These efforts can seek evolutionary change or may involve innovative change to the design.

Continuous improvement is a key aspect of BPM whereby feedback from the process and the customer are evaluated against design goals.

PROCESS ATTRIBUTES

Process Flow and Process Resources

A process flow consists of a set of connected process activities organized into a stream, sequence, course, succession, series, or progression, all based on the process input/output states, in which each process input/output defines the process flow that together performs a behavior. These process activities may connect to static resources, including business objects of various types, and to roles.

Process resources such as roles, which are represented as pools or lanes in BPM notation (BPMN) process or collaboration models, have an important role in describing work, in that they signify the allocations of responsibility and thus require consideration in the analysis and design of the work.

4. 部署/实施

在这一阶段,基于未来模型(to-be models)的流程将供业务使用。流程模型及其中的信息模型可以作为测试的基础,并可用于在实施阶段提供高水平的支持。

5. 运行/维护(监控)

此阶段涉及业务流程及其在生产环境中的促成因素的成功运行。在此阶段,必须努力保护流程,以确保其运营与设计目标保持一致。如果没有监督,流程会被局部优化或以其他方式修改,从而不必要地增加成本或降低价值。流程监控阶段是最后的阶段,也是主要任务,这最终是循环下一次迭代中分析阶段的输入。鉴于分析阶段涉及确定业务流程的动态和静态结构及其相互关系,但监控阶段只涉及一个方面:PPI 的测量,以及验证流程状态的时间、成本和质量。L. Verner[19]引用了 Gartner Group 创造的术语"业务活动监控",用来描述生成实时绩效指标以评估业务运营的速度和有效性的能力。

6. 持续改进

一旦新的业务流程投入运行,就有必要不断开展工作,通过不断努力学习和改进流程的设计来实现预期目标。这些努力可以寻求进化的变化,也可能涉及对设计的创新改变。

持续改进是 BPM 的一个关键方面,可根据流程设计目标对客户的反馈进行评估。

4.6.4　流程属性

1. 流程和流程资源

流程流由一组相互关联的流程活动组成,这些流程被组织成流、序列、过程、连续、系列或进程,所有这些都基于流程输入/输出状态,其中每个流程输入/输出定义一起执行某个行为的流程流。这些流程活动可以连接到静态资源,包括各种类型的业务对象和角色。

流程资源(如角色)在 BPM 标识(BPMN)流程或协作模型中表示为池或通道,在描述工作时具有重要作用,因为它们表示责任分配,因此在分析和设计工作时需要更加认真地考虑。

For transactional and tacit work, process resources may be either human or auto-mated via software applications.[20] Resource allocation can be useful in showing where one system connects resources to another or where there is an exchange between roles. In our business process model examples, we have identified the following resources:

- Enterprise organization (e.g., sales and distribution, marketing department, warehouse employee, etc.)
- System organizational units of ERP (e.g., client, company code, sales area, etc.)
- Information cubes (e.g., purchase order), dimensions (e.g., time, material, unit)
- Business objects of ERP (e.g., SAP purchase order BUS 2012)
- External Web services (Break Even Point)
- Data entities (e.g., customer master file, condition master, customer order)

Data Flow

For the analysis to be sound, data flow needs to be viewed separately from the process flow. A deficiency of BPMN is that it considers just the process flow and does not consider and integrate into a holistic model the separate flows of the business and information objects. Also, BPMN does not recognize that the assignment of business objects or information cubes to process activities may occur and that exposing how, where, and who views static data, information, or data flow is also useful in showing where business data structures are used in the process flow and how they change states.

Process Automation (Application)

Process automation may be supported through a number of means including a spe-cialized BPM engine. To provide a complete solution any tool used to manage pro-cesses requires the specification of the process and data flows, together with their association with the above resources.

WHY THE SUBJECT IS IMPORTANT AND THE PROBLEMS AND CHALLENGES IT WILL SOLVE

A major problem for business process professionals is the volatile environment in which they must drive change through the business process improvement life cycle. The volatility of these conditions is highlighted by the fact that "If there is one constant in the market, it is that things are always changing faster and are more dynamic,"[21] thus enforcing the idea that organizations and enterprises are under continuous pressure when optimizing their business processes and thus have to con-stantly play catch-up with their competitors.

Optimization of business processes most commonly stems from the need to solve three main business problems/strategies:

1. Those that pertain to productivity enhancement
2. Market expansion
3. The creation of new markets[22]

对于事务性和隐性工作,流程资源可以是人力也可以通过软件应用程序实现自动化[20]。在一个系统将资源连接到另一个系统的位置或角色进行交换时,资源分配将变得非常有用。在我们的业务流程模型示例中,我们确定了以下资源:

- 企业组织(如销售与渠道、市场部、仓库员工等);
- ERP系统组织单位(如客户、公司代码、销售区域等);
- 信息多维数据集(如采购订单)、尺寸(如时间、材料、单位);
- ERP的业务对象(如SAP采购订单2012);
- 外部Web服务(盈亏平衡点);
- 数据实体(如客户主文件、条件主文件、客户订单)。

2. 数据流

为了使分析合理,需要将数据流与流程流分开查看。BPMN的一个缺点是它只考虑流程流,并没有考虑并将业务和信息对象的单独流程整合到整体模型中。此外,BPMN不会将业务对象或信息多维数据集分配给流程活动,包括公开如何、在何处和何人,即可以查看静态数据、信息或数据流的人员,尽管这些信息在显示流程中使用的业务数据结构以及它们如何更改状态方面是很有用的。

3. 流程自动化(应用)

可以通过包括专用BPM引擎在内的许多手段来支持流程自动化。为了提供完整的解决方案,任何用于管理流程的工具都需要规范流程和数据流,以及它们与上述资源的关联。

4.6.5 为什么这个主题很重要,它将解决的问题和挑战

业务流程专业人员面临的一个主要问题是动荡的环境。在这种环境中,他们必须在业务流程改进生命周期中推动变革。"如果市场上有一个不变的因素,那就是事情总是变化得更快、更有活力"[21]这有助于强化组织和企业不断面临压力的理念。这一事实凸显了这些条件的波动在优化业务流程时,必须不断与竞争对手保持同步。

业务流程优化最常见的原因是需要解决三个主要的业务问题/策略:

(1)与提高生产力有关的方面;

(2)市场的扩张;

(3)新市场的建立[22]。

The goal of optimizing business processes can also be one of pure optimization, by reducing time and costs and improving quality within the organization.

The interrelations between these and other strategies or strategic goals are depicted as cause-and-effect chains within balanced scorecards being addressed within the examples of the to be models.

In a report by the Gartner Group,[23] one of the four usage scenarios driving the purchase of BPM Suites was the "Support for a continuous process improvement program," which highlights recognition of the need to optimize business processes. This change is important to enable an enterprise to overlay its application assets with a business-level representation of the end-to-end processes that are then supported by the software assets. This allows the enterprise to see and assess how applications contribute capability and enable the business. The model-driven approach is seen as one of the best ways to enable business and IT professionals to manage and change processes collaboratively to achieve these improvements. Although process-centric models have a critical role in this work, these models must be both complemented by and connected to other applicable information models. Collectively, this approach creates a unified set of models that can provide a complete picture of all phases of the process life cycle. The result is a portfolio of business-oriented models that foster a shared understanding as to how best to pursue business process management objectives.[24]

There has been a significant rate of failure of many BPM projects. The size and cost[25] of these failures expose the correlation between the need for improvement of process and information models and the need for successful completion of the BPM projects. The fact that these models must cross all levels and hierarchies of an organization creates a high level of complexity, with the consequence that many levels of decomposition/composition are required to produce useful and consistent information models,[3] and which therefore can be controlled. Often the reason for failure of the BPM projects lies with the problem that the initial process requirements were not correctly understood, formulated, or communicated throughout the design process.[26] Again, this highlights the need for methods of representation to empower the business process engineer, together with the tools and infrastructure engineers, and other contributors and stakeholders, to achieve greater success.[27]

INFORMATION MODELS WITHIN AS-IS AND TO-BE MODELS

Among the many challenges associated with process modeling, process engineering, and process architecture, questions about how to produce quality as-is and to-be process models are of great concern. The answer to these challenges is not easy because BPM and BPMN do not consider a process in its full context; it is extremely difficult to repeatedly determine the scope, level, and quality standard for processes.

Figure 3 illustrates the architectural layers that are relevant to the analysis, specification, and management of process, e.g., process modeling, as well as relevant to the context of the process architecture. This figure shows the process layer as enabled by the behavior and features of the objects in the application layer, which in turn provides access to the persistent data structure of the data layer. In addition,

优化业务流程的目标也可以是纯优化的目标之一，通过减少时间和成本提高组织质量。

这些战略与其他战略或战略目标之间的相互关系被描述为平衡记分卡中的因果关系链，并在未来的模型中加以处理。

Gartner 集团的一份报告指出[23]，推动购买BPM套件的四种使用方案之一是"支持流程持续改进计划"，这突出了对优化业务流程需求的认识。这种变化对于使企业能够将其应用资产与随后由软件资产支持的端对端流程的业务级别表示进行覆盖是很重要的。这使企业能够查看和评估应用程序如何实现功能并使业务成为可能。模型驱动的方法被视为使业务和IT专业人员能够协作管理和更改流程以实现这些改进的最佳方法之一。尽管以流程为中心的模型在这项工作中起着关键作用，但这些模型必须与其他适用的信息模型相互补充并相互连接。总的来说，这种方法创建了一组统一的模型，可以提供流程生命周期所有阶段的完整画面。最终的结果是得到一系列面向业务的模型，这些模型促进了对如何最好地实现BPM目标的共同理解[24]。

许多BPM项目的失败率很高，这些失败的规模和成本[25]暴露了流程和信息模型改进的需求与成功完成BPM项目需求之间的相关性。这些模型必须跨组织的所有级别和层次，这一事实造成了高度的复杂性，其结果是需要许多级别的分解/组合来生成有用且一致的信息模型[3]，因此必须控制这些信息模型。BPM项目失败的原因通常在于最初的流程需求在整个设计流程中没有得到正确的理解、制定或沟通[26]。这突出表明需要一种代表方法来赋予业务流程工程师及工具和基础架构工程师以及其他贡献者和利益相关者权力，从而获得更大的成功[27]。

4.6.6　现状（As-Is）和未来（To-Be）模型中的信息模型

在与流程建模、流程工程和流程体系结构相关的许多挑战中，关于如何生成现有和将来的流程模型的问题是非常值得关注的。得到这些挑战的答案并不容易，因为BPM和BPMN没有在其完整的背景中考虑流程，重复确定过程的范围、级别和质量标准也是极其困难的。

图3说明了与流程的分析、规范和管理相关的体系结构层（如流程建模），以及与流程体系结构的背景相关的体系结构层。图3显示了由应用层中对象行为和功能启用的流程层，这反过来又提供了对数据层持久性数据结构的访问。此外，

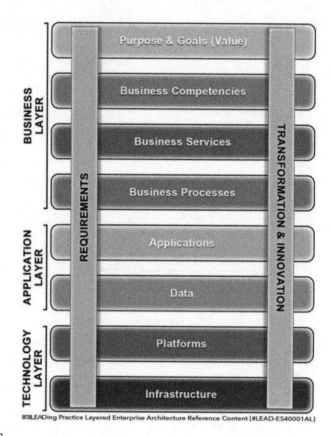

FIGURE 3

Architectural layers.

the figure shows that the need for a process to produce value, and thereby support the enterprise strategy, may only be achieved through the objects within the competency and service layer. It is the services that expose the value of the processes and the competencies that organize, contextualize, and align the processes and services to the enterprise view of value. Enterprise processes and therefore enterprise process models must be designed within and connect to this context and to the relevant objectives that reside in each. By working in this manner, we are exercising the principles behind the objects that ensure that the object of interest, in this case a process, is completely and fully defined. Furthermore, the value in assigning the objects across the layers is that within the layers the various stakeholders who have concerns about the objects view them.

It is important when defining the contextual, conceptual, logical, and physical aspects[28] It relates to a specific way of modeling process aspects of object clustering to define the correct levels of hierarchy. In our example organization, we have assigned four levels, as shown in Figure 4.

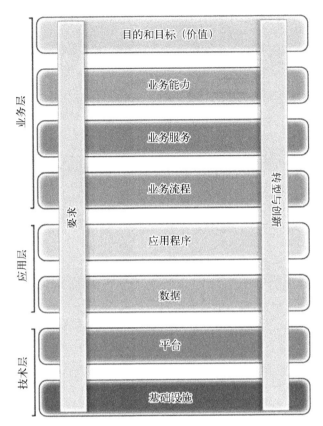

图 3　架构层

图3显示,只有通过能力和服务层内的对象才能实现一个流程来产生价值,从而支持企业战略。正是这些服务展示了流程的价值和能力,从而将流程和服务组织起来、进行继承,并使其与企业价值观保持一致。企业流程以及企业流程模型必须在此继承中设计并连接到每个背景中的相关目标。通过以这种方式工作,我们正在使用对象背后的原则,以确保感兴趣的对象(在这种情况下是一个过程)被完全和完全地定义。此外,跨层分配对象的价值在于各层中,对对象有所了解的各种利益相关者都查看了这些对象。

在定义前后关系、概念、逻辑和物理方面[28]时,为定义层次结构的正确层次,它涉及一种特定的对象聚类过程方面的建模方法。在我们的示例组织中,我们分配了四个级别。如图4所示。

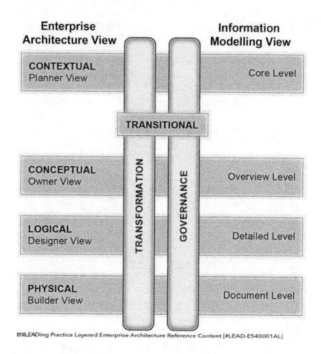

FIGURE 4

Conceptual and logical object clustering hierarchy levels.

The levels and views in Figure 4 should be understood in the following ways [29]:

- Contextual models are the perspective of the planners of the enterprise, and in creating the link between process and information models this is a core level.
- Conceptual models are the perspective of the owners of the enterprise, and in creating the link between process and information models this is the overview level.
- Logical models are the perspective of the designers of the enterprise, and in creating the link between process and information models this is the detailed level.
- Physical models are the perspective of the builders of the enterprise, and in creating the link between process and information models this is document level.

What differentiates the views and levels are not only the details, but in reality the specific models used or developed within them and subsequently different contexts in terms of purpose and goals from the models. The reason this is so important is that the different levels all have different value potential, e.g., purpose and goals, and as a result the different views and levels have their specific transformation potential and governance concept that need to be explored and interlinked throughout the layers (Figure 4). Decomposition and composition happen through the relevant objects across the views and levels and their models, an abstraction that represents and considers the process and information as a whole. As illustrated in

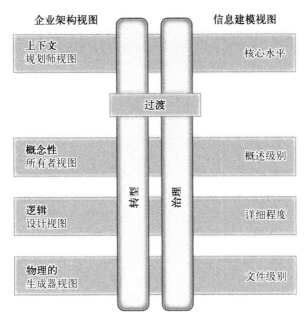

图4　概念和逻辑对象聚类层次结构级别

图4中的层次和视图应通过以下方式理解[29]：

- 前后关系模型是企业规划者的视角，在创建流程和信息模型之间的链接时，这是一个核心级别；
- 概念模型是企业所有者的视角，在创建流程和信息模型之间的链接时，这是概述级别；
- 逻辑模型是企业设计者的视角，在创建流程和信息模型之间的链接时，这是详细的层次；
- 物理模型是企业构建者的视角，在创建流程和信息模型之间的链接时，这是文档级别。

区别视图和层次的不仅是细节，而且实际上是在其中使用或开发的特定模型，以及随后在目的和目标方面不同于模型的不同前后关系。这一点之所以如此重要，是因为不同的层次都具有不同的价值潜力（如目的和目标），因此不同的视图和层次都有其特定的转换潜力和治理概念，需要在各个层次中进行探索和相互关联（图4）。分解和组合通过视图和层次及其模型中的相关对象发生，这是一个将过程和信息作为一个整体来表示和考虑的抽象。如图3所示，企业应该被视为一

Figure 3, an enterprise should be considered as a whole which subsequently includes the views and models that capture the

- Business layer, such as the resources, roles, value aspects, enterprise capabilities, functions, and services
- Application layer, representing the automated processes and thereby the application components, application modules, tasks, application services, and data components, data objects, data entities, data tables, and data services
- Technology layer, such as the platform components, platform function, platform devices, and platform services, as well as the infrastructure components, infrastructure functions, infrastructure devices, and infrastructure services

In addition to the views and levels discussed, aspects important for both information modeling and process modeling are the subject of tagging and thereby classification and categorization. Processes, information objects, and services can be tagged according to their strategy, tactics, and operational tiers. Figure 5 illustrates an example of the enterprise tiers and relevant process areas.[30]

As illustrated in Figure 5, the enterprise tiers represent tagging possibilities that link the processes, goal and objective view, decision making, and system measurement and reporting view. Therefore, classifying the process links to multiple aspects needed in the information models:

- Strategic aspects: This tier affects the entire direction of the firm. An example is the mission, vision, strategic business objectives (SBOs), and specific business performance indicators (BPIs) and business plans. The strategic tier has the long-term, complex decisions made by executives and senior management and the measurement reporting view is used for the most scorecards.
- Tactical aspects: The aspects at this tier are more medium-term, less complex decisions made mostly by middle managers. They follow from strategic decisions and aim to meet the critical success factors, the way to do this is for governance, evaluation, reports, control and monitoring and the measurement reporting view which is used for most dashboards.
- Operational aspects: At this tier day-to-day decisions are made by operational managers and are simple and routine; the measurement reporting view used is for the most cockpits.

As-Is Modeling

The purpose of as-is modeling is to explore and capture how the processes are performed today. This provides a baseline for describing the business.

Determining the Hierarchy Level

The following section describes a suitable procedure for determining the level within the hierarchy that is applicable to the analytical work being performed, the so-called decomposition or 'composition level of the information objects.[31]

个整体,包括以下内容的视图和模型:

- 业务层,如资源、角色、价值方面、企业能力、功能和服务;
- 应用层,表示自动化流程,从而表示应用程序组件、应用程序模块、任务、应用程序服务和数据组件、数据对象、数据实体、数据表和数据服务;
- 技术层,如平台组件、平台功能、平台设备和平台服务,以及基础设施组件、基础设施功能、基础设施设备和平台服务。

除所讨论的视图和级别之外,对信息建模和流程建模都很重要的方面是标记的主题,从而进行分集和分类。流程、信息对象和服务可以根据它们的策略、战术和操作层进行标记。图5展示了企业层和相关流程区域的示例[30]。

如图5所示,企业层表示链接流程、目标和目标视图、决策制定以及系统度量和报告视图的标记可能性。因此,对流程进行分类可以链接到信息模型中需要的多个方面,包括以下方面。

- 战略层面:这一层影响公司的整个方向。例如,任务、远景、SBO、特定业务绩效指标和业务计划。战略层具有由高管和高级管理人员作出的长期、复杂的决策,而大多数记分卡都使用评估报告的方式。
- 战术层面:这一层的层面更多的是中期的、不那么复杂的决策,主要由中层管理者做出。他们遵循战略决策,以满足关键的成功因素为目标。这样做的方法是用于治理、评估、报告、控制和监视,以及用于大多数仪表板的度量报告视图。
- 运营方面:在这一层,日常决策由运营经理做出,简单而常规,使用的测量报告视图适用于大多数驾驶舱。

1. 现状建模(As-Is modeling)

AS建模(As-Is modeling)的目的是探索和捕获当前流程是如何执行的,这为描述业务提供了一个基线。

2. 确定层次结构级别

以下部分描述用于确定层次结构内适用于正在执行的分析工作的级别的合适过程,即所谓的信息对象的分解或组合级别[31]。

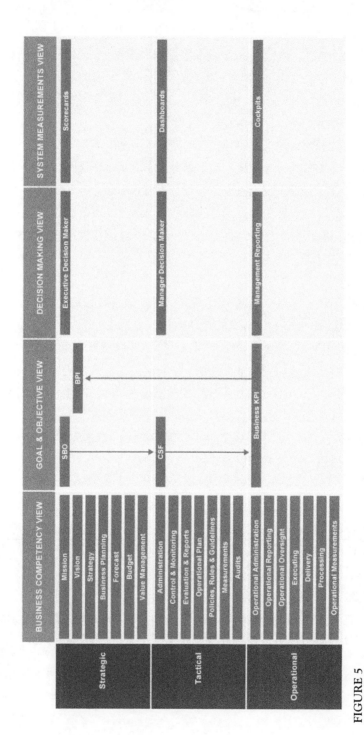

FIGURE 5

Example of enterprise tiers.

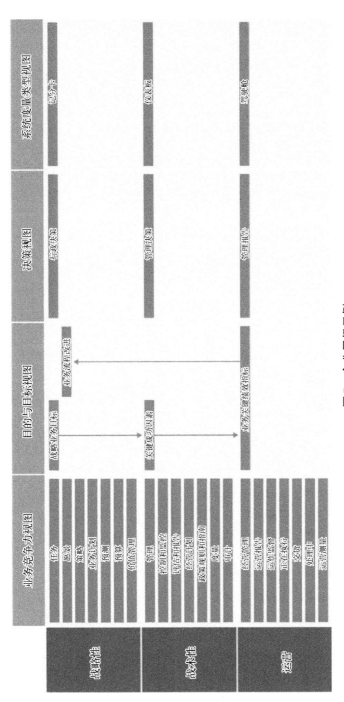

图 5　企业层级示例

Various alternative views exist of the approach to this particular problem. Figure 6 presents the principle variations: the supply chain operations reference model (SCOR)[32] and the American productivity and quality center (APQC).[33] As shown in the figure, each framework attempts to describe and populate the various levels of detail of process with authoritative process inventories. They alternatively provide for four or five levels of process decomposition. On the other hand, the SAP Business Blueprint Solution Manager (in the current version 7.1), which must implement these processes, has support only for three process levels. Unification is therefore not possible without finding another way to approach the problem.

Another related challenge is that when other models are been used, these models must also be consistent with and align with the applicable process model structure. For example, for the models used to describe an enterprise to be unified, it is critical that there be a method to connect horizontally i.e., within the same level, an information object such as resources (Business Competency layer of figure 3) to relate in a logical and coherent way with an information object from the process layer, e.g., a process activity.

A possible solution to this problem is to find and establish horizontal and vertical connections between the objects of interest so as to place the various concepts in layered structures to link the leading process layer structure from "above" to all other applicable layers while simultaneously consolidating/integrating them with their respective data layer from "below". Looking at the intersection of the different frameworks in Figure 6, it makes sense for the processes and all other layers to be set initially to three levels with a default going downward, e.g. Level 1–Business Process, Level 2–Business Step, and Level 3–Process Activity (ref. LEAD column in Figure 6). Process activities access data entities (Data layer) on the same level (horizontal navigation), with the result that finally the data table (data layer) appears (vertical navigation) at Level 4, where the associated key (used to establish and identify relations between tables), foreign keys (establish and enforce a link between

Levels of Hierarchy	APQC PCF	LEAD	SAP SolMan	SCOR
	1. Category	Process Area		
	2. Process Group	Process Group		
	3. Process	1. Business Process	1. Szenario	1. Level 1 2. Level 2
	4. Activity	2. Process Step	2. Process	3. Level 3
	5. Task	3. Process Activity	3. Process Step	4. Level 4
		4. Transaction / PPI		

FIGURE 6

Comparison of the different levels of hierarchy of the process layer.

关于解决这个特殊问题的方法,存在着各种不同的观点。图6给出了变更的原理:SCOR[32]和APQC[33]。如图6所示,每个框架都试图用权威的流程清单来描述和填充流程的各个详细级别。它们还可以提供4～5个级别的流程分解。另一方面,SAP业务蓝图解决方案(Business Blueprint Solution Manager)(版本7.1)必须实现这些流程,它只支持三个流程级别。因此,如果不找到解决这个问题的另一种方法,统一是不可能的。

另一个相关挑战是当使用其他模型时,这些模型还必须与采用的流程模型结构保持一致。例如,对于用于描述要统一的企业的模型,关键是有一种方法可以水平连接,即在同一级别中,信息对象(如图3中的业务能力层)以逻辑和一致的方式与流程层中的信息对象(如流程活动)相关联。

这个问题的一个可能的解决方案是在感兴趣的对象之间找到并建立水平和垂直的连接,以便将各种概念放在分层结构中,将领先的流程层结构从"上面"链接到所有其他适用的层,同时根据它们的方面对它们进行合并/集成。从图6中不同框架的交叉点来看,流程和所有其他层最初设置为三个级别是有意义的,默认值向下。如1级——业务流程、2级——流程步骤和3级——流程活动[参见图6中的领导列(LEAD列)]。流程活动访问同一级别(水平导航)上的数据实体(数据层),最终数据表(数据层)的结果出现在级别4(垂直导航)上,其中关联键(用于建立和标识表之间的关系)、外键(建立和实施两个表之间的链接)和数据元素找

层次结构的层次	美国生产力与质量中心分组控制功能	领导	SAP SolMan	SCOR
	1. 类别	流程领域		
	2. 流程组	流程组		
	3. 流程	1. 业务流程	1. 场景	1. 1级 2. 2级
	4. 活动	2. 流程步骤	2. 流程	3. 3级
	5. 任务	3. 流程活动	3. 流程步骤	4. 4级
		4. 交易/PPI		

图6 流程级别不同层次结构的比较

two tables), and descriptive attributes are found. These attributes are then accessed by transactions (process layer) on Level 4, representing a defined (committed) status of data input and output, creating a spine.

Another important factor that indicates the positioning of transactions at Level 4 is that performance indicators (value layer) must use this level to determine the achievement of strategic business objectives. In all other layers, this performance indicator appears on the same Level 4 e.g. Business Compliance, Service Level Agreements, Process Performance, IT Governance.

Meta Information Objects Within Information and Process Modeling

To answer the question of how to model the business and application layer meta objects, we begin by providing two summaries (Figures 7 and 8). In each case, object mapping is based on the use of four business layers (whereby components are distinguished by their contribution to value,[34] competency,[35] service,[36] and process[37]) and two application layers (which classify the components as to whether they are part of application structure[38] and behavior, or data[39]). These are brought together in a matrix. The layers are classified side by side in six columns to set all objects to a coherent set of categories or layers, and then into a hierarchy of levels (Levels 1–4) to distinguish between their areas of contribution.

Looking further, Level 3 contains the data media or data objects representing data entities and dimensions (application layer). The latter connects directly to Level 2 above, together with the information cubes. From the service group (service layer) on Level 2, individual business services connected to business objects on Level 3 can be refined (Figure 7). These in turn are used with the business objects (application layer) to encapsulate process activities and events (process layer) and the data entities (application layer) on the same level (horizontal).

To complete the picture from a business perspective, the organizational structure of the enterprise (competency layer) must now be included in and distributed across the layers. These can be seen in Levels 1–3. The business areas consist of business groups; business roles are thus assigned to both business areas and business groups (competency layer). When more than three levels of enterprise hierarchies exist, it is useful to divide these into the context of three separate process levels. At Level 3 only business roles are used.

The so-called (by SAP) system organizational unit structure of the ERP application (application layer) should be modeled on the similar-sounding but different internal departmental structure. In contrast to the organizational structure for employees in the enterprise, this structure contains the mapping of external customers and suppliers, services, stock flow, cash inflows/outputs, etc., with a process activity. The system organizational units also constitute a hierarchy of several levels. A process activity on Level 3 of the process layer can access any level in the hierarchy of the system organization. All SAP Solution Manager "compatible" information objects are highlighted blue in figure 7.

到描述性属性。然后,事务(流程层)在级别4上访问这些属性,表示数据输入和输出的已定义(提交)状态,从而创建一个脊椎图的样子。

另一个表明交易在级别4的定位的重要因素是,绩效指标(价值层)必须使用该级别来确定战略业务目标的实现。在所有其他层中,此性能指标显示在同一级别4上,如业务合规性、服务级别协议、流程性能、IT治理。

3. 信息和流程建模中的元信息对象

为了回答业务如何建模和应用层元对象的问题,我们首先提供两个介绍(图7和图8)。在每种情况下,对象映射都基于四个业务层的使用(其中组件通过它们对价值的贡献[34]、能力[35]、服务[36]和流程[37]来区分)和两个应用层(它们将组件分类为它们是否是应用程序结构[38]和行为或数据[39])的一部分)。这些以矩阵形式汇集在一起,这些图层按六列并排分类,将所有对象设置为一组连贯的类别或图层,然后设置为级别层次结构(级别1 ~ 4),以区分它们的贡献区域。

更进一步,第3层包含表示数据实体和维度(应用层)的数据媒体或数据对象。后者与信息多维数据集直接连接到上面的第2级。从级别2上的服务组(服务层)可以细化连接到级别3上的业务对象的各个业务服务(图7)。这些又与业务对象(应用程序层)一起使用,以在同一级别(水平)上封装流程活动和事件(流程层)及数据实体(应用程序层)。

为了从业务角度完成这幅图,现在必须将企业的组织结构(能力层)分布在各层中。这些可以在1 ~ 3级中看到。业务领域由业务集团组成,因此,业务角色被分配给业务领域和业务组(能力层)。当存在三个以上的企业层次结构时,将它们划分到三个单独的流程层次的前后关系中是很有用的。在第3级,只使用业务角色。

ERP应用程序(应用层)的所谓(通过SAP)系统组织单元结构应该模仿类似但不同的内部部门结构。与企业中员工的组织结构不同,这种结构包含外部客户和供应商、服务、库存流、现金流入/输出等的映射以及流程活动。系统组织单位也构成若干层次的层次结构。流程层第3层上的流程活动可以访问系统组织层次结构中的任何级别。所有SAP解决方案管理器(SAP Solution Manager)"兼容"的信息对象在图7中以蓝色突出显示。

Information Meta Objects Mapping

Layer	Business Layer				Application Layer (ERP, BI, InMemory, Mobile, SOA)	
Level of Decomp.	Value	Competency	Service	Process (BPMN)	Application	Data
1	Vision					Information Object
	Mission					Information Object
	Strategy					Information Object
	Goal					Information Object
		Business Area	Service Area	Business Process	Application Module	Information Object
		Organizational unit		Pool	Organizational Unit	Information Object
						Information Object
2	Strategy					Information Object
	Goal					Information Object
		Business Group	Service Group	Process Step/Sub Process	Application Module	Information Object
		Organizational unit		Lane	Organizational Unit	Information Object
		Revenue/CostFlow		Revenue/CostFlow		Information Object
		Information Cube		Service Group(Flow)	Information Cube	Information Object
						Information Object
						Information Object
3	Strategy					Information Object
	Goal					
	Objective					
		Business Object	Business Service	Process Activity	Application Function / Business Object	Information Object
				Screen	Transaction Code	Information Object
				Lane	System Organizational Unit	Data Entity
				Events	Business Object	Data Entity
				Lane	Business Object	Data Entity
				Lane	Dimension	Data Entity
				Lane	Data Entity	Data Media
				Data Object	Data Object	Data Rules
		Business Roles	Services Roles	Lane	Application Roles	
		Business Rules	Service Rules	Process Rules	Application Rules	Fact Table
			Service Level Agreement (SLA)	Process Performance Indicator (PPI)	System Measurements	
4	Performance Indicator	Business Compliance			IT Governance	Customizing Data Table
				Transaction	Application Task	Master Data Table
						Transaction DataTable
						Key
						Foreign Key
						Describing Attributes

FIGURE 7

Mapping meta objects.

信息元对象映射									
层	业务层				应用层(ERP, BI, InMemory, Mobile, SOA)				
分解等级	价值	能力	服务	流程（BPMN）	应用		数据		
1	愿景						信息对象		
	任务						信息对象		
	策略						信息对象		
	目标						信息对象		
		业务领域	服务区域	业务流程	应用模块		信息对象		
		组织单元		集资	组织单元		信息对象		
2	策略						信息对象		
	目标						信息对象		
		业务组	服务组别	流程步骤/子流程	应用模块		信息对象		
		组织单元		道	组织单元		信息对象		
		收入/成本流		收入/成本流			信息对象		
		信息立方体		道	信息立方体		信息对象		
				服务组（流程）			信息对象		
3	策略						信息对象		
	目标						信息对象		
	目的						信息对象		
		业务对象	业务服务	流程活动	应用功能	业务对象	信息对象		
				屏幕	交易代码		信息对象		
				车道	系统组织单位		数据实体		
				事件	业务对象		数据实体		
				道	业务对象		数据实体		
				道	尺寸		数据实体		
				道	数据实体		数据实体		
				数据对象	数据对象		数据媒体		
		业务角色	服务角色	道	应用角色		数据实体		
		业务规则	服务规则	流程规则	应用规则		数据规则		
4	绩效指标	业务合规性	服务水平协议（SLA）	流程绩效指标（PPI）	IT治理	系统测量	事实表		
							自定义数据表	主数据表	交易数据表
				交易	应用任务		关键	外键	描述属性

图7 映射元对象

Typical decomposition structures of the meta objects are found when navigating vertically downward; correspondingly, the compositions are found when navigating vertically upward. For example, a business area on Level 1 can be aligned horizontally against a business process or an application module, whereas a business area on Level 1 can be refined into a business group on Level 2. A business process consists of process steps or sub-processes. Information meta objects that have a vertical assignment to an underlying level can be aligned to the lower right side to an additional symbol that branches into one or the more appropriate models of the underlying layer shown within Figure 7. Organizational unit' (competency layer) may appear in the processes at Level 1 as BPMN pools (process layer) or in the lower hierarchy Level 2 as BPMN lanes. All system organizational units are assigned to Level 3. Their keys and foreign keys such as sales organization are assigned to Level 4.

As the engine that informs, influences, and drives all other behavior, vision, mission, strategy, and goal are assigned to the value layer at Level 1. Because they may be constrained by these larger factors, strategy and goal again appear at Levels 2 and 3. The value layer is not going to be implemented but realized, shown as information objects in the matrix of Figure 7. The same applies to the revenue/cost flow as well as the group services on Level 2 of the competency and service layer. Business services, however, are considered on Level 3, as methods of business objects. Their implementation is completed as a process activity (process layer) or application function (application layer). Finally, process activities may appear as collapsed sub-processes in BPMN diagrams at Level 2.

Information cubes (application layer) exist only on Level 2 to support the field of business intelligence. Information cubes consist of dimensions on Level 3 and fact tables on Level 4 of the data layer. On Level 4, the data layer contains master, transaction, and customizing tables as well as their associated keys, foreign keys, and descriptive attributes. The corresponding attributes feed (horizontally), e.g., the PPI of the process layer or the SLA of the service layer.

Business rules culminate in our example in process or application rules; responsibility for the integrity of the data rules (e.g., entity and referential integrity) lies with the database management system. Service rules will not be considered further in our example.

Level 3 of the process layer contains items that are considered resources respectively lanes of processes: system organizational units, business objects, dimensions, data entities, and roles. In xBPMN, data objects represent information objects and are interpreted within our example as data media (document) on Level 3 representing data entities at Level 3 or a data table on Level 4.

The information meta objects in fact have many more relationships than previously mentioned within this hierarchy; all relationships are shown in the following models in Figure 8, in which exactly one layer and one level are identified. The respective models represent more than one layer or more than one Level (e.g., hierarchical models) and therefore its information meta objects from Figure 7 can appear multiple times.

　　垂直向下导航时可以发现元对象的典型分解结构。相应地,当垂直向上导航时,可以找到组合。例如,级别1上的业务区域可以与业务流程或应用程序模块水平对齐,而级别1上的业务区域可以细化为级别2上的业务组。业务流程由流程步骤或子流程组成。对底层具有垂直分配的信息元对象可以对齐到另一个符号的右下角,该符号分支到图7中所示的更合适的底层模型之一。组织单元(能力层)可以作为BPMN池(流程层)出现在级别1的流程中,也可以作为BPMN通道出现在级别2较低的层次结构中。所有系统组织单元被分配到第3级。他们的密钥和销售组织等外键被分配到第4级。

　　作为通知、影响和驱动所有其他行为、愿景、任务、战略和目标的引擎,它们被分配到1级的价值层。因为可能会受到这些较大因素的限制,策略和目标再次出现在2级和3级。价值层不是将要被实现而是已实现,并在图7的矩阵中显示为信息对象。这同样适用于收入/成本流以及能力和服务层的第2级上的组服务。但是,业务服务在第3级被视为业务对象的方法,它们的实现是作为流程活动(流程层)或应用程序功能(应用层)完成的。最后,在第2级的BPMN图中,流程活动可能显示为折叠的子流程。

　　信息多维数据集(应用层)仅存在于第2层,以支持商业智能领域。信息多维数据集由数据层第3层的维度和第4层的事实表组成,在第4层,数据层包含主表、事务表和自定义表及其关联键、外键和描述性属性。相应的属性提要至(水平地),例如,流程层的PPI或服务层的SLA。

　　业务规则在我们的流程或应用程序规则示例中达到顶峰,数据库管理系统负责数据规则的完整性(如实体完整性和引用完整性),在我们的示例中不会进一步考虑服务规则。

　　流程层的第3级包含分别被视为资源的流程通道项,包括:系统组织单位、业务对象、维度、数据实体和角色。在xBPMN中,数据对象表示信息对象,在我们的示例中被解释为级别3上的数据媒体(文档),表示级别3的数据实体或级别4的数据表。

　　信息元对象实际上比之前在这个层次结构中提到的有更多的关系。图8中的模型显示了所有的关系,其中确定了一个层和一个级别。各自的模型表示多个层或多个级别(如分层模型),因此图7中的信息元对象可以出现多次。

Information Models Mapping

Layer	Business Layer				Application Layer	
Level of Hierarchy	Value	Competency	Service	Process	Application	Data
1	Balanced Scorecard; Objective Diagram	Organizational Chart (Business)	Function Tree	BPMN Process Diagram (Business)	Value Added Chain Diagram; Function Allocation Diagram (level 0); Function Allocation Diagram	
2	Balanced Scorecard; Objective Diagram	Organizational Chart (Business)	Function Tree	BPMN Process Diagram (Business)	Value Added Chain Diagram; Function Allocation Diagram	
3	Balanced Scorecard; Objective Diagram; KPI Allocation Diagram	Organizational Chart (Business); Business Vocabulary; Accounting Model	Function Tree; UML Class Diagram	E-Business Scenario Diagram; BPMN Collaboration Diagram (Business Rules)	eEPC; BPMN Process Diagram (Application); Function Allocation Diagram	Data Warehouse Structure Diagram (Information Cube); BPMN Process Diagram (Data); Organizational Chart (Application); ERM; Information Engineering
4	KPI Allocation Diagram	KPI Allocation Diagram	KPI Allocation Diagram	KPI Allocation Diagram	KPI Allocation Diagram; Screen Diagram (Mobile)	Document Flow; Data Warehouse Structure Diagram (Dimension); Data Warehouse Structure Diagram (Fact Table); Attribute Allocation Diagram

FIGURE 8

Mapping information models.

信息模型映射 层 级别层次结构	业务层				应用层	
	价值	能力	服务	流程	应用	数据
1	平衡计分卡 目标图	组织结构图（业务）	功能树	BPMN流程图（业务）	增值链图 功能分配图（0级）	
2	平衡计分卡 目标图	组织结构图（业务）	功能树	BPMN流程图（业务）	增值链图 功能分配图	数据仓库结构图（信息立方体）
3	平衡计分卡 目标图	组织结构图（业务） 业务词汇 会计模型	功能树 UML类图	电子商务场景 BPMN协作图（业务规则）	3PMN流程图（应用） 扩展事件驱动流程链 功能分配图	BPMN流程图（数据） 企业资源管理 文件流程 组织结构图（应用） 信息工程
4	KPI分配图 KPI分配图	KPI分配图 KPI分配图	KPI分配图	KPI分配图	KPI分配图 屏幕图（手机）	数据仓库结构图（维度） 数据仓库结构图（事实表） 属性分配图

图 8　映射信息模型

The balanced scorecard (value layer) for the organization should not only exist on Level 1 but should also be included (cascaded) to Level 2 as departmental balanced scorecards and therefore exist for each individual department (Business Group). On Level 3 we can find an employee balanced scorecard. In addition to the external customer/supplier relationship at Level 1, the related cause-and-effect chains on Level 2 should also depict the internal relationship among all departments (Business Groups). The objective diagram on Level 1 shows an objective hierarchy for each of the four perspectives of the balanced scorecard of the enterprise (Level 1) and its departments (Level 2), whereas Level 2 connects the strategic objectives (goals) of the departments of the balanced scorecard to the corresponding process steps. Via key performance indicator (KPI) allocation diagrams on Levels 3 and 4, the goals are connected to objectives that are later connected to their KPIs.

The organizational structure (competency layer) includes three levels; the corresponding department hierarchy can be mapped into a single organizational chart or broken down into hierarchies over these three levels. The individual departments are identified for reasons of simplification in our example as cost centers. Using the standard accounting model, the individual transactions will be booked according to the rules of accounting on Level 3. The KPI allocation diagram on Level 4 is used to measure business compliance.

Value-added chain diagrams describe Levels 1 and 2 of the application layer (Figure 8). Level 2 can also be represented by either an xBPMN process diagram or an e-business scenario diagram (process layer). On Level 3 both EPC and xBPMN process diagrams are used in the application layer. The connection to the business objects is represented through an UML class diagram (service layer). The KPI allocation diagrams on Level 4 cover the measurement of the service level agreements and process performance indicators (process layer).

The lowest level of the application and data layer (Level 4) covers the screen diagram and the attribute allocation diagram. The requirement and importance of fully integrating mobile workplaces into an organization's business processes is paramount in today's mobile society.[40] The key, foreign key, and attributes that describe the transactions of a screen and documents, and system organizational units are mapped to the attribute allocation diagram. The KPI allocation diagrams map the values of KPI (IT Governance) that have been identified with the fact table.

On Level 3 the data layer contains a document flow diagram, an ERM, and an information engineering model. The data warehouse information cubes are represented as star schema on Level 2, as dimensions on Level 3, and as fact tables on Level 4 of the data layer. One BPMN process diagram (data) specifies and collects the assignments of the data entities (data layer) and one BPMN process diagram (application) the system organizational units (application layer) of the function allocation diagrams, showing the process models in different complementary views for demo company Global Bike inc. (GBI).[41]

组织的平衡记分卡(价值层)不仅应该存在于1级,还应该作为部门平衡记分卡包含(级联)到2级,因此每个部门(业务组)都应该存在。在3级,我们可以找到员工平衡计分卡。除了第1级的外部客户/供应商关系外,第2级的相关因果链还应描述所有部门(业务组)之间的内部关系。级别1的目标图显示了企业(级别1)及其部门(级别2)的平衡计分卡的四个视角中的每个视角的客观层次结构,而级别2连接了平衡计分卡上部门战略目标(目标)到相应的流程步骤。通过级别3和级别4上的KPI分配图,目标与后来连接到其KPI的目标相关联。

组织结构(能力层)包括三个层次。相应的部门层次结构可以映射到单个组织结构图中,也可以分解为这三个级别的层次结构。在我们的示例中,由于简化,将各个部门标识为成本中心。使用标准会计模型,将根据第3级的会计规则来记录单个交易。级别4上的KPI分配图用于衡量业务合规性。

增值链图描述了应用层的第1级和第2级(图8)。级别2也可以由xBPMN流程图或电子化的业务方案图(流程层)表示。在级别3,在应用层中使用EPC和xBPMN流程图。与业务对象的连接通过UML类图(服务层)表示相似,第4级的KPI分配图包括服务水平协议和PPI(过程层)的度量。

应用程序和数据层的最底层(第4层)包括软件功能界面图和属性分配图。在当今的移动社会,将移动办公场所完全集成到组织的业务流程中的需求和重要性是至关重要的[40]。描述软件功能界面和文档事务以及系统组织单元的键、外键和属性映射到属性分配图。KPI分配图映射已经与事实表标识的KPI(IT治理)的值。

在第3层,数据层包含文档流程图、ERM和信息工程模型。数据仓库信息多维数据集在级别2上表示为星型模式,在级别3上表示为维度,在数据层的级别4上表示为事实数据表。一个BPMN流程图(数据)指定并收集功能分配图的系统组织单元(应用层)的数据实体(数据层)和一个BPMN流程图(应用程序)的分配,为原型(DEMO)演示公司Global Bike inc.(GBI)以不同的视角展示流程模型[41]。

EXAMPLE AS-IS MODEL (SALES AND DISTRIBUTION)

Business Process Model and Notation Model

Business process model and notation (BPMN) is a standard for graphical representation of business processes that provides a means for specifying business processes.[42] The objective of BPMN is to support business process management for both technical users and business users.

During the categorization of information models, the information meta objects have an important part in the analytic process, depending on whether the objects are types (like "employee") or instances of types (like "Sales Person 1"). The enterprise information model can be used to depict many different types of organization: for example, for a specific branch of an organization or an entire enterprise (consisting of several organizations), or for only one specific organization. The enterprise information model usually also depicts the actors (subjects) within the organization, including the entire organizational chart (usually with employee name, position, department, etc.), which can be defined either by so-called type or instance level, or sometimes mixed together. In comparison, the enterprise objects (customer, supplier, material, etc.) and services (quotation provision, sales order provision, etc.) are generally assigned only as types to a model. Eventually, during the execution phase of a single business process, only instances of all objects remain. Figure 9 shows an example of a fragment of BPMN diagram in which both types (e.g. "create sales order") and instances of types (e.g. "Sales Person 2") are used.

The BPMN collaboration diagram in Figure 9 records the as-is status of a typical sales and distribution process at Level 3 (process activity). What are expressed and can be seen are the process and the data flows of the data objects and actors involved within the operating organization. The upper black box includes activities of external customers and the exchanged documents (data objects). Both pools are located at the hierarchy Level 1, the departmental three (only "Marketing" is visible in Figure 9) lanes of the GBI on Level 2, and the seven (only three of them are visible in Figure 9) role lanes at Level 3. This allows the process to be assigned on Level 3 and thus go through many hands. The occurrence of various intermediate events wait until the process terminates; until then, numerous data objects flow back and forth between the various process activities.

The data objects flow at Level 3 can be defined by the logical/physical procedures of the organization, which require the fulfillment of certain conditions or an allocation of certain resources. The document flow in Figure 10 shows how individual documents from left to right reference each other; thus, in our example, customer payment, customer invoice, and outbound delivery are all based on a sales order. Documents such as goods issue' or customer payment are required to maintain certain business compliance such as the HGB (Handelsgesetzbuch, the German commercial law) or USGAP (United States General Accounting Principles).

4.6.7 示例"现状"模型(销售和分销)

1. BPMN 模型

BPMN 是业务流程图形表示的标准,它提供了指定业务流程的方法[42]。BPMN 的目标是支持技术用户和业务用户的 BPM。

在信息模型的分类过程中,信息元对象在分析过程中具有重要的作用,这取决于对象是类型(如员工)还是类型的实例(如"销售人员 1")。企业信息模型可用于描述许多不同类型的组织[如组织的特定分支或整个企业(由多个组织组成)],或仅针对一个特定组织。企业信息模型通常还描述组织内的参与者(主体),包括整个组织结构图(通常包括员工姓名、职位、部门等),可以通过所谓的类型或实例级别来定义,或者有时混在一起。相比之下,企业对象(客户、供应商、材料等)和服务(报价提供、销售订单提供等)通常仅作为模型的类型分配。最终,在单个业务流程的执行阶段,仅保留所有对象的实例。图 9 示出了 BPMN 图的片段的示例,其中使用了两种类型(如"创建销售订单")和类型实例(如"销售人员 2")。

图 9 中的 BPMN 协作图记录了 3 级(流程活动)典型销售和分销流程的现状。所表达和可以看到的是操作组织中涉及的数据对象和参与者的过程和数据流。上面的黑盒包括外部客户的活动和交换的文档(数据对象)。这两个池都位于层次结构级别 1 上,包括:部门级别 2 上 GBI 的三个(只有"营销"在图 9 中可见)通道,以及级别 3 上的七个(只有三个)角色通道。这允许在级别 3 上通过多种方式分配流程。各种中间事件的发生一直等到流程终止,在此之前,许多数据对象在各种进程活动之间来回流动。

级别 3 的数据对象流可以由组织的逻辑/物理流程定义,这些流程需要满足特定的条件或分配特定的资源。图 10 中的文档流显示了各个文档如何从左到右相互引用。因此,在我们的示例中,客户付款、客户发票和出站交付都基于销售订单。诸如货物发出或客户付款等文件需要维护某些业务遵从性,如德国商法(Handelsgesetzbuch,HGB)或美国通用会计原则(United States General Accounting Principles,USGAP)。

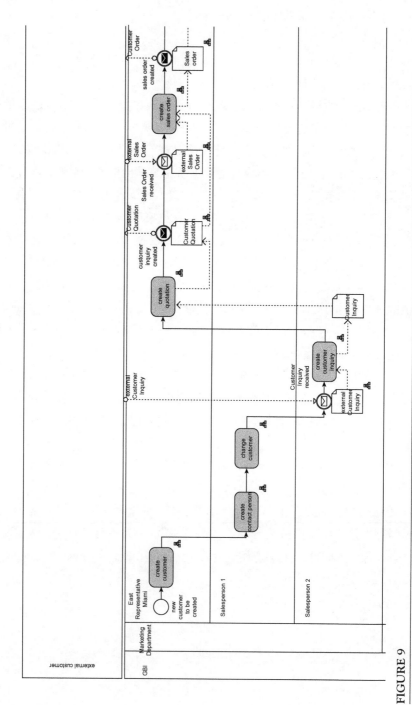

FIGURE 9

Business process model and notation collaboration diagram with process and information flow and organizational lanes (Level 3).

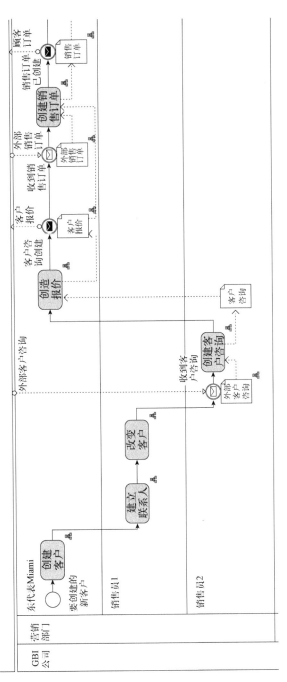

图 9　带有流程和信息流以及组织通道的 BPMN 协作图（级别 3）

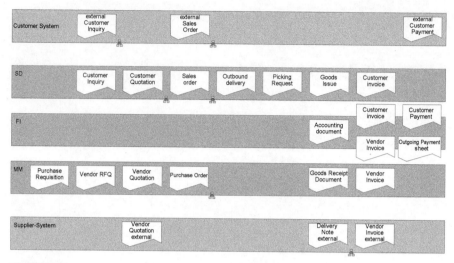

FIGURE 10

SAP document flow (Level 3).

The accounting sales model presented in Figure 11 represents the set of documents from the document with the double book entry activity necessary to execute a complete customer payment. Once the ware leaves the company, there is an effect on the balance, inasmuch as the value of the ware is missing. At the same moment, an account accrues to the customer who requested the material. A goods issue records the decreasing of material in the inventory and discharges the real account in finance. The debtor bill creates an account for the debtor. Then, in-payment bill balances the debtor bill and money gets transferred to the bank account.

Event-driven Process Chain Model

Event-driven Process Chain (EPC) diagrams are another approach to expressing business process work flows.

As shown in Figure 10, for the flows to be truly unified, three different functional areas (data objects systems within SAP) must be integrated: the customer system, sales and distribution (SD), and financial (FI) system; the external sales order should ultimately be stored as a sales order in the SD system. Because the two systems are not physically connected, until this occurs there will be a data discrepancy/media disruption between the two data objects. This is evident on the EPC in Figure 12, which depicts the processing through time from top to bottom. What the model shows is that the data (data object "sales order") of the incoming document "external sales order" must be entered manually by East Representative Miami in the screen mask VA01 Order create. This is shown

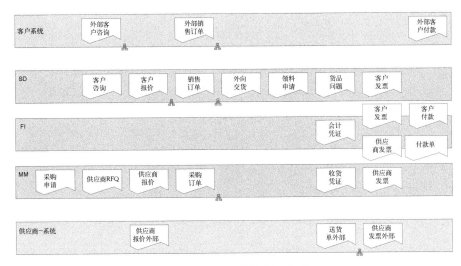

图10　SAP文档流（级别3）

图11中所示的会计销售模型表示执行完整的客户付款所需的双簿输入活动的文档中的一组文档。一旦商品离开公司,就会对平衡产生影响,因为商品的价值消失了。与此同时,对提供该材料的供应商建立一个账户,记录库存中物料的出货情况,并在财务中扣除,为债务人开立票据账户,然后,根据分期付款平衡债务人票据和资金被转移到银行账户的情况。

2. EPC模型

EPC图是表达业务流程工作流的另一种方法。

如图10所示,为了使流程真正统一,必须集成三个不同的功能软件(SAP中的数据对象系统): 客户系统、销售和分销(SD)以及财务(FI)系统。外部销售订单最终应作为销售订单存储在SD系统中。因为这两个系统没有物理连接,在此之前,两个数据对象之间将存在数据差异/媒体中断(media disruption)。这一点在图12中的EPC中很明显,它描述了从上到下的时间处理过程。模型显示的是传入文档外部销售订单的数据(数据对象销售订单)必须由East Representative Miami在screen mask VA01 order create中手动输入。图13中更详细地显示了这一点,其中

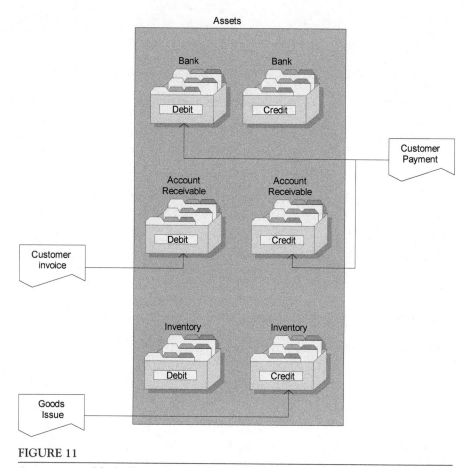

FIGURE 11

Accounting model sales/financial impact (Level 3).

on a more detailed level in Figure 13, where purchase order number is recognized as a foreign key attribute for the hierarchies on Level 4 of the model, thus providing the exact reference to the existing purchase order number of the customer.

The enterprise information models also include its customers and suppliers, such that all attributes on Level 4 refer to a unified data model (a portion of which is presented in Figure 14) on Level 3.

The focus of the BPMN process model in Figure 15 is on a portion of the persistent integration (data read/write) of the recorded sales process with the appropriate master (e.g., customer, material, and condition) and transaction data (e.g., sales order, goods issue, etc.) on Level 3 at the GBI (Level 1) in marketing (Level 2). The data are implemented and nested in Level 3 over overlapping

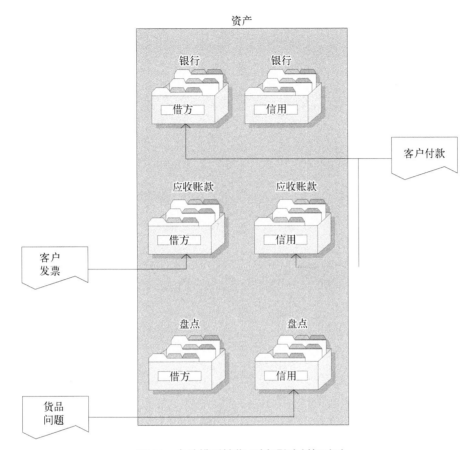

图11　会计模型销售/财务影响（第3级）

将采购订单号识别为模型第4层层次结构的外键属性,从而提供对客户现有采购订单号的准确引用。

　　企业信息模型还包括其客户和供应商,这样,级别4上的所有属性都指向级别3上的统一数据模型(部分数据模型如图14所示)。

　　图15中的BPMN流程模型的重点是持续记录(数据读/写)销售流程与市场营销(第2级)GBI(第1级)第3级的主数据(如客户、物料和条件)和交易数据(如销售订单、发货等)。这些数据在重叠车道上实现并嵌套在级别3中,并且不包含在图15中。图15显示了当识别出一个新客户时,执行业务流程以首先分配客户主

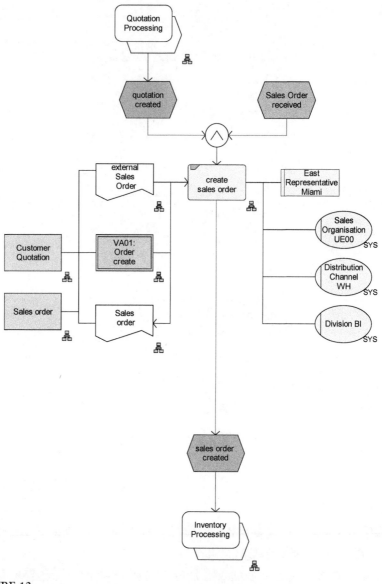

FIGURE 12

Event-driven Process Chain "create customer order" with documents, data entities, SAP screens, position and SAP system organizational units (Level 3).

lanes and are not included in this figure. Figure 15 shows that when a new customer is identified, a business process is performed to assign first the customer master data that are in turn associated with the condition master. The condition master overlaps in the upper part with the material master (not visible in the

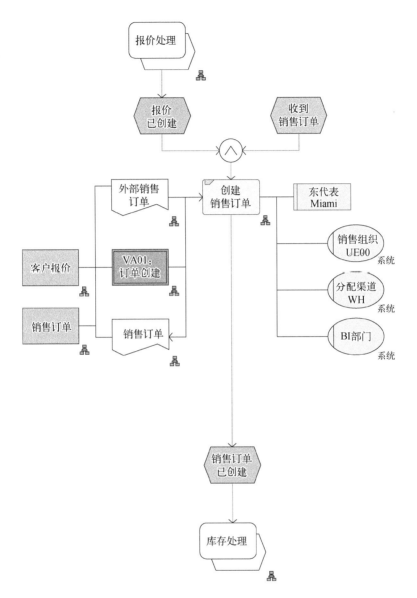

图 12　EPC "创建客户订单"，包括文档、数据实体、SAP 软件功能界面、职位和 SAP 系统组织单位（3级）

数据，该客户主数据又与条件主数据相关联。条件主数据与物料主数据在上半部分重叠（图 15 中不可见），因此，创建客户查询等第一笔交易数据也意味着生成物

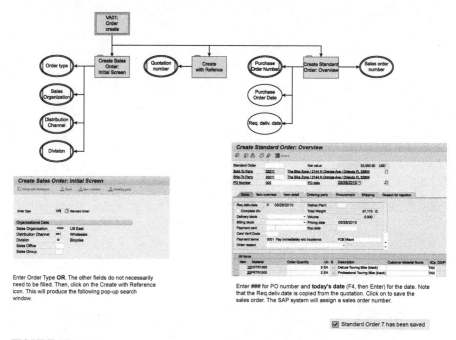

FIGURE 13

SAP screen diagram "VA01 order create" (Level 4).

FIGURE 14

Fragment of entity–relationship (ER) model customer order (Level 3).

figure); therefore, creation of the first transaction data such as customer inquiry also means that material or a combination of materials, customers, and condition master (see overlap) data is generated. Many other transaction data are based on this combination.

Entity–Relationship (ER) Model

The ER model is a method for describing the persistent data or information aspects of a business domain using properties of the data.

The fragment of the ERM associated with the example (Figure 14) shows the dependencies between the customer master data and different transaction data. A sales order leads to at least a (partial) delivery one or more outbound deliveries

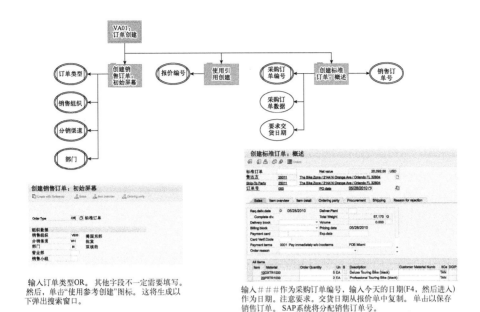

图13　SAP软件功能界面图"VA01 order create"（级别4）

图14　实体-关系（entity-relationship，ER）模型客户订单的片段（级别3）

料或物料、客户和条件主数据的组合（见重叠）。许多其他事务数据都基于这种组合。

3. ER 模型

ER模型（entity-relationship model）是一种使用数据属性描述业务域的持续数据或信息方面的方法。

与示例关联的ERM片段（图14）显示了客户主数据和不同事务数据之间的依赖关系。销售订单导致至少一个（部分）交货，一个或多个对外销售订单是从单个

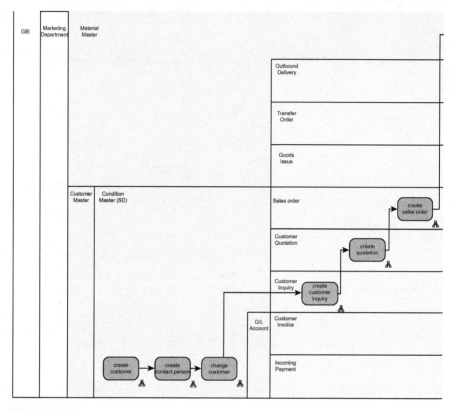

FIGURE 15

Business process management notation process diagram with process flow and data lanes (Level 3).

are created from a single sales order, whereas a (collective) delivery is associated with at least one sales order. The corresponding key or foreign key and descriptive attributes (Level 4) are not visible in this view; however, the input/output attributes (Figure 13) are shown via a 1:1 relationship, with the exception of system organization objects that are defined by customizing data entities.

The BPMN of Figure 16 shows the same process on Level 3 at the GBI (Level 1) in marketing (Level 2), this time as a function of the instances of the organizational units hierarchy of the involved system—which can exist at Level 3—similar to the BPMN collaboration diagram in Figure 15. The system organizational units are shown in Figure 16 as nested but not overlapping lanes (not visible in the example figure), which are all on the same Level 3. The system organizational units covering four extra levels are exposed within their own hierarchy in Figure 16. On the first two levels of this separate hierarchy, the company code

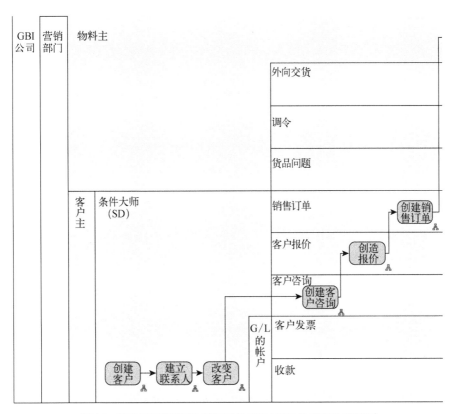

图15　带有流程流和数据通道的 BPM 符号流程图（级别3）

销售订单创建的，而（集体）交货与至少一个销售订单关联。相应的键或外键和描述性属性（级别4）在此视图中不可见。但是，输入/输出属性（图13）通过1∶1关系显示，但由自定义数据实体定义的系统组织对象除外。

　　图16的BPMN在营销（级别2）的GBI（级别1）的级别3上显示了相同的流程，是作为所涉及系统的组织单元层次结构的实例的功能——这些实例可以存在于级别3——与图15中的BPMN协作图类似。系统组织单元在图16中显示为嵌套但不重叠的路径（在示例图15中不可见），它们都在同一级别3上。图16展示了系统组织单元覆盖的四个额外的级别的各自层次结构。在这个单独的层次结构的前两个级别，公司代码US00被分配给客户端GBI，包括销售区域GBI 2.0加上工厂

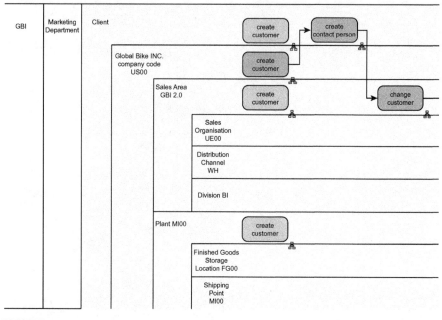

FIGURE 16

Business process management notation process diagram with process flow and system organizational unit lanes (Level 3).

US00 is assigned to the client GBI and consists of the Sales Area GBI 2.0 plus the Plant MI00, etc. A customer can be created as a "general customer (Client)", as a "sales area customer", as a "company code customer "or can be assigned to a delivery plant (plant MI 1000).

To-Be Modeling

Models can be used to describe or capture the current behavior and structure of the business. They can also be used to express possible future ways of doing business, which can then be developed. These to-be models allow decision makers to develop a shared understanding regarding how to do business and to consider design trade-offs, just as one would do with a more tangible product such as a house, a car, a toaster, or an item of clothing.

EXAMPLE OF TO-BE (BPMN) MODEL (MATERIALS MANAGEMENT)

Process automation typically focuses on the to-be status, in this case for materials management. The BPMN collaboration diagram for the purchasing process

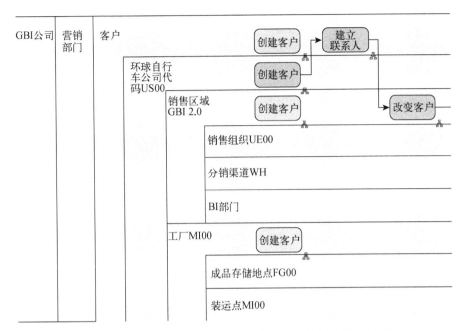

图16　带有流程流和系统组织单元通道的BPM符号流程图（级别3）

MI00等。客户可以创建为"一般客户（客户）""销售区域客户""公司代码客户"或可以分配给交付工厂（工厂MI 1000）。

4. 面向未来的模型（To-Be modeling）

该模型既可以用来描述或得到业务的当前行为和结构，也可以用来表达未来可能的商业方式，然后可以进行开发。这些所谓的模型使决策者能够就如何做生意和考虑设计权衡达成共识，就像人们对待更有形的产品，如房子、汽车、烤面包机或衣物一样。

4.6.8　未来（BPMN）模型示例（材料管理）

流程自动化通常关注于将来的状态，在本例中是物料管理。购买流程的BPMN协作图在图17中包括三个池：位于内部客户（黑盒，即不可能看到内部），低

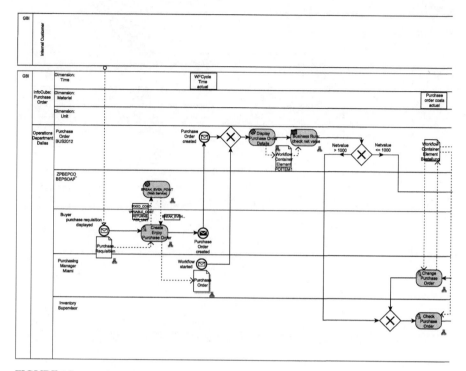

FIGURE 17

Business process management notation collaboration diagram "Purchasing Process" with process and information flow (Level 3).

in Figure 17 includes three pools, above the internal customers (black box, i.e., it not possible to see inside), below the external supplier (which is all expressed as a black box), and in the middle of the purchasing process at Level 3. The left frame for the middle and upper pool depicts the organization GBI at Level 1, which can be found next to the internal customer and the Operations Department Dallas as a lane on Level 2. These use the info cube "Purchase Order" with the three dimensions of time, material, and unit, which are also designated with lanes on Level 3. The buyer, purchasing manager Miami, and inventory supervisor use the SAP Business Object 2012 (Business Order) and the Web service Break Even Point (BEP) (both on Level 3). After displaying a purchase requisition the process is started, e.g., from a mobile work place (terminal), by the buyer. Once the break-even point has been calculated automatically, the buyer generates a purchase order. The consequence of this process is that at a predefined time, an event is automatically generated that starts a business rule, which ultimately forwards a decision to increase the inventory limits to the

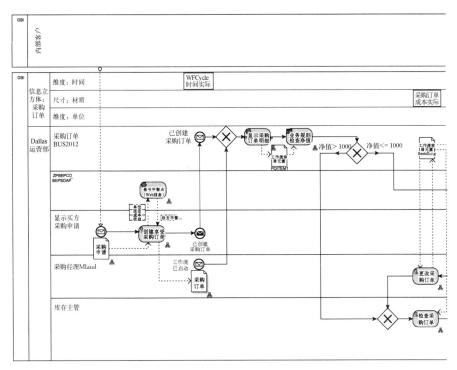

图17　包含流程和信息流（级别3）的BPM符号协作图"采购流程"

于外部供应商（都表示为一个黑盒子），以及在第3级的采购过程中。中间和上面池的左侧框架描述了级别1的组织GBI，可以在级别2的通道中找到该组织GBI，该组织GBI位于内部客户和达拉斯运营部门的旁边。它们使用信息立方体采购订单，包含时间、材料和单位的三个维度，这些维度也被指定为第3层的通道。作为买方的采购经理和库存主管使用SAP Business Object 2012（业务订单）和Web服务计算盈亏平衡点（BEP）（均在第3层）。一旦盈亏平衡点被自动计算出来，买方就会生成一个购买订单。此流程的结果是在预定义的时间自动生成一个事件，该事件启动业务规则，最终将增加库存限制的决策转发给库存管理员。或者，采购经

inventory supervisor. Alternatively, the Purchasing Manager Miami can start the workflow manually. Upon completion of the workflow, both the workflow order cycle time and the number of traversed workflows measured can then be found as PPI on Level 4. The data object flows are displayed, as well as the business documentation together with such values as purchase requisition and purchase order (see SAP Document Flow in Figure 10) technical data object flows such as the so-called workflow container flow.

Unified Modeling Language Model

The UML class diagram (Figure 18) shows a section of the business object BUS 2012 on Level 3. It displays the component together with its attributes and methods: the automated receive activity "Display Purchase Order Details ()", the user task "Change Purchase Order ()", the "Create Enjoy Purchase Order()", and the send task "Display Object ()" used by three different roles and therefore shown in three separate lanes in Figure 17. The attributes of the UML class diagrams are integrated on Level 4 with keys, foreign keys, and attributes.

Note. The UML class diagram needs to be expanded for use in Web and enterprise services.

Star Scheme

The use of the so-called star scheme or star schema is a design strategy to improve access to data for the purpose of generating complex reports. The data structure separates business process data into facts that hold the measurable quantitative data about a business, and dimensions that are foreign keys related to the fact data. This information is held in what is often referred to as a data warehouse or data mart; data are held for the purpose of reporting or analytics, so-called online analytic processing, as opposed to online transaction processing, in which data are optimized for transaction processing.

In the following example, and building on the case example, the information cube Purchase Order presented in Figure 19 is located on Level 2. It contains the three dimensions "Time","Material" and "Unit" reffering figure 17.

Information Engineering

Information engineering (IE) is an architectural method for planning, analyzing, designing, and implementing persistent data structures in an enterprise. Its aim is to enable an enterprise to improve the management of its resources, including capital, people, and information systems, to support the achievement of its business vision.

理可以手动启动工作流。在工作流完成后，可以在第4级上找到工作流订单周期时间和测量的遍历工作流数量，还将显示数据对象流、业务文档以及诸如采购请求和采购订单等值（参见图10中的SAP文档流）的技术数据对象流（如所谓的工作流容器流）。

1. UML 模型

UML类图（图18）显示了级别3上的Business Object Bus 2012的一部分。它显示组件及其属性和方法：自动接收活动"显示采购订单详细信息（　）"、用户任务"更改采购订单（　）""创建享受采购订单（　）"和发送任务"显示对象（　）"。其由三个不同的角色使用，因此在图17中以三个单独的通道显示。UML类图的属性与键、外键和属性集成在级别4上。

注意事项：UML类图需要扩展以用于Web和企业服务。

---

2. 星型模式

所谓的星型方案或星型模式是一种改进数据访问的设计策略，目的是生成复杂的报告。数据结构将业务流程数据分成包含有关业务的可测量定量数据的事实，以及与事实数据相关的外键维度。此信息保存在通常称为"数据仓库"或"数据集市"的位置，我们为了报告或分析而保留数据，即所谓的在线分析处理，而不是在线交易处理，其中数据被优化用于交易处理。

在以下示例中，图19中显示的信息立方体采购订单位于第2级。它包含时间、材料和单位三个维度，见图17。

3. 信息工程

信息工程（IE）是一种在企业中规划、分析、设计和实现持久数据结构的体系结构方法。其目的是为支持实现其业务愿景，而使企业能够改善对其资源的管理，包括：资本、人员和信息系统。

Purchase Order BUS2012
Purchase Order Number
Account Number of Vendor
BAPIMEPOHEADER
BAPIEKKO
BAPIEKPO
BAPIPARA
BAPIMEPOITEM
BAPIMEPOITEM
Display Purchase Order Details()
Create Enjoy Purchase Order ()
Display Object()
Change Purchase Order()

FIGURE 18

UML class diagram for SAP business object BUS 2012 (Level 3).

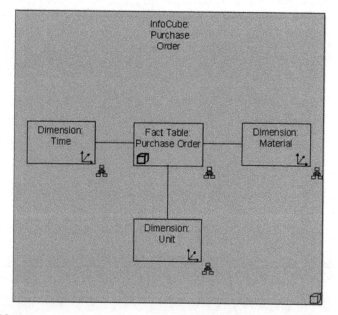

FIGURE 19

Information cube Purchase Order (Level 2).

The IE model (Figure 20) at Level 3 shows how it is possible using an in-memory database at Level 4 to accelerate access to the relevant information cube: for example, sales and distribution data. The customer and product (material master) attributes views, as well as the data foundation, include data entities used by extract–transfer–load of the

采购 订单 BUS2012
订购单号码 供应商帐号 BAPIMEPOHEADER BAPIEKKO BAPIEKPO BAPIPARA BAPIMEPOITEM BAPIMEPOITEM
显示采购订单明细（） 创建享受采购订单（） 显示对象（） 更改采购订单（）

图18　SAP业务对象BUS 2012的UML类图（级别3）

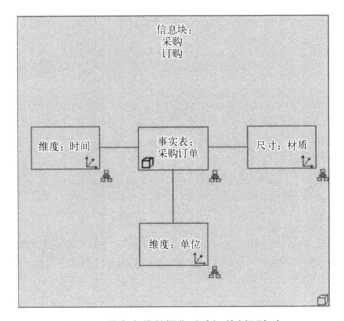

图19　信息多维数据集采购订单（级别2）

第3级的IE模型（图20）展示了如何使用第4级的内存数据库加速对相关信息多维数据集（如销售和分销数据）的访问。客户和产品（材料主人）属性视图以及数据基础，包括主数据和事务数据的提取−传输−加载所使用的数据实体，如来自

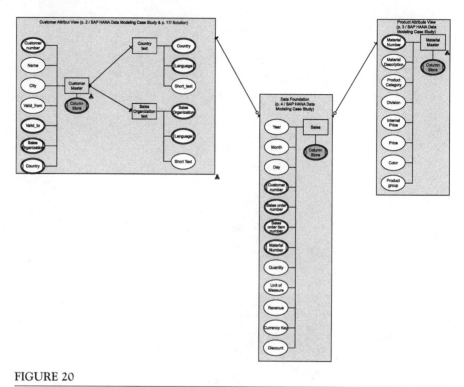

FIGURE 20

SAP High-Performance Analytic Appliance (HANA) data model analytical view (Level 3).

master and transaction data, such as material master or sales from the data warehouse, which are stored subsequently and used via column store in the in-memory database.

Current practice has evolved so that data are now stored separately in two different systems, and evolving strategy is looking toward the future and has everything implemented in one system, whereby the multiple views are combined with one primary key for the customer or material.

The details of how such a fact table (Figure 19) can be organized and its related components are shown in Figure 21. The four PPI pairs—each of as-is (actual) and to-be (plan) status—correspond to the objectives of the four perspectives of a balanced scorecard, e.g., Figure 22.

The objective diagram on Level 2 (Figure 23) shows the breakdown of the strategic objective "improve purchase order process" for the Check Purchase Order process step on the three process dimensions: quality management target, time target, and ABC target (not relevant here). These dimensions are not to be confused with the dimensions of the information cube. Arranging the strategic objective leads into a hierarchy for a KPI allocation diagram (value layer of Figure 8) on Level 3 (not shown for reasons of space), which then leads by another hierarchy to Level 4 and subsequent horizontal navigation to the data layer (Figure 8) branches in the fact table shown (Figure 21).

Level 2 depicts the process step Check Purchase Order, which can also be navigated horizontally into the e-business scenario map (Figure 24). This is shown with

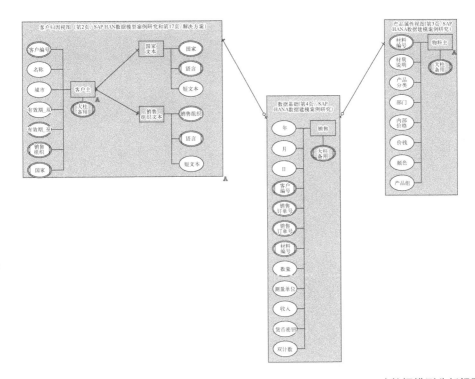

图20 SAP高性能分析工具（high-performance analytic appliance, HANA）数据模型分析视图（级别3）

数据仓库的材料主数据或销售数据，这些数据随后被存储并通过内存数据库中的列存储使用。

当前的实践已经发展到现在将数据分别存储在两个不同的系统中，并且发展战略正在朝着未来发展，且所有的东西都在一个系统中实现，从而将多个视图与客户或物料的一个主键结合在一起。

图21显示了如何组织这样一个事实表（图19）及其相关组件的详细信息。这四个PPI对——每个PPI对的现状（实际）和未来（计划）状态——对应于平衡计分卡的四个透视图的目标，如图22所示。

第2层的目标图（图23）显示了检查采购订单流程步骤的战略目标"改进采购订单流程"在三个流程维度上的细分：质量管理目标、时间目标和ABC目标（此处不相关）。这些维度不能与信息多维数据集的维度混淆。将战略目标引导到级别3上的KPI分配图（图8的值层）的层次结构中（由于空间原因未显示），然后再由另一个层次结构引导到级别4，并随后水平导航到事实表中的数据层（图8）分支（图21）。

级别2描述了流程步骤检查采购订单，也可以水平导航到电子商务场景图中（图24）。右侧显示（内部）客户，左侧显示（内部）供应商，工艺流程从上到下延伸。

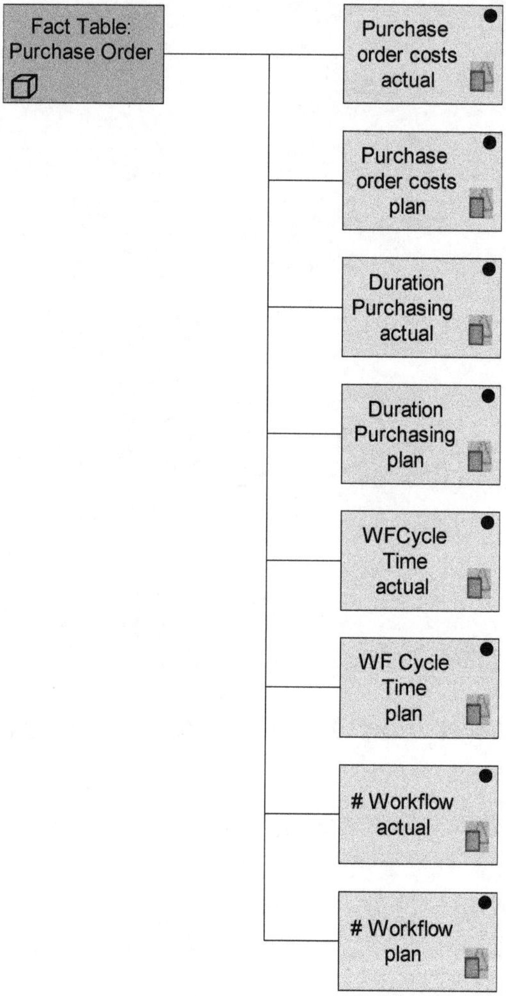

FIGURE 21

Fact table Purchase Order with PPIs (Level 4).

(internal) customers to the right, together with the (internal) suppliers to the left. The process flow extends from top to bottom. Also evident in this diagram is the revenue/cost and business service group flow and the document flow, which are actually a deeper level, at Level 3. This historically grown property is characteristic of the e-business scenario diagram during the document flow from right to left, e.g., from the purchasing department to the external supplier as a purchase order, or from left to right, e.g., from the supplier to the FI department as an external vendor invoice. The net cash flow may show only the (internal) customers toward the (internal) suppliers. In this case, they roughly correspond to the costs of the three pictured internal departments and the difference of the incoming moving/standard price minus the

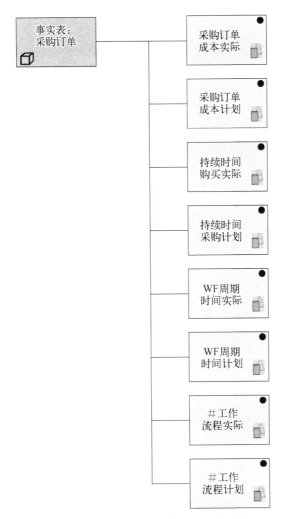

图21 带PPIs的采购订单事实表(4级)

在图24中还可以明显看到收入/成本和业务服务组流以及文档流,它们实际上是一个在第3层的更深层的层次。在文档流从右到左(例如,从采购部门作为采购订单到外部供应商)或从左到右(例如,从供应商到FI部门作为外部供应商发票)期间,这种历史上增长的属性是电子商务场景图的特征。净现金流可能只显示(内部)客户对(内部)供应商,在这种情况下,它们大致相当于图24中所示的三个内部部门的成本以及传入移动/标准价格减去实际流出的外部付款的差额。在返回业

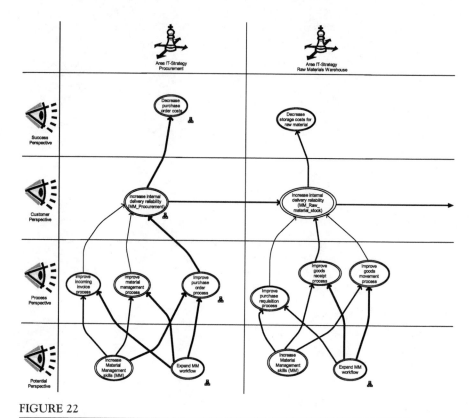

FIGURE 22

Department balanced scorecard cause-and-effect chain procurement (Level 2).

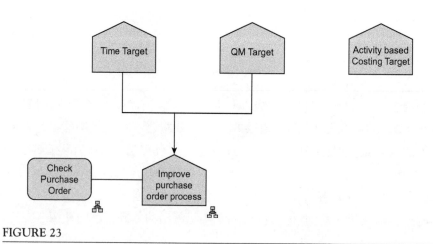

FIGURE 23

Objective diagram of process perspective of purchase department (Level 2).

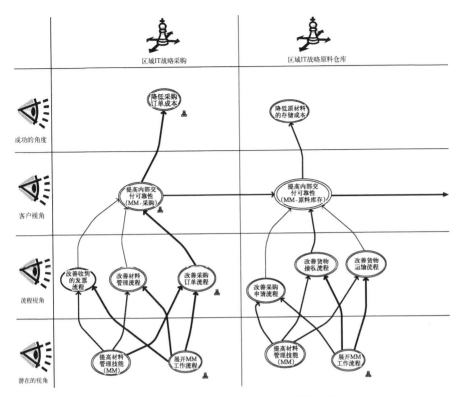

图22　部门平衡计分卡因果链采购（第2级）

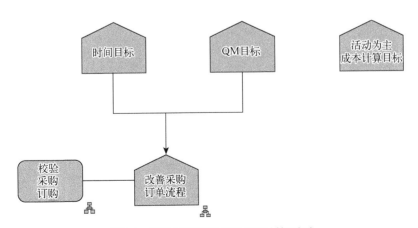

图23　采购部流程视角目标图（第2级）

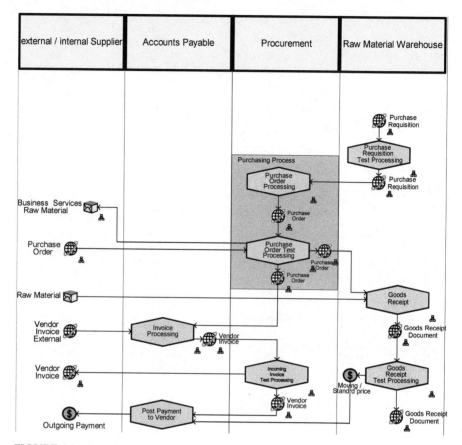

FIGURE 24

e-Business scenario diagram fragment (Level 2).

actually flowing externally outgoing payments. The services that are to be provided in return business services, e.g., to the purchasing department and another internal supplier such as an IT department run counter to the net financial flow. The BPMN shown in the Figure 17 collaboration diagram Purchasing Process (gray shaded) includes both the process Create Purchasing Order referring to RFQ and Check Purchase Order.

Balanced Scorecard Cause-and-Effect Chain

Alternatively, it is possible to navigate horizontally from the objective diagram in Figure 23 via the process objective "Improve Purchase Order Process" into the corresponding department balanced scorecard cause-and-effect chain (Figure 22). In the example considered, the two areas of procurement and raw material warehouse are be managed as cost centers. Furthermore, a procure-to-stock scenario is assumed. The strategy is based on an expansion of the existing IT resources (materials management (MM) skills and workflow system), which is also reflected

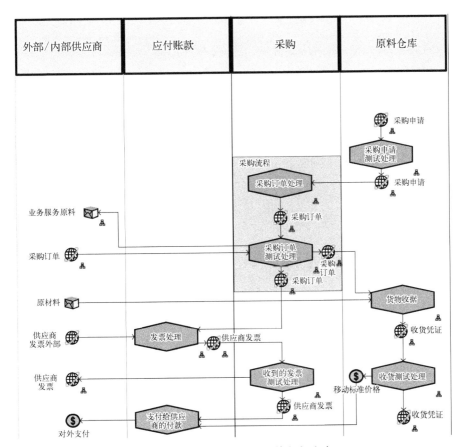

图24 业务电子化场景图片段（2级）

务服务时提供的服务,例如,向采购部门和另一个内部供应商(如IT部门)提供的服务与净财务流背道而驰。图17协作图"采购流程"(灰色阴影)中显示的BPMN既包括参考询价创建采购订单的流程,也包括检查采购订单的流程。

4.平衡计分卡因果链

或者,可以通过流程目标"改进采购订单流程"从图23中的目标图水平导航到相应的部门平衡计分卡因果链(图22)。在所考虑的示例中,采购和原材料仓库的两个区域作为成本中心进行管理。此外,假设采购到库存的策略,该策略基于对现有IT资源[物料管理(material management,MM)技能和工作流系统]的扩展,这也反映在图24的潜在透视图中。如图24所示,采购(部门)的内部客户是采购部的

in the Potential Perspective of the diagram. The internal customer of procurement (department) is, according to Figure 24, the raw material warehouse and internal customer objective of the purchasing' department, therefore, for example, an increase in the internal delivery reliability as a consequence of its internal customers, so the range of raw material warehouse is all a part of the flow. Ideally, a cost savings to the department occurs and supports the goals of the business in question and its customers, represented by a connecting line from the left to the right cause–effect chain. Another horizontal line connecting to the right emphasizes this point, where the warehouse must support the goals of its internal customer (e.g., production or sales and distribution receiving "finished goods"). Over a two-step hierarchical jump of the strategic objective, the model user vertically navigates to Level 4, where by horizontal navigation, he ultimately gets to the fact table (Figure 21) and can assign one KPI couple (actual/plan) to each of the four perspectives of the balanced scorecard.

In Figure 22 we assumed the internal customer of the finished goods warehouse department, seen as an internal supplier, could possibly be the sales and distribution department as mentioned above. We also found in Figure 24 that for every department, internal suppliers exist. If we have a closer look at the sales and distribution department balanced scorecard in Figure 25, we can see another internal supplier of the sales and distribution department: the human capital management department. Its supplier is the internal/external job market. The characteristics of the internal/external job market finally are its objectives on the potential perspective: increase general education and increase national culture. The cultural aspects might have an important role in the success of BPM in the future.[43]

LESSON LEARNED

As we have highlighted, business process modeling is a key element when aligning business processes with the requirements of an organization. With the right methodology and appropriate artifacts, it is possible to provide a clear, complete, accurate, and actionable framework for information and process modeling.

WHAT WORKED

Business process management notation process models must be complemented by and extended with information models aspects for several reasons:

1. Information modeling aspects within the process are important in any ERP implementation projects, primarily to streamline the execution of the business process and to support all report requirements
2. Reporting requirements can stem from different information aspects in an end-to-end process flow
3. Integration of more information objects of the business world (i.e., mission, vision, strategy, objectives/requirements engineering)
4. Integration of three enterprise tiers (strategic, operational, and tactical)

原材料仓库和内部客户目标,因此,例如,由于内部客户的原因,内部交货可靠性增加,因此原材料仓库的范围都是流程的一部分。

理想情况下,通过由从左到右的因果链连接线表示的那样,部门既可以节省成本,又可支持相关业务及其客户的目标,另一条连接到右边的水平线强调了这一点。在这一点上仓库必须支持其内部客户的目标(例如,接收"成品"的生产或销售和配送)。在战略目标的两步分层跳跃中,模型用户垂直导航到第4级,在那里通过水平导航,他最终到达事实表(图21),并可以为平衡计分卡的四个透视图中的每个透视图分配一个KPI对(实际/计划)。

在图22中,我们假设作为内部供应商的成品仓库部门的内部客户可能是上述的销售和配送部门。我们还在图24中发现,对于每个部门,都存在内部供应商。如果我们仔细查看图25中的销售和分销部门平衡计分卡,我们可以看到销售和分销部门的另一个内部供应商:人力资源管理部。它的供应商是内部/外部就业市场。内部/外部就业市场的特点最终是其潜在视角上的目标:加强员工教育和企业文化建设。文化方面可能在BPM未来的成功中扮演重要角色[43]。

4.6.9 经验

正如我们所强调的,当业务流程与组织的需求保持一致时,业务流程建模是一个关键元素。有了正确的方法和适当的制品,就有可能为信息和流程建模提供清晰、完整、准确和可操作的框架。

4.6.10 做的工作

BPM符号流程模型必须得到信息模型方面的补充和扩展,原因如下。

(1)流程中的信息建模方面在任何ERP实现项目中都很重要,主要是为了简化业务流程的执行并支持所有报告需求。

(2)报告需求可以来自端到端流程流中的不同信息方面。

(3)业务世界中更多信息对象的集成(如使命、抱负、策略、目标/需求工程)。

(4)三个企业层(战略、操作和战术)的集成。

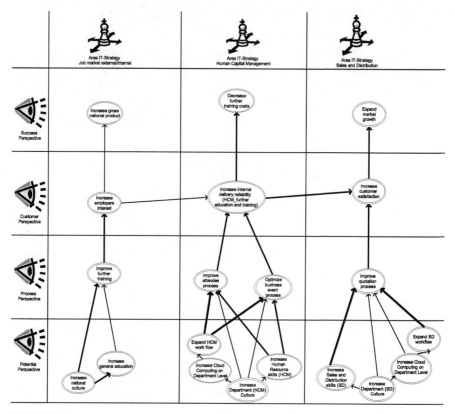

FIGURE 25

Department balanced scorecard cause-and-effect chain Human Capital Management (Level 2).

5. Performance management in an organization can be:
     a. Strategically related—measuring performance against a strategic plan
     b. Tactically related—enabling oversight, governance, evaluation, and audits
     c. Operationally related—measuring operational related activities
6. The information system will also need to be able to respond to strategic, tactical, and operational requirements, and operational requirements simultaneously
7. Different levels of abstraction (from overview to detailed level of composition/decomposition)
8. Identification and cascading of internal customer/supplier relationships (i.e., procurement/warehouse/employee)
9. Integration of dynamic (time dependent) and static information models (i.e., organizational chart, ERM)
10. Identification of more BPMN resources (i.e., system organizational units, business objects, or information cubes)

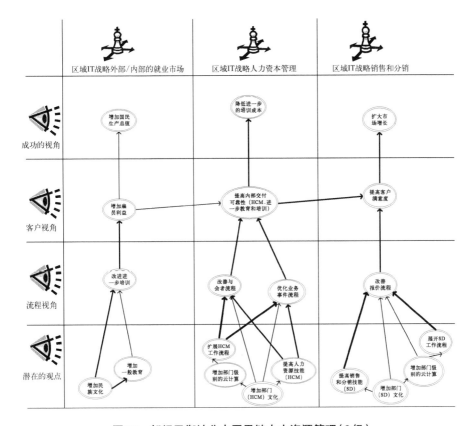

图25　部门平衡计分卡因果链人力资源管理（2级）

（5）组织中的绩效管理包括以下方面。

a. 与战略相关：根据战略计划评估战略相关绩效。

b. 与战术相关：授权监督、治理、评估和审计。

c. 与操作相关：评估操作相关的活动。

（6）信息系统还需要能够响应战略、战术需求，同时满足运营需求。

（7）不同的抽象层次（从概述到详细的集成/分解层次）。

（8）识别和关联内部客户/供应商关系（即采购/仓库/员工）。

（9）动态（依赖时间）和静态信息模型（即组织结构图、ERM）的集成。

（10）识别更多的BPMN资源（即系统组织单位、业务对象或信息多维数据集）。

11. Integration of business compliance (HGB, USGAP)
12. Integration of types and instances (such as process activities versus business department names)
13. Integration of old but content-rich information model types (EPC) with new but content-poor (BPMN) ones
14. The integrated end-to-end flow should take business, application, and technology layered requirements into consideration, thus aligning end-to-end flow process automation potential with requirements across the layers
15. Different views/layers (i.e., business and application)
16. System integration should address all of these stakeholder requirements to ensure that the correct information is available to all areas when business processes execute and afterward.
17. All related objects, in terms of business objects, information objects, and data objects, should be derived automatically in the process.
18. The purpose of the designed and integrated end-to-end flow is to maximize the level of automation by which associated business, information, and data objects in the flow through the information system are derived when a business process is executed
19. Rules are applied within the process as well as information models as traditional rule sets, rule scripts, and flow rule sets
20. Transformation potential is identified in the various process and information models. Exploiting the full innovation as well as transformation potential of the opportunities must consider both process and information models

This extension can only happen within a well-elaborated enterprise information model architecture using four or more levels of composition/decomposition, which can be found in APQC, SCOR, and other frameworks. The challenge is to transfer these levels to layers other than processes, e.g., value, competency, or application. Once this has been defined, horizontal (to get a different view) and vertical navigation (to get a more/less detailed view) between different information model and object types within one single integrated enterprise information model are possible.

This integrated enterprise information model supports the entire process life cycle from Analyze to continous improvement.

WHAT DID NOT WORK

Pure BPMN collaboration or BPMN process diagrams are not sufficient to provide all of the information needed for a successful business process implementation. The integrated enterprise information model does not yet support complete model-driven implementation. With existing BPM tools fewer than 50% of the information models (e.g., with ARIS Netweaver for SAP, SAP Solution Manager,[44] SAP Business Workflow, iGrafx, or SAP BPM) can be implemented. The reason for this deficiency is that existing tools focus on specific tiers, views, levels, model types, or information objects, and have missing or limited interfaces between the different conceptual spaces in which they reside, considering only narrow aspects of the total

（11）业务合规性集成（HGB、USGAP）。

（12）类型和实例的集成（如流程活动与业务部门名称）。

（13）将内容丰富的旧信息模型类型（EPC）与内容贫乏的新信息模型类型（BPMN）集成在一起。

（14）集成的端到端流程应考虑业务、应用程序和技术分层需求，从而使端到端流程自动化潜力与跨层需求保持一致。

（15）不同的视图/层（即业务和应用程序）。

（16）系统集成应该解决所有这些利益相关者需求，以确保在业务流程执行时以及之后正确的信息可用于所有领域。

（17）所有相关的对象，就业务对象、信息对象和数据对象而言，都应该在流程中自动派生。

（18）设计和集成的端到端流程的目的是最大限度地提高自动化水平，以便在执行业务流程时，在流程中通过信息系统派生相关的业务、信息和数据对象。

（19）在流程以及信息模型中应用规则，如传统规则集、规则脚本和流规则集。

（20）识别在各种流程和信息模型中的转化潜力，利用全面创新和转型潜力的机会考虑流程和信息模型。

这种扩展只能在一个精心设计的企业信息模型体系结构中进行，该体系结构使用可以在 APQC、SCOR 和其他框架中找到的四个或更多级别的组合/分解。挑战是将这些级别转移到如价值、能力或应用程序流程之外的其他层。一旦定义了这一点，就可以在一个单一的集成企业信息模型中，在不同的信息模型和对象类型之间进行水平（以获得不同的视图）和垂直导航（以获得更多/更少详细的视图）。

这种集成的企业信息模型支持从分析到持续改进的整个过程生命周期。

4.6.11　什么是没用的

纯 BPMN 协作或 BPMN 流程图不足以提供成功业务流程实现所需的所有信息。集成的企业信息模型还不支持完整的模型驱动实现。使用现有 BPM 工具，不到 50% 的信息模型（例如，使用 ARIS Netweaver for SAP、SAP Solution Manager[44]、SAP Business Workflow、iGrafx 或 SAP BPM）可以实现。这一缺陷的原因是，现有的工具集中于特定的层、视图、级别、模型类型或信息对象，并且在它们所在的不同概念空间之间缺少或限制了接口，只考虑到整个问题的狭窄方面，例如，专注

problem, such as focusing on automation or on transformational work, without fully capturing other forms of work.

CONCLUSIONS

Findings and Summary

In this chapter, we have elaborated on the need to interlink the process models with information model aspects and have shown how it would be done. To increase the level of understanding we have provided a comparison of the different hierarchies for the process layer. We have demonstrated how it is possible to align the different levels against each other for a number of different frameworks: APQC PCF, LEAD/ GBI, SAP Solution Manager, and SCOR (Figure 6).

Through our analysis we identified the problem of determining when using multiple information models whether it is possible to map the layers from one information model to another, e.g., LEAD to SAP solution manager, and retain a consistent process model structure. As a solution, we proposed the idea of making horizontal (layers) or vertical (levels) connections of the leading process layer structure from "above" to all the other layers, with simultaneous consolidation/integration with the data layer from "below".

The result of this solution is the matrix (Figure 7) showing how we can map business and application layer meta objects over four business layers and two application layers, whereby the layers are classified side-by-side in six columns. This allows us to detail the relationship of meta information objects to each other in different layers (horizontal integration) on different levels (vertical integration). Moreover, the meta information objects are associated with each of one or more of the specific layers (1–6) and one or more of the specific hierarchy levels (Levels 1–4), thus allowing for horizontal or vertical navigation in the matrix for each of the information meta objects.

In reality, the information meta objects have many more relationships than are detailed within the meta object mapping (Figure 7); thus, we have also provided a map of the information models (Figure 8) in which exactly one layer and one level are identified. The resulting models represent more than one layer or more than one level (e.g., hierarchical models), and therefore the information meta objects from Figure 7 can appear multiple times.

We have then taken our mapping matrices and, using a case study as detailed in Section "Process life cycle", provided validation of how they can be effective in producing business and application information models. The case study and examples detailed identify how it is possible, using our matrices and methods identified by the LEADing Practice together with a range of different modeling techniques (BSC, BPMN,[45] EPC, UML, and ER modelling), to produce useful as-is and to-be process models.

The business and application information models that we have provided detail the following:

- Integration of document flows (Figure 10) required for compliance and adherence to regulations
- Integration of user interfaces (Figure 13)
- Integration of system organizational units (second organization) (Figure 12)

于自动化或转换工作,以及不应该只抓其他形式的工作。

4.6.12　结论

发现和总结

在这一节中,我们阐述了将流程模型与信息模型方面联系起来的必要性,并展示了如何做到这一点。为了提高理解水平,我们对流程层的不同层次结构进行了比较,并演示了在许多不同的框架将不同的级别彼此对齐方式,包括:APQC PCF、LEAD/GBI、SAP 解决方案管理器和 SCOR(图 6)。

通过我们的分析,确定了在使用多个信息模型时是否有可能将层级从一个信息模型映射到另一个信息模型(例如,LEAD 到 SAP 解决方案管理器)以及保持一致的流程模型结构的问题。

这个解决方案的结果是一个矩阵(图 7),它展示了我们如何在 4 个业务层和 2 个应用程序层上映射业务层和应用层元对象,这些层在 6 列中并排分类。这允许我们详细描述不同层次(水平集成)上的元信息对象之间的关系(垂直集成)。此外,元信息对象与一个或多个特定层(1 ~ 6)和一个或多个特定层次(1 ~ 4 级)中的每一个相关联,从而允许在矩阵中为每个信息元对象进行水平或垂直导航。

实际上,信息元对象的关系比元对象映射中详细描述的关系要多得多(图 7)。因此,我们还提供了信息模型的映射(图 8),其中准确地标识了一个层和一个级别。得到的模型表示多个层或多个级别(如分层模型),因此图 7 中的信息元对象可以出现多次。

然后,我们使用映射矩阵,并使用"流程生命周期"一节中详细介绍的案例研究,验证了它们在生成业务和应用程序信息模型方面的有效性。案例研究和示例详细说明了如何使用我们的矩阵和方法,以及一系列不同的建模技术(BSC、BPMN[45]、EPC、UML 和 ER 建模),来生成有用的现有和未来的流程模型。

我们提供的业务和应用程序信息模型详细说明如下:

- 整合合规和遵守法规所需的文件流(图 10);
- UI 集成(图 13);
- 整合系统组织单位(第二组织)(图 12);

- Integration of keys, foreign keys (media break), and describing attributes
- Integration of BI (three tiers), information cubes (Figure 19), dimensions, and fact tables (Figure 21)
- In-memory (SAP HANA) analytical view (Figure 20)
- Integration of enterprise services and Web services (Figure 24)
- Enterprise and department balanced scorecard cause-and-effect chains (Figure 22, 25)

These process models identify how using meta object (Figure 7) and information models (Figure 8), matrices, and the initial ideas of the LEADing Practice have improved the quality of the process models by providing extended information modeling. Thus, we have been able to identify:

- Visible connection of strategic objectives and business processes
- Internal customer supplier relationship interfaces to other departments (flow of money, services)
- Integration of three process dimensions (quality, time, and costs)
- Integration of process and data flow

Our models demonstrate how is possible to integrate six layers (20 model types and 38 information objects) (Figures 7 and 8) and provide information models that also show how composition/decomposition can provide relevant information over four levels of six layers showing both vertical and horizontal integration/navigation. To date, the authors are not aware of another solution using such a step-by-step repeatable description that enables one to build the information models into the process landscape with a high level of detail in as-is and to-be process models.

Finally, our matrix and working examples use the definition of the process life cycle and frameworks as described in the BPM Life Cycle Chapter. It can also be found as enterprise standards[46] that are flexible, agile, and highly customizable. A further benefit in using the LEADing Practice standards is that they interlink to other frameworks, methods, and approaches such as TOGAF, Zachman, FEAF, ITIL, Prince2, COBIT, DNEAF, and many others,[47] and thus provide a powerful integrated BPM framework and enterprise architecture framework.[48]

End Notes

1. Hommes B and Van Reijswoud V, "Assessing The Quality of Business Process Modelling Techniques." in *Proceedings of the 33rd Annual Hawaii International Conference on System Sciences* 1, (IEEE, 2000), 10.
2. Bosilj-Vukšić V and Ivandić-Vidović D, "Business process change using ARIS: The case study of a Croatian insurance company. Management," *Journal of Contemporary Management Issues* 10, (2005): 77–91.
3. LEADing Practice, *The LEADing Practice eXtended BPMN Standard.* Available at: http://www.leadingpractice.com/wp-content/uploads/2013/10/LEADing-Practice-XBPMN.pdf, accessed October 14, 2014.
4. Scheruhn H, Ackermann D, Braun R, and Förster U, "Repository-based implementation of Enterprise Tiers: A study based on an ERP case study," in *Human-Computer Interaction. Users and Contexts of Use*, (Springer, 2013), 446–455.

- 集成密钥、外键（媒体中断）和描述属性；
- 集成BI（三层）、信息立方体（图19）、维度和事实表（图21）；
- 内存（SAP HANA）分析视图（图20）；
- 企业服务和Web服务的集成（图24）；
- 企业和部门平衡记分卡因果链（图22和图25）。

这些流程模型确定了如何使用元对象（图7）和信息模型（图8）、矩阵以及主要实践的初始思想通过提供扩展的信息建模来提高过程模型的质量。因此，我们能够识别：

- 战略目标和业务流程的可视连接；
- 与其他部门的内部客户–供应商关系接口（资金流、服务）；
- 集成三个流程维度（质量、时间和成本）；
- 流程和数据流的集成。

我们的模型演示了如何集成六个层（20个模型类型和38个信息对象）（图7和图8），并提供在垂直和水平集成/导航上如何在六个层的四个级别上提供相关信息的信息模型。到目前为止，作者还没有发现另一种使用这种逐步可重复描述的解决方案，这种描述使我们能够在现有和将来的流程模型中以高水平的细节将信息模型构建到流程环境中。

最后，我们的矩阵和工作示例使用BPM生命周期一节中描述的流程生命周期和框架的定义。它是灵活的、敏捷的和高度可定制的，也可以作为企业标准[46]来使用。使用领导实践标准（LEADing Practice standards）的另一个好处是，它们与其他框架、方法和方式（如TOGAF、Zachman、FEAF、ITIL、Prince2、COBIT、DNEAF和许多其他方法）相互链接[47]，从而提供强大的集成BPM框架和企业架构框架[48]。

参考文献

［1］Hommes B and Van Reijswoud V, "Assessing The Quality of Business Process Modelling Techniques." in Proceedings of the 33rd Annual Hawaii International Conference on System Sciences 1, (IEEE, 2000), 10.

［2］Bosilj-Vukšić V and Ivandić-Vidović D, "Business process change using ARIS: The case study of a Croatian insurance company. Management," Journal of Contemporary Management Issues 10, (2005): 77–91.

［3］LEADing Practice, The LEADing Practice eXtended BPMN Standard. Available at: http://www.leadingpractice.com/wp-content/uploads/2013/10/LEADing-PracticeXBPMN.pdf, accessed October 14, 2014.

［4］Scheruhn H, Ackermann D, Braun R, and Förster U, "Repository-based implementation of Enterprise Tiers: A study based on an ERP case study," in Human-Computer Interaction. Users and Contexts of Use, (Springer, 2013), 446–455.

5. Verner L, "The Challenge of Process Discovery," *BPM Trends*, (BPTrends, May, 2004). Available from: www.bptrends.com.

6. Rosenberg A, von Rosing M, Chase G, Omar R et al., *Applying Real-World BPM in an SAP Environment*, (Galileo Press, 2011).

7. Ko R. K, "A Computer Scientist's Introductory Guide to Business Process Management (BPM)," *Crossroads, the ACM student magazine* 15, no. 4 (2009): 4.

8. von Rosing M, Subbarao R, Hove M, and Preston T. W, "Combining BPM and EA in complex IT projects: (A business architecture discipline)." *Commerce and Enterprise Computing (CEC), 13th IEEE Conference on Commerce and Enterprise Computing*, (IEEE, 2011), 271–278.

9. von Rosing M, "Crash Course with the LEAD Frameworks," *Methods and Approaches*. Available at: http://www.leadingpractice.com/wp-content/uploads/2013/10/Crash-Course-to-LEAD-3.0.pdf, accessed May 20, 2014.

10. LEADing Practice, *Welcome to LEADing Practice*. Available at: http://www.leadingpractice.com/, accessed May 21, 2014.

11. Weilkiens T, Weiss C, and Grass A, *OCEB Certification Guide: Business Process Management-Fundamental Level*, (Elsevier, 2011).

12. LEADing Practice, *Decomposition & Composition Reference Content*. Available at: http://www.leadingpractice.com/enterprise-standards/enterprise-engineering/decomposition-composition/, accessed May 21, 2014.

13. Weilkiens op. cit.

14. Hommes B and Van Reijswoud V, "Assessing the quality of business process modelling techniques," *Proceedings of the 33rd Annual Hawaii International Conference on System Sciences* 1, (IEEE, 2000), 10.

15. LEADing Practice, *The LEADing Practice eXtended BPMN Standard*. Available at: http://www.leadingpractice.com/wp-content/uploads/2013/10/LEADing-Practice-XBPMN.pdf, accessed May 14, 2014.

16. http://www.leadingpractice.com/enterprise-standards/enterprise-engineering/decomposition-composition/, accessed May 21, 2014.

17. The Representation of the Business Process Captured in the Model is Implemented Directly from the Properties and Relationships of the Components Portrayed in the Model.

18. Verner L. op. cit.

19. BPM, "The Promise and the Challenge", Laury Verner, 2004.

20. Mertens P, Bodendorf F, König W, Picot A et al., *Grundzüge Der Wirtschaftsinformatik* (Springer, 2005).

21. von Rosing M, Subbarao R, Hove M, and Preston T. W, "Combining BPM and EA in complex IT projects: (A business architecture discipline)," *Commerce and Enterprise Computing (CEC), 13th IEEE Conference on Commerce and Enterprise Computing*, (IEEE, 2011), 271–278.

22. van Rensburg A, "Principles for modelling business processes," *Industrial Engineering and Engineering Management (IEEM)*, (IEEE, 2011), 1710–1714.

23. Sinur J and Hill J.B, "Magic quadrant for business process management suites," *Gartner RAS Core research note*, (Gartner, 2010), 1–24.

24. Fallon RL and Polovina S, "REA analysis of SAP HCM; some initial findings," *Dresden, Germany: ICFCA 2013: 11th International Conference on Formal Concept Analysis*, 2013.

25. Ko R. K, "A computer scientist's introductory guide to business process management (BPM)," *Crossroads, the ACM student magazine* 15, no. 4 (2009).

26. Karagiannis D and Kühn H, "Metamodelling platforms," *Proceedings of the Third International Conference EC-Web*, (Berlin, Heidelberg: Springer-Verlag, 2002),182.

［ 5 ］ Verner L, "The Challenge of Process Discovery," BPM Trends, (BPTrends, May, 2004). Available from: www.bptrends.com.

［ 6 ］ Rosenberg A, von Rosing M, Chase G, Omar R et al., Applying Real-World BPM in an SAP Environment, (Galileo Press, 2011).

［ 7 ］ Ko R. K, "A Computer Scientist's Introductory Guide to Business Process Management (BPM)," Crossroads, the ACM student magazine 15, no. 4 (2009): 4.

［ 8 ］ von Rosing M, Subbarao R, Hove M, and Preston T. W, "Combining BPM and EA in complex IT projects: (A business architecture discipline)." Commerce and Enterprise Computing (CEC), 13th IEEE Conference on Commerce and Enterprise Computing, (IEEE, 2011), 271−278.

［ 9 ］ von Rosing M, "Crash Course with the LEAD Frameworks," Methods and Approaches. Available at: http:// www.leadingpractice.com/wp-content/uploads/2013/10/CrashCourse-to-LEAD-3.0.pdf, accessed May 20, 2014.

［ 10 ］ LEADing Practice, Welcome to LEADing Practice. Available at: http://www.leadingpractice.com/, accessed May 21, 2014.

［ 11 ］ Weilkiens T, Weiss C, and Grass A, OCEB Certification Guide: Business Process Management Fundamental Level, (Elsevier, 2011).

［ 12 ］ LEADing Practice, Decomposition & Composition Reference Content. Available at: http://www. leadingpractice.com/enterprise-standards/enterprise-engineering/decompositioncomposition/, accessed May 21, 2014.

［ 13 ］ Weilkiens op. cit.

［ 14 ］ Hommes B and Van Reijswoud V, "Assessing the quality of business process modelling techniques," Proceedings of the 33rd Annual Hawaii International Conference on System Sciences 1, (IEEE, 2000), 10.

［ 15 ］ LEADing Practice, The LEADing Practice eXtended BPMN Standard. Available at: http://www. leadingpractice.com/wp-content/uploads/2013/10/LEADing-PracticeXBPMN.pdf, accessed May 14, 2014.

［ 16 ］ http://www.leadingpractice.com/enterprise-standards/enterprise-engineering/decomposition-composition/, accessed May 21, 2014.

［ 17 ］ The Representation of the Business Process Captured in the Model is Implemented Directly from the Properties and Relationships of the Components Portrayed in the Model.

［ 18 ］ Verner L. op. cit.

［ 19 ］ BPM, "The Promise and the Challenge", Laury Verner, 2004.

［ 20 ］ Mertens P, Bodendorf F, König W, Picot A et al., Grundzüge Der Wirtschaftsinformatik (Springer, 2005).

［ 21 ］ von Rosing M, Subbarao R, Hove M, and Preston T. W, "Combining BPM and EA in complex IT projects: (A business architecture discipline)," Commerce and Enterprise Computing (CEC), 13th IEEE Conference on Commerce and Enterprise Computing, (IEEE, 2011), 271−278.

［ 22 ］ van Rensburg A, "Principles for modelling business processes," Industrial Engineering and Engineering Management (IEEM), (IEEE, 2011), 1710−1714.

［ 23 ］ Sinur J and Hill J.B, "Magic quadrant for business process management suites," Gartner RAS Core research note, (Gartner, 2010), 1−24.

［ 24 ］ Fallon RL and Polovina S, "REA analysis of SAP HCM; some initial fndings," Dresden, Germany: ICFCA 2013: 11th International Conference on Formal Concept Analysis, 2013.

［ 25 ］ Ko R. K, "A computer scientist's introductory guide to business process management (BPM)," Crossroads, the ACM student magazine 15, no. 4 (2009).

［ 26 ］ Karagiannis D and Kühn H, "Metamodelling platforms," Proceedings of the Third International Conference EC-Web, (Berlin, Heidelberg: Springer-Verlag, 2002),182.

27. van Rensburg A, "Principles for modelling business processes," *Industrial Engineering and Engineering Management (IEEM)*, (IEEE, 2011), 1710–1714.

28. John A. Zachman, Henrik von Scheel, and Mark von Rosing, "The focus of Enterprise Architecture," *The Complete Business Process Handbook*, (Morgan Kaufman, 2014).

29. John A. Zachman, Henrik von Scheel, Mark von Rosing. "The focus of Enterprise Architecture," *The Complete Business Process Handbook 2*, (Morgan Kaufman, 2014).

30. LEADing Practice-Categorization & Classification Body of Knowledge, 2014.

31. LEADing Practice, *Decomposition & Composition Reference Content*. Available at: http://www.leadingpractice.com/enterprise-standards/enterprise-engineering/decomposition-composition/, accessed May 21, 2014.

32. https://supply-chain.org/f/SCOR-Overview-Web.pdf.

33. http://www.apqc.org/.

34. Business Objects that Capture the Scope and Value of the Business.

35. Business Objects that Capture and Describe the Essential Organizations Skill And Knowledge Needed to Fulfil the Scope And Purpose of the Business.

36. Business Objects that Realize Behaviour.

37. Business Objects Necessary to Execute Work and Create Value.

38. Business Objects Necessary to Describe the Structure and Behaviour of Software that Enables Work.

39. Business Objects Necessary to Describe the Persistent Information Used within the Software.

40. Gumpp A and Pousttchi K, "The 'Mobility-M'-framework for application of mobile technology in business processes." *INFORMATIK 2005-Informatik LIVE - Jahrestagung der Gesellschaft für Informatik e V (GI) 2*, 2005, 523–527.

41. Scheruhn H, Sicorello S, Weidner S, Repository-based ERP case studies: A study about chances and benefits of agile case study development in Witold Abramowicz, John Domingue, Krzysztof Wecel (Eds.): *Business Information Systems Workshops - BIS 2012 International Workshops and Future Internet Symposium, Vilnius, Lithuania, May 21-23, 2012 Revised Papers*. Springer 2012.

42. http://www.bpmn.org/.

43. Zhao F, Scheruhn H and von Rosing M, "The Impact of Culture Differences on Cloud Computing Adoption" In: *Human-Computer Interaction*, (Springer 2014), 776–785.

44. Scheruhn H, Ackermann D, Braun R and Förster U, "Repository-based implementation of Enterprise Tiers: A study based on an ERP case study," In: *Human-Computer Interaction. Users and Contexts of Use*, (Springer, 2013), 446–455.

45. LEADing Practice, *The LEADing Practice eXtended BPMN Standard*. Available at: http://www.leadingpractice.com/wp-content/uploads/2013/10/LEADing-Practice-XBPMN.pdf, accessed May 14, 2014.

46. APQC and von Rosing M, "Crash Course with the LEAD Frameworks," *Methods and Approaches*. Available at: http://www.leadingpractice.com/wp-content/uploads/2013/10/Crash-Course-to-LEAD-3.0.pdf, accessed May 20, 2014.

47. LEADing Practice. *Interconnects with Existing Frameworks*. Available at: http://www.leadingpractice.com/about-us/interconnects-with-main-existing-frameworks/, accessed May 26, 2014.

48. LEADing Practice. *Welcome to LEADing Practice*. Available at: http://www.leadingpractice.com/, accessed May 21, 2014.

[27] van Rensburg A, "Principles for modelling business processes," Industrial Engineering and Engineering Management (IEEM), (IEEE, 2011), 1710−1714.

[28] John A. Zachman, Henrik von Scheel, and Mark von Rosing, "The focus of Enterprise Architecture," The Complete Business Process Handbook, (Morgan Kaufman, 2014).

[29] John A. Zachman, Henrik von Scheel, Mark von Rosing. "The focus of Enterprise Architecture," The Complete Business Process Handbook 2, (Morgan Kaufman, 2014).

[30] LEADing Practice-Categorization & Classification Body of Knowledge, 2014.

[31] LEADing Practice, Decomposition & Composition Reference Content. Available at: http://www.leadingpractice.com/enterprise-standards/enterprise-engineering/decompositioncomposition/, accessed May 21, 2014.

[32] https://supply-chain.org/f/SCOR-Overview-Web.pdf.

[33] http://www.apqc.org/.

[34] Business Objects that Capture the Scope and Value of the Business.

[35] Business Objects that Capture and Describe the Essential Organizations Skill And Knowledge Needed to Fulfil the Scope And Purpose of the Business.

[36] Business Objects that Realize Behaviour.

[37] Business Objects Necessary to Execute Work and Create Value.

[38] Business Objects Necessary to Describe the Structure and Behaviour of Software that Enables Work.

[39] Business Objects Necessary to Describe the Persistent Information Used within the Software.

[40] Gumpp A and Pousttchi K, "The 'Mobility-M'-framework for application of mobile technology in business processes." INFORMATIK 2005-Informatik LIVE - Jahrestagung der Gesellschaft für Informatik e V (GI) 2, 2005, 523−527.

[41] Scheruhn H, Sicorello S, Weidner S, Repository-based ERP case studies: A study about chances and benefits of agile case study development in Witold Abramowicz, John Domingue, Krzysztof Wecel (Eds.): Business Information Systems Workshops - BIS 2012 International Workshops and Future Internet Symposium, Vilnius, Lithuania, May 21−23, 2012 Revised Papers. Springer 2012.

[42] http://www.bpmn.org/.

[43] Zhao F, Scheruhn H and von Rosing M, "The Impact of Culture Differences on Cloud Computing Adoption" In: Human-Computer Interaction, (Springer 2014), 776−785.

[44] Scheruhn H, Ackermann D, Braun R and Förster U, "Repository-based implementation of Enterprise Tiers: A study based on an ERP case study," In: Human-Computer Interaction. Users and Contexts of Use, (Springer, 2013), 446−455.

[45] LEADing Practice, The LEADing Practice eXtended BPMN Standard. Available at: http://www.leadingpractice.com/wp-content/uploads/2013/10/LEADing-PracticeXBPMN.pdf, accessed May 14, 2014.

[46] APQC and von Rosing M, "Crash Course with the LEAD Frameworks," Methods and Approaches. Available at: http://www.leadingpractice.com/wp-content/uploads/2013/10/Crash-Course-to-LEAD-3.0.pdf, accessed May 20, 2014.

[47] LEADing Practice. Interconnects with Existing Frameworks. Available at: http://www.leadingpractice.com/about-us/interconnects-with-main-existing-frameworks/, accessed May 26, 2014.

[48] LEADing Practice. Welcome to LEADing Practice. Available at: http://www.leadingpractice.com/, accessed May 21, 2014.

The BPM Way of Implementation and Governance

Mark von Rosing, Henrik von Scheel, August-Wilhelm Scheer

INTRODUCTION

Most process initiatives today include some sort of business process automation, whereas implementation and governing of complex processes in this complex setting can be a daunting thing.

The reality is that 72% of process automation (IT) projects fail to deliver on time, on budget, or on value.[1] Most IT projects fail during the implementation/deployment phase.[2] Research indicates that only 25% of failed process automation (IT) projects occur because of unsolvable technical issues, whereas 75% of all failures are due to a complex mixture of problems including missing process leadership, missing employee process skills, bad communication and so on. This clearly suggests at least part of the gap is between the "as-is" process landscape, the 'to-be' process design, and the actual business transformation sought in and through the process implementation.

In Part V, we focus on the "Way of Implementing" and "Way of Governing", spanning from agile way from process design, to process implementation, BPM change management, process outsourcing, holistic process governance, project, program, portfolio, and BPM Governance as well as BPM Alignment.

We will outline the "Way of Implementing" or the approach, you, the practitioner, follow to apply the way of working and modeling into the physical and thereby the process execution and concrete relevant aspects. In the "Way of Governing," we outline the approach the practitioner follows to steer and govern what exists. It consists both of a holistic BPM Governance approach as well as a separate governance process that spans across the BPM Life Cycle, for example, process analysis, process design, process implementation, and run, monitor, and optimize the existing process.

End Notes

1. The Standish Group, CHAOS (2009)
2. Trad, Antoine; Kalpi, Damir & Trad, Hiam, The Selection and Training Framework for Managers in Business Innovation Transformation Projects, (June 2013)

The Complete Business Process Handbook. http://dx.doi.org/10.1016/B978-0-12-799959-3.00026-4

第五部分

5.1　BPM实施和治理方式

Mark von Rosing, Henrik von Scheel, August-Wilhelm Scheer

介绍

当今大多数流程活动都包含某种业务流程的自动化,而在此复杂环境中实现和管理复杂流程可能是一件令人生畏的事情。

现实情况是72%的流程自动化(IT)项目不能按时、按预算或按价值交付[1],大多数IT项目在实现/部署阶段失败[2]。研究表明,只有25%的流程自动化(IT)项目的失败是由无法解决技术问题造成,而75%的失败归因于复杂的混合问题,包括:缺少流程领导、员工流程技能缺失、沟通不畅等。这清楚地表明,至少部分差距存在于现有流程、未来流程设计和流程转换过程这三者之间。

在第五部分中,我们聚焦于"实现路径"和"治理方法",从敏捷方法到流程设计、流程实现、BPM变革管理、流程外包、整体流程治理、项目、程序、组合、BPM治理以及BPM一致性。

我们将概述"实施方式"或方法,您和其他从业者应该遵循将此工作方式和建模方法自然地应用于流程执行等相关方面。在治理方式中,我们概述从业者应该遵循的方法,既包括整体BPM治理方法,也包含跨越BPM生命周期的单独治理流程,如流程分析、流程设计、流程实施以及运行维护等方面,进而监控和优化现有流程。

参考文献

[1] The Standish Group, CHAOS (2009).
[2] Trad, Antoine; Kalpi, Damir & Trad, Hiam, The Selection and Training Framework for Managers in Business Innovation Transformation Projects, (June 2013).

Applying Agile Principles to BPM

Mark von Rosing, Joshua von Scheel, Asif Qumer Gill

INTRODUCTION

The term "Agile" has attracted significant attention across industry and academia.[1] Agile is not new. The history of agile concepts can be traced back to 1930s. It has its foundation in iterative and incremental approaches. Many ways exist in which agile concepts can be applied across various disciplines and industry verticals, such as agile software development, agile project management, agile supply chain, agile manufacturing, agile service management, agile enterprise, and the list goes on. Similarly, agile concepts can also be applied to business process management (BPM) planning, analysis, architecture, design, implementation, operation, monitoring, and improvement. However, before jumping on the bandwagon of Agile BPM, it is important to understand what is meant by "Agile." What are the building blocks or principles underlying agile? What does it mean to use agile principles? What is the difference between agile and traditional non-agile ways of working? Why do we need to be agile? How is an Agile BPM capability established? The purpose of this chapter is to provide the precise and practical answers to these fundamental questions. This chapter is organized as follows. Firstly, it describes the agile thinking and its origin. Secondly, it describes the agile characteristics, values and principles. Thirdly, it describes the agile practices or ways of working. Fourthly, it describes the difference between the agile and traditional ways of working. Fifthly, it describes the application of agile ways of working to BPM and defines the Agile BPM. Sixthly, it discusses how to establish an Agile BPM capability by using the agility adoption and improvement model. Finally, it concludes the chapter with key take away points.

WHAT IS AGILE?

Although the basic "agile" term comes from the Latin word *agilis* and means to drive, do, and see. The basic meaning of agile is to move quickly, lightly, and easily. In the 1930s, the automobile industry introduced the first agile concepts through the introduction of optimization concepts and work splitting. Further, agile concepts have been applied within the lean manufacturing/lean consumption paradigms. With Agile's growing popularity, other industry segments started realizing that agile principles are not limited to any specific industry segment or functional group. Most relevant to this discussion, over the past decade the software industry has successfully adopted agile principles, and Agile has become a popular software project and product development methodology. Agile methods and practices can be traced back to the incremental software development methods as far back

The Complete Business Process Handbook. http://dx.doi.org/10.1016/B978-0-12-799959-3.00027-6

5.2 将敏捷原则应用于BPM

Mark von Rosing, Joshua von Scheel, Asif Qumer Gill

5.2.1 介绍

敏捷一词已经引起了企业界和学术界的广泛关注[1]。敏捷并不是全新的概念，它的历史可以追溯到20世纪30年代，迭代和增量的方法是它的基础。敏捷概念可以通过多种方式应用于不同的学科和行业垂直领域，如敏捷软件开发、敏捷项目管理、敏捷供应链、敏捷制造、敏捷服务管理、敏捷企业等，并且该领域清单还在增加。同样，敏捷概念也可以应用于BPM的规划、分析、架构、设计、实现、操作、监控和改进。然而，在追逐敏捷BPM的潮流之前，了解敏捷的含义是很重要的，包括：什么是敏捷的基础构件或原则？使用敏捷原则意味着什么？敏捷和传统的非敏捷工作方式有什么区别？为什么我们需要敏捷？如何建立敏捷的BPM能力？本节的目的是为这些基本问题提供精确而实际的答案。本节组织如下：第一，介绍敏捷思维及其起源；第二，阐述敏捷的特点、价值观和原则；第三，介绍敏捷实践或工作方式；第四，阐述敏捷与传统工作方式的区别；第五，描述敏捷工作方式在BPM中的应用，并定义敏捷BPM；第六，讨论如何利用敏捷采用和改进模型建立敏捷的BPM能力。最后，本节以关键的知识点作为结束。

5.2.2 敏捷是什么？

敏捷术语来自拉丁语"agilis"，意思是驾驶、做和看。敏捷的基本含义是快速、轻而易举地移动。在20世纪30年代，汽车工业通过引入优化概念和工作分解，首先引入了敏捷概念。此外，敏捷概念已应用于精益制造、精益消费模式中。随着敏捷越来越受欢迎，其他行业部门开始意识到敏捷原则并不局限于任何特定的行业部门或功能领域。与这一讨论最相关的是，在过去的十年中，软件行业已经成功地采用了敏捷原则，而敏捷已经成为一种流行的软件项目和产品开发方法。敏捷方法和实践可以追溯到早在1957年[2]的增量软件开发方法，之后因重量级的瀑布法

as 1957[2] before falling out of favor for the heavyweight waterfall method. More recently, the agile movement began to come back when, in 1974, a paper by E. A. Edmonds introduced an adaptive software development process.[3] Concurrently and independently, the same methods were developed and deployed by the New York Telephone Company's Systems Development Center under the direction of Dan Gielan. Also in the early 1970s, the concepts of Evolutionary Project Management (EPM), which has evolved into Competitive Engineering, got their start. These were followed with the so-called lightweight agile software development methods, which evolved in the mid-1990s as the carminative reaction against the waterfall-oriented methods, which were characterized by their critics as being heavily regulated, regimented, and micromanaged, and having overly incremental approaches to development. Proponents of these newer, lightweight agile methods contend that they are returning to development practices, which were present early in the history of software development.[4] Compared to traditional software engineering, agile development is mainly targeted at complex systems and projects with dynamic, "undeterministic", and nonlinear characteristics, in which accurate estimates, stable plans, and predictions are often hard to get in early stages, and big upfront designs and arrangements will probably cause a lot of waste, that is, not economically sound. These basic arguments and precious industry experiences learned from years of successes and failures have helped shape agile's flavor of adaptive, iterative, and evolutionary development.[5]

Early implementations of agile methods include Rational Unified Process (1994), Scrum (1995), Crystal Clear, Extreme Programing (1996), Adaptive Software Development, Feature Driven Development (1997), and Dynamic Systems Development Method (DSDM) (1995). After the Agile Manifesto[6] was published in 2001,[7] these have since been referred to collectively as "agile methodologies."

Although Agile is now being applied and discussed around software development, the core of Agile is also about the ability to structure organizations in such a way that they can embrace change and adapt quickly to service the customers in their ever-changing needs. However, taking a big-bang approach to Agile is not really a viable option for many organizations, as most successful adoptions of Agile are tailored to the strengths and limitations of the specific organization.

Like any other change, Agile adoption is not always welcomed right away and faces resistance. Organizations observe many types of frictions that reduce the momentum during Agile implementation. These frictions absorb energy because of the resistance at various levels. Friction is not a fundamental force but occurs because of the turbulence caused by the change. Three main types of frictions apply to the strategy linkage, organization, processes, and technical agility. In this way, Agile is referred to as a mindset, change, flexibility, nonfunctional requirement (link to strategy and goals), culture, and the ways of working, approach, or philosophy. This section discusses the basic definition of agility and introduces the agile features—the characteristics, values, principles, and practices of which Agile is composed.

则（waterfall methodology）的出现而不再受青睐。直到1974年，E. A. Edmonds在当年的一篇论文中介绍了一个适应性软件开发过程，敏捷思想开始回归[3]。同时，同样的方法也由纽约电话公司的系统开发中心在Dan Gielan的指导下开发和部署。同样在20世纪70年代初，渐进式项目管理（evolutionary project management，EPM）的概念已经演变为竞争工程，并开始了它们的发展。紧随其后的是所谓的轻量级的敏捷软件开发方法，该方法在20世纪90年代中期演化为针对面向瀑布法则方法的逆向工程，这些瀑布法则方法的批评者认为它们的特点是受到严格的监管、管制和微观管理，并且开发方法过于笨重。而敏捷方法的支持者声称这些新的、轻量级的软件开发方法在历史的早期就已经出现了[4]，它们正在回归到开发实践中。与传统的软件工程相比，敏捷开发主要针对具有动态的、不确定的和非线性特征的复杂系统和项目。这主要是在项目的早期阶段很难获得准确的估计、稳定的计划和预测，而复杂的前期设计和规划可能会造成大量浪费，也就是说在操作上可行性较低。综上，以往的研究和从多年的成功和失败中吸取的宝贵的行业经验帮助塑造了敏捷的适应性、迭代和进化开发风格[5]。

　　敏捷方法的早期实现包括：统一软件开发过程（1994年，rational unified process）、迭代式增量软件开发过程（1995年，scrum）、水晶项目管理体系（Crystal clear）、极限编程（1996年，extreme programing）、自适应软件开发（Adaptive software development）、特征驱动开发（1997年，feature driven development）和动态系统开发方法（1995年，dynamic system development method，DSDM）。在《敏捷宣言》（Agile Manifesto）[6]于2001年发表之后[7]，这7种方法被统称为"敏捷方法论"。

　　虽然敏捷现在正在围绕软件开发进行应用和讨论，但敏捷的核心也在于能够以这样的方式构建组织，即他们可以接受变化并快速适应客户以满足不断变化的需求。然而，对于许多组织而言，对敏捷采取大规模方法并不是一个可行的选择，因为大多数成功采用敏捷的方法都是针对特定组织的优势和局限性而定制的。

　　与其他任何变化一样，敏捷应用并不总是立即受到欢迎并会面临阻力。组织会遇到许多类型的冲突，这些冲突降低了实施敏捷过程中的动力。这些冲突不是一种基本力量，而是由变化引起的湍流而产生的。三种主要的摩擦类型适用于战略链接、组织、流程和技术敏捷性。通过这种方式，敏捷被称为思维定式、变化、灵活性、非功能性需求（与战略和目标的链接）、文化以及工作方式、方法或哲学。本节讨论了敏捷的基本定义，并介绍了敏捷的特点、价值观、原则和实践，敏捷就是由这些特征组成的。

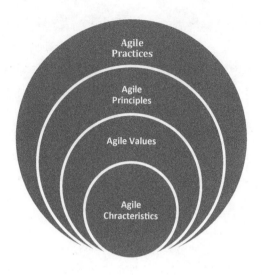

FIGURE 1

What is agile?

Figure 1 shows the conceptual relationship between agile features. At the core, and by far the most critical to the nature of Agile, are its characteristics; slightly less important are the values that are employed when Agile is practiced. This is followed by the agile principles that guide how Agile is applied, and then finally are the agile practices that form the basis for work within an agile setting.

Qumer and Henderson-Sellers (2008) provide the following precise definitions of agility and agile methods.

"Agility is a persistent behavior or ability of a sensitive entity that exhibits flexibility to accommodate expected or unexpected changes rapidly, follows the shortest time span, uses economical, simple and quality instruments in a dynamic environment, and applies updated prior knowledge and experience to learn from the internal and external environment."[8]

"A software development method is said to be an agile software development method when a method is people focused, communications oriented, flexible (ready to adapt to expected or unexpected change at any time), speedy (encourages rapid and iterative development of the product in small releases), lean (focuses on shortening time frame and cost and on improved quality), responsive (reacts appropriately to expected and unexpected changes), and learning (focuses on improvement during and after product development)."[9]

AGILE CHARACTERISTICS

The agility definition highlighted the five fundamental agile characteristics: responsiveness, flexibility, speed, leanness, and learning. These five characteristics can be used to describe and measure the agility of an object or entity.

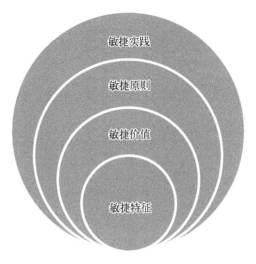

图1　什么是敏捷?

　　图1显示了敏捷特性之间的概念关系。敏捷本质的核心也是目前为止最关键的是它的特性,稍微不那么重要的是在实践敏捷时所采用的价值观。接下来是指导如何应用敏捷的敏捷原则,最后是形成敏捷工作基础的敏捷实践。

　　Qumer和Henderson-Sellers(2008年)对敏捷性和敏捷方法给出了以下精确的定义。

　　"敏捷是表现出灵活性,能够快速适应预期或意外的变化,遵循最短的时间跨度,在动态环境中使用经济、简单和优质的仪器,应用更新的先前知识和经验向内部和外部环境学习的敏感实体的一种持久行为或能力[8]。"

　　"当一种方法体现为以人员为中心、以通信为导向、灵活(随时准备好适应预期或意外的变化)、快速(鼓励小版本产品的快速和迭代开发)、精益(侧重于缩短时间框架和成本)以及对改进的质量、响应性(对预期和意外的变化做出适当的反应)和学习(关注产品开发期间和之后的改进),我们就将该软件开发方法称为一种敏捷的软件开发方法[9]。"

5.2.3　敏捷特性

　　敏捷定义强调了五个基本的特性:响应性、灵活性、速度、精益和学习。这五个特征可以用来描述和度量对象或实体的敏捷性。

- *Responsiveness*: is the ability of an object or entity to scan and sense the external and internal opportunities; and form an appropriate response according to the situation at hand.
- *Flexibility*: is the ability of an object or entity to accommodate expected or unexpected changes.
- *Speed*: is the ability of an object or entity to provide a speedy or quick response to expected or unexpected changes.
- *Leanness*: is the ability of an object or entity to provide a speedy and flexible response with optimal or minimal resources without compromising the quality.
- *Learning*: is the ability of an object or entity to learn through continuously managing and applying up-to-date knowledge and experience.[10]

AGILE VALUES

Similarly, the six agile values provide fundamental statements that describe agile preferences:

1. Individual and interactions over processes and tools
2. Working software over comprehensive documentation
3. Customer collaboration over contract negotiation
4. Responding to change over following a plan
5. Keeping the process agile
6. Keeping the process cost-effective

The agile values one to four were provided by the Agile Manifesto (2001). The fifth agile value "keeping the process agile" was provided by Koch in 2005.[11] The sixth value of "keeping the process cost-effective" was provided by Qumer and Henderson-Sellers.[12]

AGILE PRINCIPLES

Agile Software development is based on 12 guiding principles, which are set out in the Agile Manifesto[13]:

1. Our highest priority is to satisfy the customer through early and continuous deliver of valuable software.
2. Welcome changing requirements, even late in development. Agile processes harness change for the customer's competitive advantage.
3. Deliver working software frequently, from a couple of weeks to a couple of months, with a preference to the shorter timescale.
4. Business people and developers must work together daily throughout the project.
5. Build projects around motivated individuals. Give them the environment and support they need, and trust them to get the job done.

- 响应性：是一个对象或实体体会、感知外部和内部机会的能力，并根据现有的情况形成适当的响应。
- 灵活性：对象或实体适应预期或意外变化的能力。
- 速度：对象或实体对预期或意外变化提供快速响应的能力。
- 精益：是一个对象或实体在不影响质量的情况下，以最佳或最小的资源提供快速、灵活的响应的能力。
- 学习：是一个对象或实体通过不断管理和应用最新知识和经验来学习的能力[10]。

5.2.4 敏捷价值

同样，敏捷的六个价值观体现了敏捷性：

（1）个人和流程与工具之间的互动；

（2）工作软件优于全面的文档；

（3）客户合作重于合同谈判；

（4）响应变化而不是遵循计划；

（5）保持流程敏捷；

（6）保持流程的成本效益。

敏捷价值观（1）到（4）由敏捷宣言（2001年）提供。Koch于2005年提供了第五个敏捷价值"保持流程敏捷"[11]。Qumer和Henderson-Sellers提供了第六个价值"保持流程的成本效益"[12]。

5.2.5 敏捷原则

敏捷软件开发基于12条指导原则，这些原则在敏捷宣言[13]中有所阐述。

（1）我们的首要任务是通过尽早和持续交付有价值的软件来满足客户。

（2）欢迎改变需求，甚至在开发后期。敏捷流程利用变化来实现客户的竞争优势。

（3）频繁地交付可工作的软件，从几周到几个月，优先考虑更短的时间尺度。

（4）业务人员和开发人员必须每天在整个项目中一起工作。

（5）围绕有动力的个人建立项目。为他们提供所需的环境和支持，并相信他们能够完成工作。

6. The most efficient and effective method of conveying information to and within a development team is face-to-face conversation.
7. Working software is the primary measure of progress.
8. Agile processes promote sustainable development. The sponsors, developers, and users should be able to maintain a constant pace indefinitely.
9. Continuous attention to technical excellence and good design enhances agility.
10. Simplicity—the art of maximizing the amount of work not done—is essential.
11. The best architectures, requirements, and designs emerge from self-organizing teams.
12. At regular intervals, the team reflects on how to become more effective, then tunes and adjusts its behavior accordingly.

As the 12 guiding principles make clear, they are software centric, to apply in the BMP context. We will show later how they can be tailored to apply in a different setting with great effect.

AGILE PRACTICES

A number of agile methods exist (e.g., XP, Scrum, and Lean). These methods provide concrete agile practices that adhere to the agile characteristics, values, and principles. The scope of each of the methods is slightly different from the others. For instance, XP focuses on employing technical software development practices such as Refactoring, "Pair Programming", Automated Testing, Continuous Integration and so on. Scrum focuses on project management practices and the use of "Sprints" to deliver functionality. Generally, agile development is supported by a bundle of concrete practices covering areas that may include the full range of product development from requirements, design, modeling, coding, testing, project management, process, quality, and so on. The result is that we learn two things: first, the differences indicate that no standard single agile method is available, which may be applied or adopted off-the-shelf; and second, the best practices from different agile methods can conceivably be combined to create a situation-specific agile method. What is important to note here is that the key to being agile is to focus on harnessing agile characteristics, values, and principles underlying the specific agile practices.

AGILE VERSUS TRADITIONAL WAYS OF WORKING

Agile and traditional waterfall methods are two distinct ways of developing software. The Waterfall model can essentially be described as a linear model of product delivery. Like its name suggests, waterfall employs a sequential set of processes as subsequently indicated in Figure 2. Development flows sequentially from a start point to the conclusion, the delivery of a working product, with several different stages along the way, typically: requirements, high-level design, detailed implementation,

（6）将信息传递给开发团队和在开发团队中最有效和最高效的方法是面对面的对话。

（7）工作软件是衡量进展的主要方法。

（8）敏捷过程促进可持续发展。发起人、开发人员和用户应该能够无限期地保持恒定的步调。

（9）持续关注卓越的技术和良好的设计能够提高灵活性。

（10）简单性（将未完成工作量最大化的艺术）是至关重要的。

（11）最好的架构、需求和设计来自组织的团队。

（12）每隔一段时间，团队就应该思考如何提高效率，然后相应地调整自己的行为。

正如12条指导原则所阐明的那样，它们以软件为中心适用于BMP环境。稍后我们将展示如何定制它们，以在不同的环境中应用并产生巨大的效果。

5.2.6 敏捷实践

存在许多敏捷方法（如XP、Scrum和精益），这些方法提供了遵循敏捷特征、价值和原则的实践形式。每种方法的作用域与其他方法略有不同，例如，XP侧重于采用重构、"结对编程"、自动化测试、持续集成等技术软件开发实践；Scrum侧重于项目管理实践和使用"Zoho Sprints"（一种敏捷开发项目管理工具）来交付功能。通常，敏捷开发是由一组具体的实践来支持的，这些实践覆盖需求、设计、建模、编码、测试、项目管理、过程、质量等的整个产品开发范围。结果是，我们学到了两件事：首先，差异表明没有标准的单一敏捷方法可用，可以应用或采用现成的敏捷方法；其次，可以将不同敏捷方法的最佳实践组合起来创建特定于场景的敏捷方法。这里需要注意的是，敏捷的关键是集中精力利用特定敏捷实践背后的敏捷特征、价值和原则。

5.2.7 敏捷与传统的工作方式的对比

敏捷和传统的瀑布方法是两种不同的软件开发方法。瀑布模型本质上可以描述为产品交付的线性模型。就像它的名字所暗示的那样，瀑布使用一组顺序的流程，如图2所示。为交付一个工作产品，开发从开始点到结束点依次进行，过程包括需求、高阶设计、详细实施、验证、部署和客户验证等几个不同的阶段，之后通常

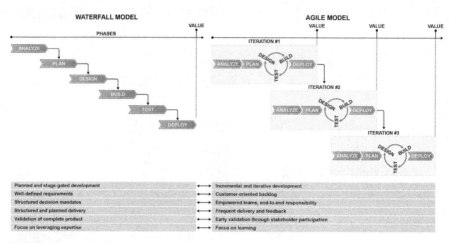

FIGURE 2

Agile versus traditional waterfall.

verification, deployment, and customer validation, often followed with stages to cover the running/maintenance of the product, and to address the need for continuous improvement.

The emphasis of Waterfall is on the project plan and managing all work against the plan. For this reason, a clear plan and a clear vision should exist before beginning any kind of development. Because the Waterfall method requires upfront, extensive planning, it permits the launch of a known feature set, for an understood cost and timeline, which tends to please clients.

Furthermore, Waterfall development processes tend to be more secure because they are so plan oriented. For example, if a designer drops out of the project, it isn't a huge problem, as the Waterfall method requires extensive planning and documentation. A new designer can easily take the old designer's place, seamlessly following the development plan. As described above, Agile offers an incredibly flexible design model, promoting adaptive planning and evolutionary development. Agile might be described as freeform software design. Workers only work on small packages or modules at a time. Customer feedback occurs simultaneously with development, as does software testing and deployment. This has a number of advantages, especially in project environments in which development needs to be able to respond to changes in requirements rapidly and effectively.

By way of comparison, instead of a big-bang waterfall product delivery, Agile focuses on delivering early value or product features in small increments, which is referred to as a minimum viable product or as having minimum marketable features. An agile project is organized into small releases, in which each release has multiple iterations. Within each iteration just enough work is pulled off the stack, planned, analyzed, designed, developed, tested, integrated, and then deployed in the production or a production-like staging environment. During and following

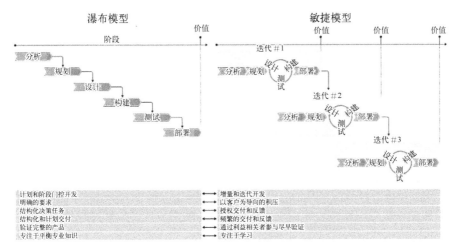

图2　敏捷vs传统瀑布法则

是涵盖产品运行/维护的阶段,以及满足持续改进的需要。

　　瀑布法则的重点是项目计划和根据计划管理所有工作。因此,在开始任何形式的开发之前,都应该有一个清晰的计划和远景。因为瀑布方法需要预先的、广泛的计划,它允许以一个可理解的成本和时间线启动一个已知的特性集,这往往会取悦客户。

　　此外,瀑布式开发过程是面向计划的,因此也更安全。例如,如果一个设计师退出了项目,由于瀑布方法需要大量的计划和文档,一个新的设计师可以轻而易举地取代旧的设计师,无缝地遵循开发计划,这不是一个大问题。如上所述,敏捷提供了一个难以置信的灵活设计模型,促进了适应性规划和进化开发。敏捷可以被描述为自由形式的软件设计,工作人员每次只能在小包装或模块上工作,并且客户反馈与开发同时发生,软件测试和部署也是如此。这具有许多优点,特别是在项目环境中,其中开发需要能够快速有效地响应需求的变化。

　　相比之下,敏捷关注于以小增量交付早期价值或产品特性,而不是大瀑布式的产品交付,这被称为最低可行产品或具有最低可销售特性。敏捷项目被组织成小版本,每个版本有多个迭代。在每个迭代中,只需从堆栈中抽出足够的工作,进行计划、分析、设计、开发、测试、集成,然后在生产或类似生产的阶段环境中进行部署。

the iteration the product is demonstrated to concerned stakeholders for feedback and commitments. Each iteration also involves retrospective activity, which is aimed at identifying and addressing the issues of the agile practices. In each iteration, different developers may work on different modules or requirements (also known as user stories) throughout the development process and then work to integrate all of these modules together into a cohesive piece of working-software release. In summary, this can be seen as a process, which consists of analysis and planning stages, followed by a rapid design, build, and test cycle, all of which then ends with deployment.

Experience with the agile approach has shown that it can be especially beneficial in situations in which it is not possible to define and detail the project requirements, plan, and design upfront. Agile is also an excellent option for experimental circumstances. For example, if you are working with a client whose needs and goals are a bit hazy, it is probably worthwhile to employ the agile method. The client's requirements will likely gradually clarify as the project progresses, and development can easily be adapted to meet these new, evolving requirements. Agile also facilitates interaction and communication—collaboration is more important here than doing design in isolation. Because interaction among different designers and stakeholders is key, it is especially conducive to teamwork-oriented environments.

Figure 2 compares and contrasts key elements of Agile and Waterfall Development. In this figure, we see graphically the life cycle of each development model. Below each type of life cycle are listed the key properties of each method and how they relate to the equivalent properties of the alternative method.

AGILE BPM

Although Agile is not a silver bullet that can be applied to all problems, however, it does provide ways of working that could be suitable to the circumstances in which frequently changing business and customer requirements or other conditions of uncertainty force the organization to pursue quick wins for developing capabilities, services, or systems. As Agile is about making complex things simple or simpler, this section of the chapter will highlight how the agile concepts can be applied to enable BPM in all the various areas and disciplines as defined in by Qumer and Henderson-Sellers.[8] We must, however, keep in mind that agility of process is not in and of itself Agile BPM and that to incorporate agility into BPM actually requires a fundamental shift in the strategy, operations, and tactics of the way BPM works and how modeling is carried out in an organization. This section tackles this up-to-date subject in the context of:

1. The benefits and limitations of Agile and how to apply it to BPM
2. An Agile BPM method
3. A firmly defined terminology
4. A concept to develop agile capabilities in the BPM Center of Excellence (CoE).

在迭代期间和之后,产品向相关的利益相关者展示反馈和承诺。每个迭代还包括回顾活动,其目的是识别和解决敏捷实践的问题。在每个迭代中,不同的开发人员可能在整个开发过程中处理不同的模块或需求(也称为用户故事),然后将所有这些模块集成到一个具有内聚性的工作软件版本中。总之,这可以看作是一个过程,它由分析和计划阶段组成,然后是一个快速的设计、构建和测试周期,所有这些都以部署作为结束的标志。

敏捷方法的经验表明,在无法预先定义和详细说明项目要求、计划和设计的情况下,敏捷方法尤其有用。敏捷也是测试环境下一个很好的选择。例如,如果您正在与需求和目标有点模糊的客户端合作,那么使用敏捷方法可能是值得的。随着项目的进展,客户的需求可能会逐渐清晰,开发可以很容易地适应这些新的、不断发展的需求。敏捷还促进了交互和交流——在这里,协作比单独进行设计更重要,因为不同的设计人员和利益相关者之间的交互是关键,所以它特别有利于面向团队合作的环境。

图2比较了敏捷开发和瀑布开发的关键元素。在图2中,我们可以看到每个开发模型的生命周期。在每种类型的生命周期下面列出了每种方法的关键属性,以及它们与替代方法的等效属性之间的关系。

5.2.8 敏捷BPM

尽管敏捷不是一个可以应用于所有问题的灵丹妙药,但是它确实提供了一种工作方式,这种工作方式可以适应频繁变化的业务和客户需求或其他条件的不确定性迫使组织追求快速赢得发展、服务或系统。由于敏捷是将复杂的事情变得简单或更简单,本节的这一部分将重点介绍如何应用敏捷概念在Qumer和Henderson-Sellers[8]定义的所有不同领域和规程中启用BPM。然而,我们必须记住:流程的敏捷性本身并不是敏捷的BPM,而是要融入到企业中。对BPM的敏捷性实际上需要在BPM工作方式的战略、操作和战术以及如何在组织中执行建模方面进行根本性的转变。本节在以下背景下讨论最新主题:

(1)敏捷的好处、局限性以及如何将其应用于BPM;

(2)一个敏捷BPM方法;

(3)一个明确定义的术语;

(4)在BPM CoE中开发敏捷功能。

The Benefits and Limitations of Agile and How to Apply It to BPM

We have seen that Agile offers several benefits (e.g., value to customer, organization, staff, and community) over traditional ways of working. We have seen, for example, that Agile focuses on developing a minimally marketable or viable product or service features, which will provide value to customers and community. In contrast to the traditional waterfall approach, it focuses on delivering value early to customers and community in short increments, which range in duration from anywhere between a few weeks to months. This seems helpful for the organizations and staff seeking to improve time to market and quality while reducing the cost of production and failure. Clearly then, agile ways of working not only help delivering value early, but they also seem appropriate in recognizing the risks and failure early to mitigate their impact.

A part of exploring the potential around Agile BPM also includes understanding the traditional problems and challenges when adapting a new concept. As with so many things, resistance or friction impede adaptation of a new way of thinking and working with agile concepts. For an organization adopting Agile BPM concepts, numerous challenges are possible. However, the most common challenges we have encountered are as follows:

- *Static Friction:* The force that must be overcome before agile concepts can be implemented in a nonagile organization, for example, friction observed before piloting first Agile BPM project.
- *Dynamic Friction:* The force that must be overcome to maintain uniform agile motion and the friction encountered when people don't see immediate results after a new Agile BPM project. It is important for the Agile BPM leader to constantly communicate value of "inspect and adapt." Once the BPM CoE and the organization learn to manage incremental value driven by agile process, dynamic friction starts diminishing by itself.
- *Political Friction:* The force resisting agile progress because of politics that can come from the BPM CoE or the organization itself. A good Agile BPM leader can influence negative politics by persuasive communication in Agile's favor.
- *Knowledge Friction:* The force that must be overcome due to the BPM CoE and the organizational lack of competencies and resources who understand Agile and its precepts, workings, and value. Most organizations use external consultants or hire an Agile BPM specialist to train, coach, and mentor employees so that they gain an agile knowledge base.

Once a solid agile knowledge base is in place in the BPM CoE, this friction will generally start to diminish.

These frictions limit the ability of an organization to maximize the use of Agile BPM in an optimized Way of Working, Modeling, and Governing. Table 1: Indicates the typical friction factors across organizational areas.

Friction is not the only challenge organizations will face when moving from traditional BPM to an Agile BPM way of thinking. Other influencing factors cause Agile BPM to fail or to deliver significantly lower value than expected:

- *Innovation is only done from the process perspective*: As it is in the IT world, the current view of Agile is very much defined by the Software/Application

1. 敏捷的好处和局限性,以及如何将其应用于BPM

我们已经看到,与传统的工作方式相比,敏捷提供了许多优势(如对客户、组织、员工和团体的价值)。例如,我们已经看到,敏捷专注于开发最低限度的可销售或可行的产品或服务功能,这将为客户和团体提供价值。与传统的瀑布式方法不同,它专注于以较短的增量提前向客户和社区交付价值,其持续时间从几周到几个月不等。这似乎有助于组织和员工在减少生产成本和失败的同时,寻求提高上市效率和质量。显然,敏捷的工作方式不仅有助于及早实现价值,而且在及早识别风险和失败以减轻其影响方面似乎也很合适。

探索敏捷BPM的潜力还包括在适应新概念时理解传统问题和挑战。与许多事情一样,阻力或冲突阻碍了对新思维方式和敏捷概念的适应。对于采用敏捷BPM概念的组织,可能存在许多挑战。但是,我们遇到的最常见挑战如下。

- 静态冲突:这是在非敏捷组织中实施敏捷概念之前必须克服的力量,例如,在试用第一个敏捷BPM项目之前观察到的冲突。
- 动态冲突:在新的敏捷BPM项目之后,当人们看不到立即的结果时,为了保持敏捷的统一运动而必须克服的力量,以及当人们在一个新的敏捷BPM项目后没有立即看到结果时所遇到的冲突。对于敏捷BPM领导者来说,不断传达"检查和适应"的价值是非常重要的。一旦BPM CoE和组织学会了管理敏捷流程驱动的增量价值,动态摩擦就会自行减少。
- 政治冲突:由于可能来自BPM CoE或组织本身的政治而抵制敏捷进步的力量。一个好的敏捷BPM领导者可以通过有利于敏捷的说服性沟通来影响消极政治。
- 知识冲突:由于BPM CoE和组织缺乏理解敏捷及其规则、工作和价值的能力和资源,必须克服的力量。大多数组织使用外部顾问或聘请敏捷的BPM专家来培训、训练和指导员工,从而获得敏捷知识库。

一旦在BPM CoE中建立了坚实的敏捷知识库,这种冲突通常就会开始减少。

这些冲突限制了组织以优化的工作、建模和管理方式最大限度地使用敏捷BPM的能力。表1显示了跨组织领域的典型冲突因素。

当组织从传统的BPM转向敏捷的BPM思维方式时,冲突并不是唯一的挑战。其他影响因素导致敏捷BPM失败或交付的价值明显低于预期。

- 创新只能从流程的角度进行:正如IT领域一样,当前的敏捷观点很大程度上是由软件/应用程序和底层技术的角度定义的,这在一定程度上导致了

Table 1 *Example of Friction Factors Across Organizational Areas*

	Static	Political	Dynamic	Knowledge
Organizational	We are unique	Agile BPM versus non agile BPM	Not yet getting value for agile	Waterfall versus agile
Process	Why change?	Change control vs embrace change	Agile process is too fluid	Fear of the unknown, only know BPM water-fall methods
Technical	Where to start?	BPM CoE and process architecture committee vs community of practice	Not enough resources for process and software automation projects	Resources lack agile competencies.

and underlying technology perspective; this gives rise to a degree of vague-ness of requirements, especially within the business layer. We also see this in the process community, in which they limit business innovation and transformation to what they can see the process can do rather than work-ing in its context. Often this is based on a traditional BPM focus around optimizing the existing processes. However, this view limits Agile BPM concepts from enabling true business agility. The reason is that the current IT Agile methods only have feedback loops between the Plan and Deploy phases for a BPM project, placing an emphasis on the feedback from what is possible in the process and creating a disjointedness loop back to the business. Resulting in Agile BPM teams not having gone through the mul-tiple agile business iterations capturing the value and performance aspects relevant in the Analyze and Plan phase (the Business Layer Context). Therefore, Agile BPM needs a better business requirement loop, which is elaborated on later in this chapter.

- *Multiple changes at once:* Changing both value and performance expecta-tions as well as changing business requires the organization to introduce far too many changes during the iteration's Design, Build, and Test phases. This makes it very difficult for the Agile BPM teams to complete the process analysis, process alignment, process changes, process design, process automa-tion, and so on, in the required 2-4 week sprint cycle. Agile BPM, therefor,e needs to build a better requirement and execution approach into the overall approach.

- *Users are not sure of what they can get versus what they want:* Business and process users are not always sure of what they can get from BPM initiatives. As Steve Jobs said, "It's really hard to design products by focus groups. A lot of times people don't know what they want until you show it to them."[14]

- "Having a developer who has a deep understanding of the business is often better than an inexperienced business person with no understanding of how

表1 跨组织领域的冲突因素的例子

	静 态	政 治	动 态	知 识
组织	我们是独一无二的	敏捷BPM与非敏捷BPM的比较	还没有从敏捷中获得价值	瀑布和敏捷
流程	为什么要改变	变革控制vs拥抱变革	敏捷过程过于流畅	害怕未知,只知道BPM瀑布方法
技术	从哪里开始	BPM CoE和流程架构委员会vs实践社区	没有足够的资源用于流程和软件自动化项目	资源缺乏敏捷能力

需求的模糊性,尤其是在业务层中。我们在流程社区中也看到了这一点,在这个社区中,他们将业务创新和转换限制在他们能够看到流程可以做什么,而不是在其背景中工作。通常,这是基于围绕优化现有流程的传统BPM重点。然而,这种观点限制了敏捷的BPM概念实现真正的业务敏捷性。原因在于,当前的IT敏捷方法只在BPM项目的计划和部署阶段之间有反馈循环,强调流程中可能的反馈,并创建返回业务的分离循环,导致敏捷BPM团队没有经历多个敏捷业务迭代捕获相关的价值和性能方面的分析和计划阶段(业务层背景)。因此,敏捷BPM需要更好的业务需求循环,本节后面将对此进行详细阐述。

- 同时进行多个变更:变更价值和效果期望以及更改业务都要求组织在迭代的设计、构建和测试阶段引入太多的变更,这使得敏捷的BPM团队很难在所需的2 ~ 4周的周期内完成流程分析、流程协调、流程更改、流程设计、流程自动化等。因此,敏捷的BPM需要在整体方法中构建更好的需求和执行方法。

- 用户不确定他们能得到什么,也不确定他们想要什么:业务和流程用户并不总是确定他们能从BPM计划中得到什么。正如 Steve Jobs(史蒂夫·乔布斯)所说,"按用户设计产品真的很难。很多时候人们不知道自己想要什么,直到您把它展示给他们。"[14]

- "拥有一个对业务有深刻理解的开发人员,通常要比一个缺乏经验、不了解技术如何实现需求工作的业务人员好。"这也涉及我们之前讨论过的挑

technology can enable work about "requirements." This also touches upon the challenge we previously discussed in which many process experts limit the business innovation and transformation aspects to what they can see that the process can do. The result is that value derived from the process as well as the execution of the innovation cycle, therefore, for the most part, comes too short and does not deliver the desired result or value.

- *Limited executive sponsorship equals limited agility*: Agility requires sponsorship at the highest level. It requires dedicating top performers and empowering them to challenge the status quo. That is, not just automating the existing siloed approach; executives need to resolve innovation and transformation blockers rapidly and with a focus on the final goal.
- *Agile Focus is operational and not strategic*: Agile teams tend to focus on operational accomplishments and report it to tactical level, however, at the expense of strategic business objectives. In BPM, this is tragic, for missing the big picture can lead to long-term failure at the expense of apparent success in the short term.
- *Agile has too little governance*: Agile teams often lack sufficient checks and balances; if this occurs, they can cause lots of damage in a very short time. Agile BPM must interlink with BPM Governance (see chapter BPM Governance).
- *Giving up on quality*: Because of the high demands or urgency that is placed on process deployment, an Agile BPM team often falls back on the crutch of checking only for process pain points/process defects instead of maintaining a high level of quality of overall BPM changes. (see the chapters on BPM Change Management and BPM Governance)

Some of these failure points are indicated in the following Agile figure, which for many organizations represents the current agile method with its failures and problems.

In Figure 3 we see the Agile development method laid out to show the Analyze and Plan stages, followed by the design, build, and test cycle, all of which then ends with the deployment stage with the key criticisms, or weaknesses, mapped to the applicable points in the method.

An Agile BPM Method

To overcome the challenges of the agile method and enable an organization to adapt Agile, BPM must enable strategic alignment and provide the necessary link to performance and value expectations, requirement management, coordination with business impact and changes, better quality, and thereby value creation and realization. For this, we need to augment the traditional agile approach to incorporate a stronger requirement management and an agile feedback loop in the analysis phase. This loop should consider all layers of the enterprise, that is, business, application, and technology, thus allowing the use of these

战,在这个挑战中,许多流程专家将业务创新和转换方面限制在他们能够看到流程所能做的事情上。其结果是,在大多数情况下,从创新的流程中获得的价值无法实现用户期望的结果。

- 有限的高管支持等于有限的敏捷:敏捷需要最高层的支持,它需要最优秀的人才的加入,并赋予他们挑战现状的能力。也就是说,不仅要使现有的孤立方法自动化,管理者还需要快速解决创新和转型障碍,并将重点放在最终目标上。
- 敏捷关注的是操作性而非战略性:敏捷团队倾向于关注运营成果,并将其报告给战术层面,然而,这是以牺牲战略业务目标为代价的。但在BPM中错过了大局可能会导致长期失败,而在短期内显然会失败,这将是一个悲剧。
- 敏捷的管理太少:敏捷团队经常缺乏足够的制衡,如果发生这种情况,它们会在很短的时间内造成很大的损害。敏捷BPM必须与BPM治理相互关联(参见BPM治理一节)。
- 放弃质量:由于流程部署中的高需求或紧迫性,敏捷的BPM团队通常只会检查流程痛点/流程缺陷,而不是保持整个BPM变更的高质量(请参阅有关BPM变更管理和BPM治理的章节)。

下面的敏捷图显示了其中的一些失败点,对于许多组织来说,这代表了当前敏捷方法的失败和问题。

在图3中,我们看到了展示分析和计划阶段的敏捷开发方法,接下来是设计、构建和测试周期,所有这些阶段都以部署阶段结束,并将关键的不足和弱点反映到方法中。

2. 敏捷 BPM 方法

为了克服敏捷方法的挑战并使组织能够适应敏捷,BPM必须与战略一致,并提供必要的链接以实现绩效和价值预期、需求管理、与业务影响和变更的协调、更好的质量,从而达到价值的创建和实现。为此,我们需要增强传统的敏捷方法,以便在分析阶段实现更强大的需求管理和敏捷反馈循环。此循环应该考虑包括业务、应用程序和技术在内的企业的所有层级,从而允许在设计构建和测试阶段以敏

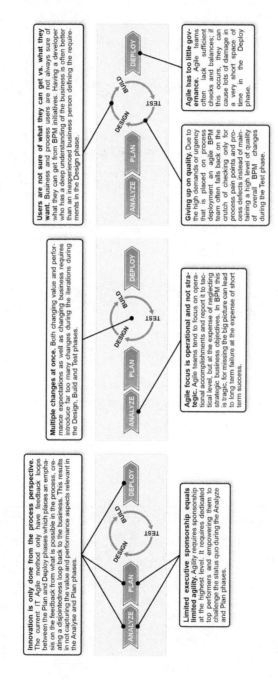

FIGURE 3

Agile weakness point indicators.[15]

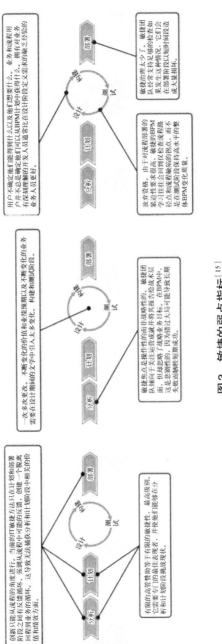

图 3 敏捷的弱点指标 [15]

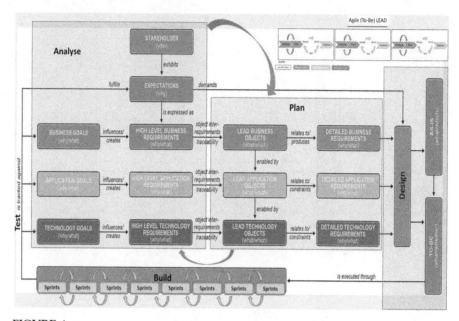

FIGURE 4

Details of agile BPM way of working.[16]

requirements in an Agile Way through the design Build and Test phase and to assess testing against the requirements prior to deployment. This is shown in Figure 4.

Agile Analysis

Agile analysis, in the context of Agile BPM, suggests active collaboration with the stakeholders to identify the requirements with necessary details at the release and iteration levels, instead of trying to get the complete detailed requirements up-front. The key difference (compared to traditional process analysis) is that the Agile BPM focuses on the relevant value and performance drivers and analyses in which and how they can be executed. A process in scope can be identified, modeled, analyzed, and decomposed into subprocesses for the Agile BPM project. Within each subprocess, a set of requirements is documented at high level in terms of process user stories in collaboration with the stakeholder at the beginning of the project. The identified user stories or requirements within each subprocess are estimated and prioritized. The prioritized process requirements are organized into short iterations and releases. A set of high-level process requirements or user stories for a given iteration can be further clarified, detailed, and confirmed (signed off) just before the start of the iteration (Zero Iteration). For instance, user stories planned for any given iteration can be detailed and signed-off beforehand. The detailed signed-off user stories can be made ready (just-in-time documentation) one iteration in advance before the start

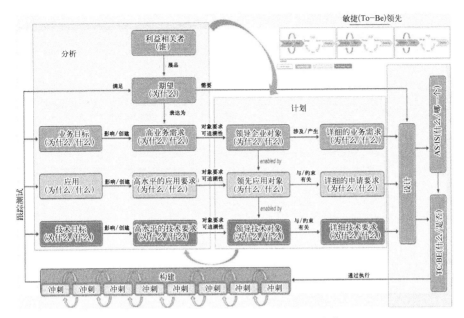

图4　敏捷BPM工作方式的详细信息[16]

捷的方式使用这些需求,并在部署之前根据需求评估进行测试。如图4所示。

1)敏捷分析

在敏捷BPM的背景下,建议将敏捷分析与利益相关方紧密合作。要在发布和迭代过程中确定需求和必要的详细信息,而不是试图预先得到完整的详细需求。关键差异(与传统流程分析相比)在于敏捷BPM关注相关的价值和性能驱动因素,以及在哪些方面及如何执行它们。范围内的流程可以被识别、建模、分析并分解为敏捷BPM项目的子流程。在每个子流程中,在项目开始时与利益相关者合作,确定每个子流程中已识别的用户需求或要求,并确定其优先级,优先处理和发布被组织确定为短迭代的项目。在迭代开始之前(零迭代),可以进一步澄清、详细确认(签署)给定迭代的一组高级流程要求或用户需求。例如,为任何给定迭代计划的用户故事都可以预先详细描述和签署。在下一次迭代中开始开发这些用户故事的

Requirement#	Who/Whom Specification For Example, Stakeholder/ Owner	Where Specification For Example, Layer, Objects, Area (Process, Service, Data, Infrastructure, etc.)	What Specification: High-Level Requirements	What Specification: Detailed Requirements
#				
#				
#				
#				
#				
#				

Table 2 *Example of an Agile BPM Template Developed in the Analysis and Planning Phase*

of the process development of those user stories in the next iteration. It is important to note here that high-level user stories should only provide enough details that are necessary for estimation and prioritization, and should not lock in unnecessary low-level details, which may hinder the adaptability of the Agile BPM project.

Agile Planning

Traditional ways of BPM planning focus on the detailed up-front planning. Agile BPM ways of working require planning at project, release, iteration, and day level. Agile BPM focuses on initial high-level project plan that outlines number of project releases, resources, risks, and cost and benefits estimates. Out of the high-level and detailed requirements a project plan is developed to outline when and which requirements can be meet throughout as the project progresses in small releases. Table 2 illustrates an example of such a template/artifact, used to relate the captured components relevant in the plan phase aspects that is, Stakeholder/ Process Owner, relevant objects and the high-level (nonfunctional) requirements and detailed (functional) requirements. Such a template typically is in the form of a map, which can start as a simple row, and when information is added produces a catalog of rows. Because a release plan only focuses on the release in hand and the first two or three iterations for that release, such a template has the purpose of building an inventory or index list of the relevant stakeholders, objects, and requirements from the different relevant architectural layers that from the analysis phase can be used and tracked against in the planning phase.

Table 2 is the template of a map that captures requirements in a high-level and detailed form and indicates both who has an interest in the requirement and where within the layers and business objects each requirement resides. This enables the agile practitioner to use the release plan to track the project progress in small iterations. An iteration plan focuses on the iteration that will start next. It provides the detailed information about the time-boxed (2–4 weeks iteration) short-iteration activities and schedule such as additional analysis, the

表2　在分析和计划阶段的敏捷BPM开发模板的示例

需求#	谁/谁的规范,如相关者/所有者	哪里的规范,如层、对象、区域(流程、服务、数据、基础设施等)	什么规格:高级需求	什么规格:详细要求
#				
#				
#				
#				
#				

过程之前,可以提前一次迭代准备详细的已签署用户故事(即时文档)。这里需要注意的是,用户的高层管理者应该只提供评估和确定优先级所需的足够详细信息,不应该锁定不必要的低级细节,因为这可能会妨碍敏捷BPM项目的适应性。

2)敏捷规划

传统的BPM规划方法侧重于详细的前期规划,而敏捷BPM工作方式则需要在项目、发布、迭代和日常级别进行规划。敏捷BPM侧重于初始的高级项目计划,该计划概述了项目发布、资源、风险以及成本和收益估算的数量。项目计划的制订基于高层次和详细的需求,以概述随着项目的小规模、分阶段进展,什么时候和哪些需求可以得到满足。表2举例说明了一个模板/组件的示例,即将制订的与计划阶段方面相关的利益相关者/流程所有人、相关对象和高级(非功能性)需求以及详细(功能性)需求关联起来的例子。这样的模板通常是以地图的形式出现的,地图可以从一个简单的行开始,当添加信息时,将生成一个行目录。由于发布的计划仅是当前版本以及该版本的前两次或三次迭代,这个模板的目的是从不同的相关架构层构建一个相关的相关者、对象和需求的清单列表,从分析阶段、规划阶段可以再次使用和跟踪这些清单列表。

表2是一个地图的模板,它以高级和详细的形式获得需求,并指出谁对需求感兴趣,以及每个需求在层和业务对象中的位置。这使得敏捷实践者能够使用发布计划在小迭代中跟踪项目进度。迭代计划关注下一次开始的迭代。它提供了关于时间框(2～4周的迭代)的短期迭代活动和日程安排的详细信息,如额外的分析、设计、构建阶段的开发和测试方面等,它还包括迭代的日期和时间,并显示用例

design, the development in the build phase, and the test aspects, and so on, It also includes date and time of iteration, and shows cases and retrospectives. Daily planning in each iteration is achieved via daily stand-up meetings, in which team members discuss what they did yesterday, what will they do today? Are there any impediments? Agile project, release, iteration, and daily planning enable Agile BPM.

Agile Architecture and Design

Agile design for BPM can kick off by reviewing the existing As-Is process model and identified requirements for the target To-Be process model. Instead of a detailed up-front design, a high-level design for the To-Be process can be developed at the start of the project. This high-level design will then emerge with more details in short releases and iterations. Hence, a high-level design can be built on the identified requirements and objects relevant for the target To-Be situation. A high-level design will set the foundation for the Agile BPM project choices and options that enable the detailed design in each iteration (Design Phase-Product Backlog) wanted and specified by the stakeholders within their expectations. It is important to note here that, instead, a target final To-Be process can be achieved via small Transition states. Each project release and iteration should focus on developing a stable Transition state linked to the overall final To-Be state. Once defined, linked to relevant objects and approved To-Be requirements in the execution Build Phase fall under change control. In the Agile BPM way of thinking and working, this means additions to the product backlog will be made if additional requirements are identified in the ongoing Analyze, Plan, and Design Phases. Iteration Build and Testing enable tracking of build completion and quality against the requirements identified in the Analyze phase. This is crucial to ensure that both Backward and Forward Traceability of requirements have been achieved in the preceding phases so that Value generated through the Phases is not lost. This is highlighted in Figure 5 as follows.

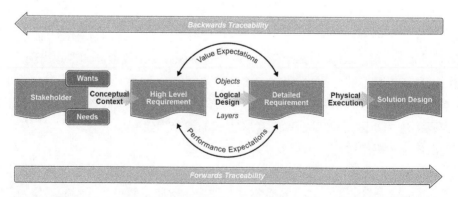

FIGURE 5

Agile BPM backward and forward traceability ensuring value generation.[17]

和回馈。每个迭代中的每日计划都是通过"每天站立会议"来实现的,在这些会议中,团队成员讨论他们昨天做了什么、他们今天将做什么、有什么障碍。总之,敏捷项目、发布、迭代和日常计划支持敏捷BPM。

3)敏捷架构和设计

BPM的敏捷设计可以通过审查"现有流程状况模型"并确定"未来流程状况模型"来开始。与详细的前期设计不同,在项目开始时为将来的流程开发一个高级设计。这个高级的设计将在简短的版本和迭代中提供更多的细节。因此,可以根据确定的需求和与将来的目标情况相关的对象构建高级设计。高水平的设计将为敏捷BPM项目的选择和选项奠定基础,这些选择和选项支持相关者期望和指定的每个迭代(设计阶段-产品待办事项列表)中的详细设计。这里需要注意的是,可以通过小的过渡状态来实现最终的目标流程。每个项目发布和迭代都应该关注于开发一个稳定的过渡状态,该状态与最终的状态相关联。定义后链接到相关对象,并在执行构建阶段中批准待办事项需求,这些需求将置于更改控制之下。在敏捷BPM的思维和工作方式中,这意味着如果在正在进行的分析、计划和设计阶段中确定了额外的需求,那么将向产品待办事项列表添加内容。迭代生成和测试可根据分析阶段确定的需求跟踪任务完成情况和质量,这对于确保在前面的阶段中实现需求的向后和向前追溯至关重要,这对通过这些阶段产生的价值不会丢失是非常重要的。图5突出显示了这一点,如下所示。

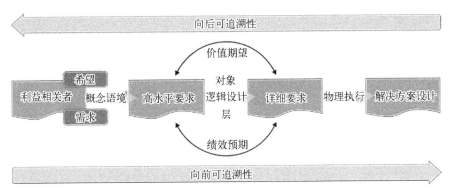

图5　确保价值生成的敏捷BPM向后和向前追溯[17]

Agile Build

Traditional ways of working focus on big-bang product or service development in the build phase. Agile ways of working focus on building the product or service minimum marketable or viable features in small iterations based on the just-in-time user stories or requirements. Agile ways of development focus on delivering value early. The focus shifts from documentation to delivering working product or service features right from the beginning. So that Agile BPM initiatives can link strategy, identify value aspects, and focus on the relevant objects, the Agile BPM initiatives in the Build phase use process architecture concepts. There is a misunderstanding that the process architect role and process architecture artifact are not needed in the agile environment. However, various BPM project focused and isolated user stories may overlook the holistic picture of the enterprise architecture and underlying business process and technology assets. To be agile in the build phase, the process user stories need to be connected and classified as described in the Process Architecture chapter.

The impact of the process user stories needs to be considered through the lens of the relevant process architectural categorizations and classifications. Agile process architecture integrated with user stories would not only result in better identification of value, performance, rules, monitoring, organizational change management, risk, and implementation strategy, but it would also provide a shared vision of the enterprise to guide the agile teams working in the distributed Agile BPM development environments. Hence, agile ways of working require process architecture principles. This enables the project architecture, at the beginning of the project, to be linked to the holistic enterprise architecture, and then the details of the architecture will evolve as the BPM project progresses in small iterations. In the specific BPM work, we suggest applying Service-oriented Architecture (SOA) and Process-oriented Architecture (POA) principles. A repeatable process or a part of the process can be developed as a "service". A business process or workflow can be managed through the choreography and orchestration of services. As Qumer and Henderson Sellers[18] point out, the agile service-oriented process is developed in small iterations.

The agile build, test, and deploy phases use the To- Be design identified in the Design phase as the defined Product backlog to work with in the Build phase. The Agile BPM way of thinking and working in these phases uses the standard agile activities:

1. Defining the Product Backlog
2. Sprint Planning Meeting
3. Defining the Sprint Backlog
4. Sprint
5. Interrogating and Testing
6. Demo Release
7. Client Feedback Meeting
8. Retrospective
9. Refactoring
10. System Changes
11. System Testing

4）敏捷构建

传统的工作方式侧重于在构建阶段进行大爆炸式的产品或服务开发，而敏捷的工作方式则关注于在用户实时场景或需求的小型迭代中构建产品或服务的最小可销售或可行的特性。敏捷的开发方式关注于尽早交付价值。重点从一开始就从文档转移到提供工作产品或服务功能。因此，构建阶段的敏捷BPM计划使用流程架构，可以将战略联系起来，识别价值等方面，并专注于相关对象、概念。有一种错误的观点是：在敏捷环境中不需要流程架构师角色和流程架构组建。然而，各种关注于BPM项目的和独立的用户场景可能会忽略企业架构、底层业务流程和技术资产的整体情况。为了在构建阶段保持敏捷，流程用户场景需要像流程体系结构一节中描述的那样进行连接和分类。

流程用户场景的影响需要通过相关流程架构分类和分类的视角来考虑。将敏捷过程体系结构与用户案例相结合，不仅可以更好地识别价值、性能、规则、监控、组织变更管理、风险和实现策略，还可以提供企业的共享愿景，以指导在分布式敏捷BPM开发环境中工作的敏捷团队。因此，敏捷的工作方式需要流程架构原则，这使得在项目开始时，项目体系结构能够与整个企业体系结构相关联，然后随着BPM项目在小迭代中不断发展，体系结构的细节也将不断演进。在具体的BPM工作中，我们建议使用SOA和"面向流程的体系结构（POA）原则"，基于此，可以将可重复流程或流程的一部分开发为服务。可以通过对服务的编排来管理业务流程或工作流。正如Qumer和Henderson Sellers[18]所指出的那样，面向服务的敏捷流程是在小迭代中开发的。

敏捷构建、测试和部署阶段使用在设计阶段确定的未来目标设计作为在构建阶段中使用的已定义产品待办事项。敏捷BPM在这些阶段的思维和工作方式使用标准的敏捷活动。

（1）定义产品待办事项列表；

（2）冲刺计划会；

（3）定义冲刺待办事项；

（4）冲刺；

（5）询问和测试；

（6）样稿演示；

（7）客户反馈会议；

（8）回顾；

（9）重构；

（10）系统更改；

（11）系统测试；

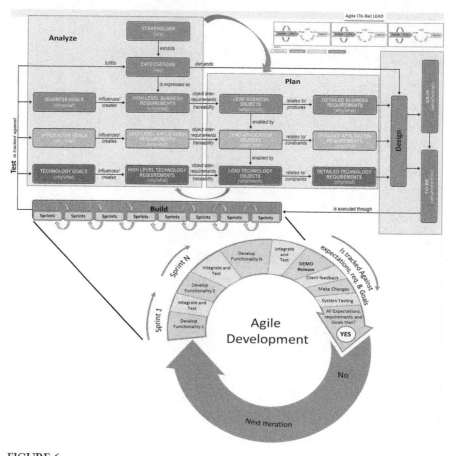

FIGURE 6

The Agile BPM build phase.[19]

12. Decision Point: Are expectations, requirements, and goals of Application and Technology Completed? If YES, then Deploy. If NO, then a new iteration is started.

The build process is presented in context in Figure 6, showing how each set of "build" sprints draws on the design to be executed.

Agile Testing

Although traditional ways of working around testing first do the testing once the whole product or service is developed, agile ways of working focus on testing the product or service minimum marketable or viable features in small iterations while the development is in progress. Agile ways of development focus on automating the testing practices, such as automated unit testing, acceptance testing, integration testing, and so on, Right from the beginning, the focus shifts from testing documentation to actually testing the

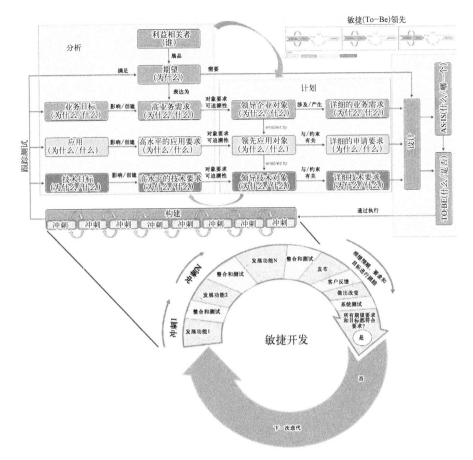

图6   敏捷BPM构建阶段[19]

（12）决策要点：应用和技术的期望、需求和目标是否完成？如果是，则部署；如果没有，则开始新的迭代。

构建过程在图6中展示，显示了构建过程中冲刺的执行步骤。

5）敏捷测试

尽管传统测试工作的方法是在整个产品或服务开发之后再进行测试，但敏捷的工作方法采用在开发过程中进行小迭代中测试产品或服务，以达到最低可销售或可行的特性。敏捷的开发测试主要采用自动化测试方法，如自动化的单元测试、验收测试、集成测试等，从一开始的重点就从测试文档转移到以满足需求为目标的实际测试工作中的产品或服务。这样做不仅是为了确保工作产品或服务满

working product or service features against requirements. This is done not only to ensure that the working product or service meets the customer expectation but also to identify any risks or blockers. In each case, whether within the business, application, or technology layer, the way in which 'testing' the requirements is performed is to simulate and compare the As-Is and target behaviors so as to expose possible defects for gaps between that which is desired and what is to be provided. This is achievable as requirements and designs are now addressed through structured models, in which every design object can be traced to its requirement and the various goals it executes. Testing in terms of tracking may occur, not just at the application and technology layers, but within the business layer and within the work system that binds the project choices and options to the specific To-Be solution design that enables the innovation and transformation expressed, for example, demanded by stakeholder expectations. These can also be 'tested' against the specific goals that relate to the requirements as well as the full 'testing' of solution design into the "work system". This in turn leads to the ability to pragmatically consider design options to fulfill and thereby meet expectations and to verify the quality of the product prior to Deployment.

It is important and now possible to resolve any test-related issues during the relevant iteration. If an issue is not resolved in a given time-boxed iteration, then do not extend the duration of the iteration, rather move the issue to the next iteration, and record and prioritize it on the product or test backlog. A user story related to a product or service feature is considered done when it has passed all the acceptance tests. If a minor issue arises, then it is fine to let the user story pass and fix the issue in the next iteration.

Agile Deployment

The process models, end-to-end flows, and or process changes can be deployed into production either after each iteration or at the end of release. An individual product or service release deployment can be combined with other releases for different products. Organizations may have their own local release cycle. Agile BPM ways of deployment requires tracking the testing and changes. For these reasons, Agile BPM in the deployment phase focuses on collaborative and communication-oriented shared responsibility, accountability, and business value-oriented change and governance. As discussed earlier, as within Agile BPM, the requirements for the product or service features are the responsibility and accountability of the owning stakeholder. Senior management is responsible and accountable for funding, empowering, and supporting the managers. For instance, traditional BPM project governance uses a gated approach to release and monitor the fixed up-front project funding and outcomes. Agile BPM, however, decomposes the project into short releases. Project funding is released based on each successful release of a project. Therefore, agile managers and agile teams are mainly responsible and accountable for the delivery of a valuable quality product or service features to the customer. Therefore, in reality, the empowered agile managers, the Agile BPM team, and customer collaborate for the value co-creation. It is important to note that Agile BPM consequently requires empowering the BPM CoE managers and the downstream business managers.

足客户的期望,而且也是为了识别任何风险或阻碍因素。在每种情况下,无论是在业务层、应用程序层还是技术层中,执行需求测试的方式都是模拟和比较预期行为与目标行为,以揭示实际与预期之间可能存在的缺陷。这是可以实现的,因为需求和设计现在是通过结构化模型来处理的,在这个模型中,每个设计对象都可以跟踪到其需求和它执行的各种目标。跟踪方面的测试可能不仅发生在应用程序层和技术层,也可能发生在业务层和工作系统中,这些系统将项目选择和选项绑定到特定的待解决方案设计中,以实现创新和转换,如满足利益相关者的期望。这些也可以针对与需求相关的特定目标进行测试,以及将解决方案设计完全测试到工作系统中。这反过来又导致了为满足预期而需要考虑实际设计选项的能力,并在部署之前验证产品的质量。

重要的是,现在可以在相关迭代期间解决任何与测试相关的问题。如果某个问题在给定的时间限制的迭代中没有得到解决,那么不要延长迭代的持续时间,而是将该问题移动到下一个迭代,并在产品或测试积压工作中记录并确定其优先级。与产品或服务特性相关的用户场景在通过所有验收测试后即被视为已完成。如果出现一个小问题,那么建议是让用户先通过并在下一次迭代中修复这个问题。

6)敏捷部署

流程模型、端到端流程或流程变更可以在每次迭代之后或在发布结束时部署到生产环境中。单个产品或服务版本部署可以与不同产品的其他版本相结合。组织可能有自己的本地发布周期。敏捷BPM的部署方式需要跟踪测试和更改。由于这些原因,部署阶段的敏捷BPM侧重于面向协作和通信的共享责任、问责制以及面向业务价值的更改和治理。如前所述,在敏捷BPM中,对产品或服务功能的要求是拥有利益相关方的责任和义务。高级管理层负责并为管理人员提供资金、授权和支持。例如,传统的BPM项目治理使用门控方法(gated approach)来发布和监控固定的前期项目资金和结果。然而,敏捷BPM将项目分解为小版本。项目资金的发放基于项目的每一个成功。因此,敏捷经理和敏捷团队主要负责向客户交付有价值的高质量产品或服务特性,因此,在现实中,被授权的敏捷经理、敏捷BPM团队和客户协作以共同创造价值。需要注意的是,敏捷BPM需要授权给BPM CoE经理和下游业务经理。

Agile Terminology

Even though agile principles can be applied to Enterprise Architecture, Agile is not an Enterprise Architecture discipline, and hence no direct Objects or Meta Objects apply. However, as shown in the above text, Agile does bring a set of new concepts that are critical to being able to comprehend any discussions on what Agile is, how it works, and how it can be applied within BPM. Having standardized terms provides a structural way of thinking and enables having common terminology in the execution of Agile BPM. It enables the organization of terms around the viewpoints associated with Agile BPM (see chapter What BPM can learn from Enterprise Architecture).

As such, terminology is used with various existing delivery frameworks, methods, and approaches that exist within the BPM CoE and the Project Management Offices (PMO); it is vital when developing such a set of standardized terminology that it be 100% vendor neutral and agnostic from various vendor solutions.

Although the terms are based on a collection of best and leading practices around how to work with Agile BPM within an organization, we do not claim that these terms are all-comprehensive, but rather want to use the terminology that is most common in agile circles and apply to the Agile BPM work (Table 3).

Table 3 *The Most Common Terms that Would Be Used Within Agile BPM*[20]

Term	Definition
Agile coach	A Person responsible for supporting and improving the capability of an organization to deliver in an agile way.
Agile driver and forces (external/internal)	Pressures that arise from outside or inside a system triggering agile approaches.
Backlog	A Prioritized list of requirements that are waiting to be worked on.
Bug	An error, flaw, mistake, failure, or fault in the process models, process rules, or process design that produces an incorrect or unexpected result, or causes it to behave in unintended ways.
Burndown chart	A Visual representation that shows work remaining over time.
Burnup chart	A Visual representation that shows work completed over time.
Business capability	An abstraction that represents the abilities and the quality of being capable, intellectually (logical) and or physically. Agile enterprise developments must be able to specify the aptitude that may be developed for the enterprise and how it will perform a particular function, process, or service.
Business change	Changes in the way an organization functions brought about through a project or other initiative.
Business resource/actor	A Specific person, system or organization internal or external that is part of or afected by the agile development to the enterprise. This can include that the agile development will influence or impact the resource/actor-defined functions and activities.

3. 敏捷术语

即使敏捷原则可以应用于企业架构,敏捷也不是企业架构的规程,因此没有直接的对象或元对象可以应用。然而,如上文所示,敏捷确实带来了一组新概念,它们对于理解关于敏捷是什么、如何工作以及如何在BPM中应用敏捷的讨论非常重要。标准化术语提供了一种结构化的思维方式,并支持在敏捷BPM的执行中使用通用术语。它支持围绕与敏捷BPM相关的观点组织术语(请参阅"BPM可以从企业架构中学到什么"一节)。

因此,术语与现有的各种交付框架、方法和方式一起使用,这些交付框架、方法和方式存在于BPM CoE和项目管理办公室(PMO)中。在开发这样一组标准化术语时,至关重要的一点是,它100%与供应商无关,并且与各种供应商解决方案无关。

尽管这些术语是基于关于如何在组织内使用敏捷BPM的最佳和领导实践的集合,但我们并不认为这些术语都是全面的,而是希望使用敏捷圈中最常见的术语,并应用于敏捷BPM工作(表3)。

表3　敏捷BPM中最常用的术语[20]

术　　语	定　　义
敏捷教练	组织中负责支持和改进敏捷方式交付能力的人
敏捷驱动和力量 (外部/内部)	触发系统外部或内部敏捷方法而产生的压力
待办事项列表	等待处理的需求的优先级列表
错　　误	流程模型、流程规则或流程设计中的缺陷、失败或错误,或导致不正确或意外的结果,或导致其以非预期运行的方式
燃尽图	显示剩余工作时间的可视化图表
燃耗图	显示随时间完成的工作的可视化表示
业务能力	一种抽象概念,代表能力、智力(逻辑)或身体的能力和品质。敏捷企业开发必须能够指定可能为企业开发的能力,以及它将如何执行特定的功能、流程或服务
业务变化	组织职能通过项目或其他计划而发生的变化
业务资源/ 参与者	内部或外部的特定的人、系统或组织,是影响企业敏捷开发的一部分的资源、参与者定义的功能和活动

Table 3 *The Most Common Terms that Would Be Used Within Agile BPM*
—Cont'd

Term	Definition
Business service	Agile concepts applied to business concepts will impact the change and development of business services. In terms of the externally visible ("logical") deed, or effort performed to satisfy a need or to fulfil a demand, meaningful to the environment.
Business workflow	A Business workflow involved in the agile development, impacting and/or changing the stream, sequence, course, succession, series, progression, as well as order for the movement of information or material from one business function, business service, business activity (worksite) to another.
Continuous integration	When individual process models are combined in, for example, an entire end-to-end process flow and tested as soon as they are produced.
Cross-functional team	A Group of people with different skills and expertise working toward a common goal.
Defect trend	A Report that shows a rolling average of the number of problems (bugs) the team has opened, resolved, and closed.
Definition of "done"	An increment of a product that is ready for continual use by the end user. Can also be referred to as "done, done."
Deployment	All of the activities that make the process models ready for use and implementation.
Elaborate	When the delivery team adds detail to high-level business requirements.
Function	These are sometimes called epic stories or epics. Functions represent large sets of functionality, for example, Accounts receivable, Accounts payable, month-end close, etc.
Functionality	The behaviors that are specified to achieve.
Information radiator	A large, highly visible display that gives a picture of progress and key issues relating to an area of work.
Iteration	A Short time period in which a team is focused on delivering an increment of a product that is useable.
Kanban board	A visual board in which columns represent a state in which a user story can reside, for example, planned, blueprinting, realization, testing, done. Stories are arranged on the Kanban board and moved from one column to another as progress is being made. Many teams build physical Kanban boards by using tape and Post-it notes. Digital Kanban boards are another alternative.
Lean	Techniques to streamline processes and eliminate any activities that do not add value to the user.
Non-functional requirements	Describe how the process models or BPM projects should operate, as opposed to functional requirements that describe how it should behave. Typical examples would be: wished behavior, process security, accessibility, usability, availability, response times, etc.

Continued

（续表）

术　语	定　义
业务服务	应用于业务概念的敏捷概念将影响业务服务的更改和开发。根据外部可见的（逻辑的）行为，为满足需求而进行的努力
业务工作流程	敏捷开发中涉及的业务工作流，影响或更改流程、序列、过程、继承、系列、进展，以及将信息或材料从一个业务功能、业务服务、业务活动（工作站点）移动到另一个业务功能、业务服务、业务活动（工作站点）的顺序
持续集成	将各个流程模型组合在一个完整的端到端流程中，并在生成后立即进行测试
跨职能团队	朝着一个共同的目标努力的一群拥有不同技能和专业知识的人
缺陷统计	显示团队打开、解决和关闭的问题（bug）数量趋势的平均值的报告
完成的定义	为最终用户持续使用而准备好的产品增量。也可以称为"完成了，完成了"
部　署	为流程使用和实现做好准备的所有活动
详尽说明	当交付团队向高级业务需求添加细节时所准备的资料
功　能	功能代表一系列的性能或效用，如应收账款、应付账款、月末结算等
功能性	指定要实现的行为
信息屏	一种大的、高度可见的显示屏，展示工作进展和与工作领域相关的关键问题
迭　代	是一种不断用变量的旧值递推新值的过程
看　板	一种可视板，其中列表示用户情景可以驻留的状态，如计划的、蓝图的、实现的、测试的、已完成的。过程被安排在看板上，随着进展情况的推进，从一列移动到另一列。许多团队使用便签构建物理看板。数字看板是另一种选择
精　益	简化流程和消除不为用户增加价值的任何活动的技术
非功能性需求	描述流程模型或BPM项目应该如何操作，而不是描述其行为方式的功能需求。典型的例子有：期望的行为、流程安全性、可访问性、可用性、响应时间等

Table 3 *The Most Common Terms that Would Be Used Within Agile BPM* —Cont'd

Term	Definition
Owner	The person who is ultimately responsible for prioritization and acceptance of delivered features on a given process or project.
Performance expectations	Although for the most tagged and classified as non-functional requirements, the performance expectations are more as they specify the desire for the manner in which, or the efficiency with which, something reacts or fulfils its intended purpose as anticipated by a specific stakeholder. It will give an input to non-functional requirements, however the performance expectations will also be used in the early validation and thereby be the baseline against performance testing.
Release	Each release is associated with some type of go-live in which a number of processes or BPM projects are moved to roll out/production. For example, a "big bang" BPM program could have just a single large release. A more phased approach could lead to many releases within a single program.
Release plans	A Plan that sets out the order in which user requirements will be released into live service.
Retrospective	A Retrospective is a focused session in which your team looks back at how the current agile approach is working and which areas can be improved. Many agile teams conduct retrospectives at set intervals of time (every 8 weeks or at the end of every sprint).
Rework	Components of a project that will need to be revisited to correct bugs or altered to meet new requirements.
Show and tell	When the delivery team demonstrates how the product or service works at the end of each iteration to elicit feedback.
Sprint	A Sprint is typically a predetermined period of time (2 weeks, 4 weeks, 6 weeks, etc.) within which a set of identified user stories needs to be complete. Alternatives to sprints are to use a Kanban variation of agile project management methodology.
Stand-up	A Short meeting conducted standing up to report progress, share impediments, and make commitments.
Task	We generally try to avoid tracking detailed tasks, but sometimes we need to breakdown a single user story into multiple tasks and assign those to different people. For example, this can be handy for tracking specific process design and development tasks. Each task belongs to one and only one user story. It is the lowest-level entity that we track.
Technical debt	Poor process design in overall process architecture. The consequence of this is that more time is needed later on in the project to resolve process issues
Testing	A set of actions undertaken to assess whether a process or process model behaves as expected.

（续表）

术 语	定 义
负责人	最终对给定过程或项目中交付特性的优先级和验收负责的人
绩效预期	虽然对于标记最多且被归类为非功能性要求，但性能期望更多，因为它们指定了对特定利益相关者所预期的某种反应或实现其预期目的的方式或效率的期望。它将为非功能性需求提供输入，但性能预期也将用于早期验证，从而成为性能测试的基准
发 布	每个版本都与某种类型的上线相关联，其中许多流程或BPM项目被移动到推出/生产。例如，大爆炸式的BPM程序可能只有一个大型版本。更分阶段的方法可能会导致单个程序中的许多版本
发布计划	一种计划，它规定了将用户需求发布到实时服务中的顺序
回 顾	回顾是一个集中的会议，在这个会议中，您的团队回顾当前敏捷方法是如何工作的，以及哪些领域可以改进。许多敏捷团队在规定的时间间隔内（每8周或在每个迭代结束时）进行回顾
返 工	需要重新访问项目的组件以纠正错误或进行修改以满足新的需求
演 示	交付团队在每次迭代结束时演示产品或服务如何工作以获得反馈
冲 刺	冲刺通常是预先确定的一段时间（2周、4周、6周等），在这段时间内需要完成一组确定的用户描述。冲刺的替代方案是使用敏捷项目管理方法的看板变体
站立会议	为报告进展、分担障碍和作出承诺而举行的简短会议
任 务	我们通常会尽量避免跟踪详细的任务，但有时我们需要将单个用户功能分解为多个任务并将其分配给不同的人。例如，这对于跟踪特定的流程设计和开发任务非常方便。每个任务都属于一个用户故事。它是我们跟踪的最低级别实体
技术陷阱	整体流程架构中的流程设计较差。这样做的结果是在项目的后期需要更多的时间来解决流程问题
测 试	用于评估流程或流程模型的行为是否如预期那样的一组操作

Table 3 *The Most Common Terms that Would Be Used Within Agile BPM* —*Cont'd*

Term	Definition
Test coverage	The proportion of a process model that has been assessed.
Time box	A fixed time frame, usually to undertake an intense increment of work
User story	Each user story describes a particular business requirement and is assigned to a single function. In many ways, a user story is the next level down in terms of detail after a function. The standard question approach is followed: "WHO is needed, WHAT we do, WHY we do it". User stories are business-centric, not technology-centric. They do not capture HOW something will be accomplished (that comes later).
Value expectation	Tagging and classifying value expectations as nonfunctional requirements is a part of specifying the anticipated benefits that are of worth, importance, and significance to a specific stakeholder. It will give an input to nonfunctional requirements, however the value expectations will also be used in the customer orientation feedback loop, relating back to the specific stakeholders value expectations.
Value proposition?	A key principle of Agile is its recognition that during a project the customers can change their minds about what they want and need (often-called requirements churn), and that unpredicted challenges cannot be easily addressed in a traditional predictive or planned manner. As such, Agile BPM concepts need to adopt an empirical approach, accepting that the problem cannot be fully understood or defined, focusing instead on maximizing the team's ability to respond to changing value and/or performance expectations and thereby emerging requirements and create specific value proposition to the new need/want.
Velocity	The rate at which a team completes work.

Building Agile Capabilities in the BPM CoE

The establishment of an agile capability in an existing BPM CoE is a challenging task. This is partly because the CoE and teams have a different way of thinking, working, modeling, and implementing. It would, therefore, be appropriate to gradually establish agile capability by introducing agile roles, practices, and tools. One way to do is to use the Agility Adoption and Improvement Model (AAIM). This model (Figure 7) was first developed and published in 2007[21] and then was updated in 2010 as "AAIM Version 2.0".[22] AAIM V2.0 has been developed based on the intensive research in agile adoption at a large scale. The AAIM can be used as a roadmap or guide for agile transformation. Organizations or teams can adopt and improve agile environment to achieve specific agile level(s).

（续表）

术　语	定　义
测试范围	已评估的流程模型的比例
时间框	一个固定的时间框架,通常承担紧张的工作增量
用户故事	每个用户故事描述一个特定的业务需求,并被分配到单个功能。在很多方面,用户故事是功能之后的下一层细节。标准的提问方式是:"需要谁,我们做什么,我们为什么这么做"。用户故事以业务为中心,而不是以技术为中心。它们没有捕捉到某件事将如何完成(稍后会实现)
价值期望	将价值预期作为非功能性需求进行标记和分类,是指定对特定利益相关者具有价值、重要性和有意义的预期收益的一部分。它将为非功能性需求提供输入,但是价值期望也将在面向客户的反馈循环中使用,与特定的涉众价值期望相关
价值主张	敏捷的一个关键原则是:它认识到在项目中客户可以改变他们想要和需要的想法(通常称为需求变更),并且不可预测的挑战不能以传统的预测或计划的方式轻易解决。因此,敏捷的BPM概念需要采用一种经验方法,接受不能被完全理解或定义问题,而是专注于最大限度地提高团队应对不断变化的价值和业绩期望的能力,从而满足新的需求,并为新的需求创造具体的价值主张
速　度	团队完成工作的速度

4. 在 BPM CoE 中构建敏捷功能

在现有 BPM CoE 中建立敏捷能力是一项具有挑战性的任务。这部分是因为 CoE 和团队有不同的思维、工作、建模和实现方式。因此,通过引入敏捷角色、实践和工具来逐步建立敏捷能力是合适的。一种方法是使用敏捷应用和改进模型(AAIM)。这个模型(图7)首先在2007年[21]开发并发布,然后在2010年更新为 AAIM 2.0 版本[22]。AAIM 2.0 版本是在大规模深入研究敏捷应用的基础上开发的。AAIM 可以用作敏捷转换的路线图或指南。组织或团队可以采用和改进敏捷环境来实现特定的敏捷级别。

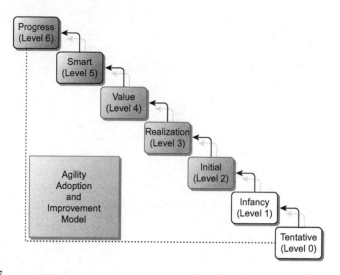

FIGURE 7

The agility adoption and improvement model V2.0.[23]

AGILITY ADOPTION AND IMPROVEMENT MODEL

The AAIM is structured into white, green, and black blocks and six levels (from 0 to level 6—Tentative to Progress). The colors indicate the levels, in which the white blocks are those levels at which initial experience in critical aspects of agility is garnered, whereas the green blocks are the levels at which the agile practices are established and entrenched, and the black blocks show those levels at which agile disciplines become universal. Each agile level has a name and specifies the lean agile principles to follow to achieve the particular level. Continuous improvement is integral to each level. The achieved lean agile level shows the lean agile maturity of an organization or team. The following section discusses the AAIM in the context of Agile BPM capability establishment.

- Tentative (Level 0)
  This level focuses on establishing an experimental environment whereby experience can be gained by BPM teams with some of the agile roles, practices, and tools. Based on this initial experience, BPM teams can communicate the perceived advantages of agile ways of working to the senior management to seek their support to begin with the further systematic establishment of the Agile BPM capability.
- Infancy (Level 1)
  This level focuses on adopting a basic elementary set of agile principles, roles, practices, and tools to support the iterative and incremental test-driven BPM development (evolutionary environment). This level provides the foundation for the further establishment of the Agile BPM capability.

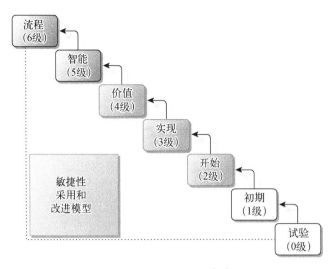

图7 AAIM 2.0版本[23]

5.2.9 AAIM

AAIM分为白色、绿色和黑色块以及6个级别（从0级到6级——暂定到进度）。颜色表示级别，其中：白色块是获得敏捷性关键方面的初始体验的级别，而绿色块是敏捷实践建立和根深蒂固的级别，黑色块显示敏捷学科变得普遍。每个敏捷级别都有一个名称，并指定要实现特定级别的精简敏捷原则。持续改进是每个级别的组成部分，实现的精益敏捷级别显示了组织或团队的精益敏捷成熟度。以下部分讨论敏捷BPM功能建立背景下的AAIM。

- 开始（0级）

这个级别的重点是建立一个实验性环境，BPM团队可以通过使用一些敏捷角色、实践和工具来获得经验。基于这一初始经验，BPM团队可以向高级管理人员传达敏捷工作方式的可感知优势，以寻求他们的支持，从而开始进一步系统地建立敏捷BPM功能。

- 婴儿期（1级）

这个级别的重点是采用一组基本的敏捷原则、角色、实践和工具来支持迭代和增量测试驱动的BPM开发（演化环境）。此级别为进一步建立敏捷BPM功能提供了基础。

- Initial (Level 2)

  This level focuses on establishing a collaborative BPM environment by adopting agile principles, roles, practices, and tools to support active communication and collaboration among the team members and internal and external shareholders.

- Realization (Level 3)

  This level focuses on establishing a simple result-focused Agile BPM capability by adopting agile principles, roles, practices, and tools to support the production of the executable BPM artifacts with minimal or reduced documentation. This is an advanced agile level, and the teams who are not accustomed to working with less documentation would find it challenging. This level could only be achieved if a well-established communication-oriented culture exists in the organization (e.g., established at level 2).

- Value (Level 4)

  This level focuses on establishing self-organizing BPM teams by adopting agile principles, roles, practices, and tools. Critical here is the fact that self-organization requires working knowledge and experience of Agile.

- Smart (Level 5)

  This level focuses on establishing a knowledge-focused Agile BPM capability by adopting agile principles, roles, practices, and tools to support knowledge management and innovation beyond the scope of an individual Agile BPM project and team.

- Progress (Level 6)

  This level focuses on establishing a continuous improvement by sustaining and continuously improving Agile BPM capability.

Some lessons learned around the Agility Adoption and Improvement journey include the realization that no single agile method or approach is a silver bullet. An organization should not focus too much on mechanically adopting only one agile framework end to end, such as Scrum or XP, but rather should focus on establishing and harvesting an agile mind set, values, principles, thinking, practices, roles, tools, and culture by using some kind of road map or adoption model of progress. The essential lesson is to let agile teams Assess, Tailor, Adopt, and Improve their own agile method suitable to their needs, context, or project, and focus on more "Facilitating and Guiding" teams with appropriate "Reward and Incentive Program" in their agile transformation journey while avoiding the imposition of "Agile" on teams.

CONCLUSION

Traditional BPM ways of working focus on detailed up-front planning, requirements analysis, process analysis, process design, process implementation, and continuous improvement to adjust changes. In other words, traditional takes a waterfall approach to BPM. Here, the assumption is that all the requirements for the process work are fixed, known, or complete. A lot of time and resources are spent up front

- 初级（2级）

此级别关注于通过采用敏捷原则、角色、实践和工具来建立协作BPM环境，以支持团队成员以及内部和外部股东之间的积极沟通和协作。

- 实现（3级）

此级别的重点是通过采用敏捷原则、角色、实践和工具来支持使用最少或减少的文档生成可执行BPM制品，从而建立一个简单的以结果为中心的敏捷BPM功能。这是一个高级敏捷级别，不习惯使用较少文档的团队会发现它具有挑战性。只有在组织中存在一种良好的以沟通为导向的文化（如在第2级建立的文化），才能达到这一层次。

- 价值（4级）

此级别关注于通过采用敏捷原则、角色、实践和工具来建立自组织BPM团队。关键是，自组织需要敏捷的工作知识和经验。

- 灵活（5级）

此级别的重点是通过采用敏捷原则、角色、实践和工具来支持超出单个敏捷BPM项目和团队范围的知识管理和创新，从而建立以知识为中心的敏捷BPM功能。

- 进展（6级）

这个级别的重点是通过维持和持续改进敏捷BPM功能来建立持续改进。

在敏捷应用和改进过程中获得的一些经验教训包括：认识到没有一种单一的敏捷方法或方式是新技术（silver bullet喻指新技术）。一个组织不应该太过注重机械地采用端到端敏捷框架（如Scrum和XP），而是应该关注建立和收获敏捷的思维方式、价值观、原则、思考、实践、角色、工具和文化通过使用某种路线图或采用模型的进展。必要的教训是让敏捷团队评估、定制、采用并改善自己的敏捷方法，使之适合他们的需求、环境或项目，并在敏捷转型过程中关注更多的"促进和指导"团队，提供适当的"奖励和激励计划"，同时避免敏捷的实施团队。

5.2.10　结论

传统的BPM工作方式侧重于详细的前期计划、需求分析、流程分析、流程设计、流程实现和持续改进以调整更改。换句话说，传统的BPM采用瀑布式方法。这里的假设是，流程工作的所有需求都是固定的、已知的或完整的。大量的时间和资源被预先用于实现一个固定的或明显完整的需求和计划列表的假象，而没有实

for achieving this illusion of a fixed or apparently complete list of requirements and plans, without actually delivering a single feature of a working product or service. By the time requirements are completely defined, signed off, and developed, business focus and market competition may have already been changed in response to an always-changing business environment, or changing performance or value expectations. Organizations need to be agile in response to such changing business environments and expectations. We have, therefore, focused in this chapter on the question of why we need to be agile, when, and how Agile BPM could be applied, as well as how to establish an agile capability. Applying the Agile BPM way of thinking and working will ensure that the BPM CoE teams work in a faster way and apply Kaizen principles of continuous improvement directly in their way of working. In that, they learn from what they do and how, Agile adapts and reshapes their manner of delivering the project and involves stakeholders in a new way, for example, as coparticipants in the process.

End Notes

1. Larman Craig., Agile and Iterative Development: A Manager's Guide. Addison-Wesley, (2004) p. 27.
   Ambler Scott., Agile Modeling: Effective Practices for Extreme Programming and the Unified Process. John Wiley & Sons, (12 April 2002) pp. 12.
   Boehm, B. R., Turner. Balancing Agility and Discipline: A Guide for the Perplexed. Boston, MA: Wesley, (2004).
   Sliger, M., Broderick, S., The Software Project Manager's Bridge to Agility. Addison-Wesley, (2008).
   Rakitin, S.R., "Manifesto Elicits Cynicism: Reader's letter to the editor by Steven R. Rakitin", IEEE (2001).
   Geoffrey Wiseman (July 18, 2007). "Do Agile Methods Require Documentation?" InfoQ.
   Abrahamsson, P., Salo, O., Ronkainen, J., & Warsta, J., Agile Software Development Methods: Review and Analysis. VTT Publications, (2002) 478.
   Guide to Agile Practices, the Agile Alliance.
   Aydin, M.N., Harmsen, F., Slooten, K. V., & Stagwee, R. A., An Agile Information Systems Development Method in Use. Turk J Elec Engin, (2004) 12(2), 127–138.
2. Gerald M., Weinberg, As quoted in Larman, Craig; Basili, Victor R. (June 2003). "Iterative and Incremental Development: A Brief History". Computer 36(6): 47–56, doi:10.1109/MC.2003.1204375, ISSN 0018–9162.
3. Edmonds, E.A., "A Process for the Development of Software for Nontechnical Users as an Adaptive System". General Systems, (1974) 19: 215–218.
4. See note 559 above.
5. Larman, C., Agile and Iterative Development: A Manager's Guide. Addison-Wesley, (2004) p. 27. ISBN 978-0-13-111155-4.
6. Agile Manifesto., Manifesto for Agile Software Development, (2001) http://agilemanifesto.org/.
7. Ibid.

际交付一个工作产品或服务的单一特性。在需求完全定义、签署和开发之前,业务重点和市场竞争可能已经随着不断变化的业务环境、性能或价值预期的变化而发生了变化。组织需要敏捷地响应这些不断变化的业务环境和期望。因此,在这一节中,我们关注的问题是:为什么我们需要敏捷,何时及如何应用敏捷BPM,以及如何建立敏捷能力。应用敏捷BPM思维和工作方式将确保BPM CoE团队以更快的方式工作,并在其工作方式中直接应用持续改进的改善原则。在这种情况下,他们从他们所做的事情以及敏捷如何适应和重塑他们交付项目的方式中学习,并以一种新的方式让涉众参与进来,如作为参与过程的参与者。

参考文献

[1] Larman Craig., Agile and Iterative Development: A Manager's Guide. Addison-Wesley, (2004) p. 27.Ambler Scott., Agile Modeling: Effective Practices for Extreme Programming and the Unified Process. John Wiley & Sons, (12 April 2002) pp. 12.Boehm, B. R., Turner. Balancing Agility and Discipline: A Guide for the Perplexed. Boston, MA: Wesley, (2004).Sliger, M., Broderick, S., The Software Project Manager's Bridge to Agility. Addison-Wesley, (2008).Rakitin, S.R., "Manifesto Elicits Cynicism: Reader's letter to the editor by Steven R. Rakitin", IEEE (2001).Geoffrey Wiseman (July 18, 2007). "Do Agile Methods Require Documentation?" InfoQ.Abrahamsson, P., Salo, O., Ronkainen, J., & Warsta, J., Agile Software Development Methods: Review and Analysis. VTT Publications, (2002) 478.Guide to Agile Practices, the Agile Alliance.Aydin, M.N., Harmsen, F., Slooten, K. V., & Stagwee, R. A., An Agile Information Systems Development Method in Use. Turk J Elec Engin, (2004) 12(2), 127−138.

[2] Gerald M., Weinberg, As quoted in Larman, Craig; Basili, Victor R. (June 2003). "Iterative and Incremental Development: A Brief History". Computer 36(6): 47−56, doi:10.1109/MC.2003.1204375, ISSN 0018−9162.

[3] Edmonds, E.A., "A Process for the Development of Software for Nontechnical Users as an Adaptive System". General Systems, (1974) 19: 215−218.

[4] See note 559 above.

[5] Larman, C., Agile and Iterative Development: A Manager's Guide. Addison-Wesley, (2004) p. 27. ISBN 978-0-13-111155-4.

[6] Agile Manifesto., Manifesto for Agile Software Development, (2001) http://agilemanifesto.org/.

[7] Ibid.

8. Qumer, A., & Henderson-Sellers, B.,"A framework to support the evaluation, adoption and improvement of agile methods in practice", Journal of Systems and Software, (2008) vol. 81, no. 11, pp. 1899–1919.

9. Ibid.

10. Gill, A.Q., "Towards the Development of an Adaptive Enterprise Service System Model", Americas Conference on Information Systems, Chicago, USA, August 2013 in Americas Conference on Information Systems (AMCIS 2013), Shim, J.P et al., ed., AIS, USA (2013).

11. Koch, A.S., "Agile Software Development: Evaluating the Methods for Your Organization", Artech House, Inc, London (2005), pp. 1–272.

12. Quner, A., Henderson-Sellers, B. and McBride, T., (2007). Agility Adoption and Improvement Model, EMCIS 2007.

13. Agile Manifesto (2001), Manifesto for Agile Software Development, http://agilem anifesto.org/.

14. Source: Business Week, (May 12 1998).

15. LEADing Practice Business Process Reference Content [#LEAD-ES20005BP].

16. LEADing Practice Agile Reference Content #LEAD-ES30006ES.

17. Ibid.

18. Qumer, A., & Henderson-Sellers, B., "ASOP: an agile service-oriented process", International Conference on Software Methods and Tools, Rome, Italy, November 2007 in New Trends in Software Methodologies, Tools and Techniques. Proceedings of the sixth SoMeT_07, H. Fujita and D. Pisanelli, ed., IOS Press, Amsterdam, The Netherlands, (2007) pp. 83–92.

19. See note 16 above.

20. Taken from the LEADing Practice Agile Reference Content LEAD-ES30006ES.

21. See note 18 above.

22. Qumer, A., A Framework to Assist in the Assessment and Tailoring of Agile Software Development Methods, PhD Thesis, (2010) UTS.

23. Gill, A.Q., Bunker, D., "SaaS Requirements Engineering for Agile Development" in Xiaofeng Wang, N. Ali, I. Ramos, R. Vidgen ed., Agile and Lean Service-Oriented Development: Foundations, Theory, and Practice, IGI, USA, (2013) pp. 64–93.

［ 8 ］ Qumer, A., & Henderson-Sellers, B., "A framework to support the evaluation, adoption and improvement of agile methods in practice", Journal of Systems and Software, (2008) vol. 81, no. 11, pp. 1899−1919.

［ 9 ］ Ibid.

［ 10 ］ Gill, A.Q., "Towards the Development of an Adaptive Enterprise Service System Model", Americas Conference on Information Systems, Chicago, USA, August 2013 in Americas Conference on Information Systems (AMCIS 2013), Shim, J.P et al., ed., AIS, USA (2013).

［ 11 ］ Koch, A.S., "Agile Software Development: Evaluating the Methods for Your Organization", Artech House, Inc, London (2005), pp. 1−272.

［ 12 ］ Qumer, A., Henderson-Sellers, B. and McBride, T., (2007). Agility Adoption and Improvement Model, EMCIS 2007.

［ 13 ］ Agile Manifesto (2001), Manifesto for Agile Software Development, http://agilemanifesto.org/.

［ 14 ］ Source: Business Week, (May 12 1998).

［ 15 ］ LEADing Practice Business Process Reference Content［ #LEAD-ES20005BP ］.

［ 16 ］ LEADing Practice Agile Reference Content #LEAD-ES30006ES.

［ 17 ］ Ibid.

［ 18 ］ Qumer, A., & Henderson-Sellers, B., "ASOP: an agile service-oriented process", International Conference on Software Methods and Tools, Rome, Italy, November 2007 in New Trends in Software Methodologies, Tools and Techniques. Proceedings of the sixth SoMeT_07, H. Fujita and D. Pisanelli, ed., IOS Press, Amsterdam, The Netherlands, (2007) pp. 83−92.

［ 19 ］ See note 16 above.

［ 20 ］ Taken from the LEADing Practice Agile Reference Content LEAD-ES30006ES.

［ 21 ］ See note 18 above.

［ 22 ］ Qumer, A., A Framework to Assist in the Assessment and Tailoring of Agile Software Development Methods, PhD Thesis, (2010) UTS.

［ 23 ］ Gill, A.Q., Bunker, D., "SaaS Requirements Engineering for Agile Development" in Xiaofeng Wang, N. Ali, I. Ramos, R. Vidgen ed., Agile and Lean Service-Oriented Development: Foundations, Theory, and Practice, IGI, USA, (2013) pp. 64−93.

BPM Change Management

Maria Hove, Marianne Fonseca, Mona von Rosing, Joshua von Scheel, Dickson Hunja Muhita

INTRODUCTION

In this chapter, we will focus on what Business Process Management (BPM) Change Management is, why it is needed, when it can be applied, and the benefits of applying it. We believe that the principles of BPM Change Management are relevant to any organization, independent of industry, business model, or one's operating model. Change is a challenge faced by all organizations; it not only impacts the strategic aspects, the business models, the employees, and the way an organization utilizes technology. The degree of outside change influences the organization's ability of being able to keep in control the way it works in terms of its activities. When an organization manages the changes befalling the organization, it must also actively manage the change of its business processes. Therefore, in this chapter we will also discuss lessons learned in terms of what works and what does not around BPM Change Management and how the concepts can be developed in the BPM Center of Excellence (CoE), and applied in the BPM Life Cycle method.

LESSONS LEARNED AROUND BPM CHANGE MANAGEMENT

Like many other organizations as well as practitioners working with change management, we make good use of Elisabeth Kübler-Ross's work[1] on stages people go through as they deal with organizational change and transformation. The reason this is important is the phenomenon referred to as "the valley of despair." That, unless tackled rightfully with change management aspects, the organization will suffer, and the benefits wished by the organization will be far less than expected (for further information, read the BPM innovation and transformation chapter). Without a strong change management approach, outcomes can be poor, to say the least (Figure 1).

Many benefits and gains may be obtained by managing change, among them is that it prepares the organization for the "valley of despair" and how the organization can react. There are also Kotter's well-known recommendations for change[2]:

1. Establish a sense of urgency
2. Form a powerful guiding coalition
3. Create a vision

The Complete Business Process Handbook. http://dx.doi.org/10.1016/B978-0-12-799959-3.00028-8

5.3　BPM变革管理

Maria Hove, Marianne Fonseca, Mona von Rosing, Joshua von Scheel, Dickson Hunja Muhita

5.3.1　介绍

在本节中,我们将重点讨论什么是BPM变革管理,为什么需要它,什么时候可以应用它,以及应用它的好处。我们相信BPM变革管理的原则与任何组织都相关,并且独立于行业、业务模型或个人的操作模式。变革是所有组织都要面临的挑战,它不仅影响战略执行、商业模式以及员工和组织使用技术的方式。外部变化的程度还会影响组织掌控其经营活动的能力。当变革降临到组织头上时,组织必须积极地应对其业务流程的变革。因此,在这一节中,我们还将讨论从BPM变革管理中获取的经验和教训,包括如何在BPM CoE以及BPM生命周期方法中开发和应用这些概念。

5.3.2　关于BPM变革管理的教训

像许多其他组织以及从事变革管理的实践者一样,我们充分利用了Elisabeth Kubler-Ross关于人们在处理组织变革和转型时所做的研究[1]。变革管理之所以如此重要,是因为出现了一种被称为"绝望之谷"的现象,基于这种现象,除非正确地处理变革管理方面的问题,否则组织将遭受损失,并且组织得到的收益将远远少于预期(要获得更多信息,请阅读BPM创新与转型章节)。至少可以说,如果没有强大的变革管理方法,结果可能很糟糕(图1)。

通过管理变革可以获得许多好处和收益,其中之一就是它为组织准备了应对"绝望之谷"的方法。此外,J. Kotter还对变革提出了一些建议[2]:

（1）树立紧迫感;

（2）形成强大的指导联盟;

（3）创造愿景;

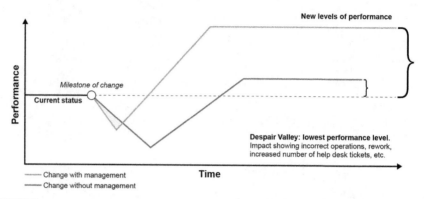

FIGURE 1

The valley of despair based on Elisabeth Kübler-Ross's work around stages of change.[1]

4. Communicate the vision
5. Empower others to act on the vision
6. Plan for and create short-term wins
7. Consolidate improvements and produce more change
8. Institutionalize new approaches

Although this is understood in Organizational Change Management (OCM), this is not really applied the same way to BPM. Although the role of technology, process design, process modeling, and all other factors in the business process management field should not be de-emphasized, this is an intuitively true statement that the major impediment to the successful implementation of a business process change is frequently the poor change management approach adopted.[3]

Therefore, analyzing and understanding the impact of change to the organization must be an integral part of a BPM CoE and the BPM process cycle of continuous improvement. Identifying the process and drawing it in BPM Notations, and then publishing, will not enable change. It can enable standardization, also integration, but not change. As Michael Hammer[4] said in 1993, "Coming up with ideas is the easy part, but getting things done is the tough part. The place where these reforms die is …down in the trenches." Moreover, who 'owns' the trenches? You and I do, and all the other people working with processes. Change imposed on the 'trench people' will not succeed without being part of the evolution or revolutionary process of change".[5] One needs to consider the change drivers in the various business areas and groups, the organizational competencies impacted, then the flow of the end-to-end business processes—those that span business functions and organizational boundaries. Determine how the change affects discrete business processes as well as how changes in various processes affect related processes and business functions.[6] Some organizations are better than others working with change and adapting it into their BPM work. As Paul Harmon wrote "Process Management and Change Management are tightly coupled".[7]

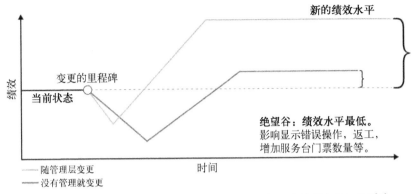

图1　基于Elisabeth Kubler-Ross围绕变革阶段研究的"绝望之谷"[1]

（4）传达愿景；

（5）授权他人按照愿景行事；

（6）计划并创造短期胜利；

（7）巩固改进之处并应对更多变化；

（8）使新方法制度化。

尽管这在组织变革管理（OCM）中是可以理解的，但是这并没有以相同的方式应用于BPM。尽管技术、流程设计、流程建模以及BPM领域中所有其他因素的作用都不应该被忽视，但阻碍成功实现业务流程变革的原因往往是采用糟糕的变革管理方法[3]。

因此，分析和理解变革对组织的影响必须是BPM CoE和持续改进的BPM流程周期的一个组成部分。一旦使用BPM的符号识别、绘制、发布流程完毕，就不应该再更改它。此后，它可以实现标准化，也可以集成到其他方面，但就是不能再更改。正如M. Hammer[4]在1993年所说："提出想法是容易的，但完成事情是困难的。这些变革将死在战壕中。"此外，谁拥有战壕？答案是：您和我，以及所有其他处理流程的人。因此，如果不把变革或变革过程的一部分施加到拥有战壕的人身上，变革就不会成功[5]。我们需要考虑各种业务领域和组织中的变革驱动因素、组织能力受到影响的程度，然后是端到端业务流程——那些跨越业务功能和组织边界的流程。确定变革如何影响单个的业务流程，以及不同流程中的更改如何影响到相关流程和业务功能[6]。有些组织能够比其他组织更好地处理变革，并将其应用在他们的BPM工作中。正如P. Harmon所写，"流程管理和变革管理应该紧密结合在一起"[7]。

LESSONS LEARNED OF THE OUTPERFORMERS AND UNDERPERFORMERS

It is generally recommended that an overall change management approach be built, and then link the BPM program of work to the change management tools that the approach has made available.[8] Although these aspects are common knowledge, it is also common knowledge that this is not common practice. In the following we will explore what practices we could learn from leading organizations (outperformers) that know how to go about this, as well as what are bad practices or not the ideal way of working (underperformers in an area).

In the following Table 1 we have listed some differences between the Underperformers' and Outperformers' way of doing BPM Change Management:

Table 1 *List Underperformer and Outperformers Ways of Doing BPM Changes*	
Underperformers	**Outperformers**
No formal method or approach for BPM change management. The underlying thought is that BPM models in process notations and process owners publish them to enable change.	Established and documented BPM change management method and approaches.
Capacity (mis)management • Projects often demand huge BPM CoE change capacity and services to deliver. • Change assessment is done by individual project, not integrated to global BPM CoE capacity assessments. • No clear delineation of new and run processes. Roles and responsibilities must be better defined, understood, and tracked for development and change/maintenance.	There is ONE perspective on BPM changes for BPM CoE, including new change and changing to existing, so that only one prioritization process is built into the BPM CoE portfolio management and governance process.
Multiple project owners, change owners. Various process change request documentation. Multiple change management processes and procedures exist today, preventing a proactive/steady state and an integrated, end-to-end change process from requirements to implementation.	Going to integrated, best practices-based BPM CoE change handling, with specific roles and responsibilities, end-to-end processes supported by a single-source of truth.
BPM technology choices is based on: • functional needs • technology requirements • process modeling capabilities Other service organizations involvement not mapped • Current process does not clearly distinguish between BPM CoE services and interlinked business and IT services • Requirement for changes are siloed in various business departments.	BPM technology choices is based on: • business drivers • change enablers Changes scope and governance are defined; entry and exit points are mapped and related coordination, control, and measurements mechanisms are described.

5.3.3　成功者和失败者的经验教训

一般建议构建一个全面的变革管理方法,然后将BPM程序变革管理工具相链接[8]。虽然这些方面是常识,但却没有被经常实践。在接下来的文章中,我们将探讨我们可以从那些知道如何做到这一点的领先组织(表现优异的组织)中学到的实践经验,以及从那些落后组织(在某一领域表现不佳的组织)中学到的教训或不理想的工作方式。

在下面的表1中,我们列出了BPM变革管理中表现不佳者和表现优异者之间的一些差异。

表1　BPM变革表现不佳者和表现优异者的方法清单

表现不佳者	表现优异者
BPM变革管理没有正式的套路或方法。其基本思想是用BPM模型作为流程符号,以及流程责任人能够发布和更改流程	建立并记录BPM变革管理套路和方法
能力(管理信息系统)管理: • 项目通常需要巨大的BPM CoE变革能力和服务以实现交付 • 变革评估由单个项目完成,而不是集成到总体的BPM CoE能力评估中 • 对新流程和运行的流程没有清晰地描述。角色和职责必须更好地定义、理解和跟踪,以便进行开发和更改/维护	对于BPM CoE,有一个关于BPM变革管理的统一视角,包括新的更改和对现有更改。以便在BPM CoE项目组合管理和治理过程中,只构建一个优先级流程
多个项目负责人、多个变革负责人 各种流程变更请求文档 目前存在多种变革管理流程和程序,从而阻止了从需求到实施的主动/稳定状态和集成的端到端变革流程	转向集成的、基于最佳实践的BPM CoE变革处理,具有特定的角色和职责,由单一事实来源支持的端到端流程
BPM技术的选择依据: • 功能需要 • 技术要求 • 流程建模能力 没有其他涉及的服务组织参与: • 当前流程没有明确区分BPM CoE服务和相互链接的业务和IT服务 • 变革的需求被隔离在各个业务部门内	BPM技术的选择依据: • 业务驱动因素 • 变革推动者 定义了变革范围和治理、规定了入口和出口,并描述了相关的协调、控制和度量机制

Enterprise architecture engagement gaps • EA guidelines, parameters, and standards impacting change management are not formalized or integrated into BPM change concepts • Changes identified by the BPM teams are not synchronized with EA teams.	EA change management is formalized, and elements from its various domains affecting BPM change management are integrated throughout the BPM CoE process life cycle.
No standards or interlinks of the various teams and their work around the change concepts, that is, change analysis/strategy, design, creation, deployment, monitoring, and link to continuous improvement.	BPM change management is integrated throughout the process life cycle.

LESSONS LEARNED AROUND BENEFIT AND VALUE REALIZATION

Some of the benefits of doing integrated and standardized BPM Change Management in the BPM CoE among others are:

• Transparency of changes throughout the life cycle—Reducing the number of unauthorized changes, leading to fewer BPM CoE service disruptions, and reduced change-related incidents.
• Enable Organizational Change Management (OCM), minimize the severity of any impact and disruption.
• Delivering change based on change drivers (value and performance drivers) maximizes value and reduces cost of change.
• Not only specifying, but establishing link between performance/value drivers and process changes, and ensuring higher project success rate, reducing failed changes, and, therefore, business disruptions, defects, and rework.
• Implementing changes that meet the clients' agreed value expectation while optimizing costs.
• Reducing the average time to align the process to the right way of working (from misalignment to alignment).
• Contributing to better portfolio, program, and project estimates of the quality, time, and cost of change.
• Contributing that change to meet governance, legal, legislative, contractual, and regulatory requirements by providing auditable evidence of change management activity.
• Ensure that all stakeholders receive appropriate and timely communication about the change so they are aware and ready to adopt and support the change.

LEADING PRACTICE SUGGESTIONS ON WHAT REALLY WORKS WELL

Developing an effective change road map that is integrated into the business process management life cycle and the BPM CoE change and issue management is imperative for change effectiveness, if your initiative is to avoid the "valley of

（续表）

表现不佳者	表现优异者
企业架构参与缺口： ● 影响变革管理的EA指导方针、参数和标准没有被形式化或集成到BPM变革概念中 ● BPM团队标识的更改没有与EA团队同步	EA变革管理是形式化的，来自影响BPM变革管理的各个领域的元素被集成到整个BPM CoE流程生命周期中
在变革分析/策略、设计、创建、部署、监控和持续改进方面，不同团队及其工作围绕变革概念没有标准或相互衔接	BPM变革管理集成在整个流程生命周期中

5.3.4　关于利益和价值实现的经验教训

在BPM CoE中实现BPM变革管理集成和标准化的一些好处包括：

● 生命周期中变更的透明性——减少未授权变更的数量，减少BPM CoE服务中断的次数，减少与变更相关的突发事件；

● 启用OCM，将影响和中断的严重性降到最小化；

● 基于变革驱动因素（价值和性能驱动因素）完成变更，实现价值最大化并降低变更成本；

● 不仅指定性能/价值驱动因素和流程变更，而且建立它们之间的联系，确保更高的项目成功率，减少失败的变更，从而减少业务中断、缺陷和返工；

● 在优化成本、满足客户约定价值预期的同时实施变更；

● 减少将流程与正确工作方式相结合的平均时间（从未结合到结合）；

● 有助于实现更好的项目组合、项目规划以及质量、工期和成本变更的评估；

● 通过提供变更管理活动的可审计证据，促进变革以满足政府、法律、立法、合同和监管要求；

● 确保所有相关者都能及时、恰当地得到关于有关变更的沟通，以便他们意识到并准备好应用和支持变革。

5.3.5　什么才是真正有效的指导建议

设计有效的变革路线图，并将其集成到业务流程生命周期管理、BPM CoE变革和问题管理之中，这对于您避免"绝望之谷"以及确保变革有效性来说是必要

despair". The purpose and objectives of BPM Change Management is to respond to both process change requests as well as the BPM client's changing requirements, while maximizing value and reducing incidents, disruptions, and rework. We feel it is vital to point out that BPM CoE Change Management and BPM Change Management are different. Although both interact, they have different purposes:

- *BPM CoE Change Management*—The resulting concept that combines the vision, strategic considerations, and requirements for Portfolio Change Management within the BPM CoE. This is done through the BPM CoE Change Board and includes alignment of programs and projects ensuring overall value creation for the BPM CoE clients.
- *BPM Change Management*—The set of activities/procedures/processes that, although supporting the BPM CoE, is about the execution of change management. Therefore, it would be the more tactical and operational change management piece of the BPM CoE. The changes are executed through the various tasks and roles in the BPM CoE.

As already pointed out, and although they are different, five common key areas compose BPM CoE Change Management as well as BPM Change Management:

1. Requirements Management
2. Planning with link to process portfolio, program, and project management
3. Value and Performance Management
4. BPM Governance
5. BPM Continuous Improvement feedback loop in terms of degree of change (low, medium, or high)

Out of scope of the above mentioned would be changes with significantly wider impacts than BPM Service Changes or BPM Project Changes. These include:

- Business organizational changes that need to be channeled through the Process Portfolio Management channel
- Structural changes that need to be channeled through the BPM CoE management
- The people-side of change that needs to be channeled through the business change management group

We realize that managing change effectively implies using processes to manage change, whereas it would mean that the BPM CoE would have to take their own medicine, we also realize that this would be a fundamental change for most organizations. It would mean that all BPM changes are recorded and evaluated, and that authorized changes are prioritized in a BPM CoE Change Management procedure. This makes the process life cycle ideal to execute the five key areas around BPM CoE Change Management and enables change in any phase as well as empowering the continuous improvement feedback loop (see Figure 2).

Integrating BPM Change Management into the process life cycle includes both changes to new and existing processes. It increases transparency of changes

的。BPM变革管理的目的和目标是响应流程变更请求和BPM客户端变更需求,同时最大化价值,并减少事件、中断和返工。虽然BPM CoE变革管理和BPM变革管理相互作用,但我们认为有必要指出他们在目的等方面的不同。

- BPM CoE变革管理——由此产生的概念结合了BPM CoE中的项目组合变革管理的愿景、战略考虑和需求。包括协调计划和项目在内都是通过BPM CoE变革委员会完成的,这能够确保通过BPM CoE为客户创造整体价值。
- BPM变革管理——一组活动/过程/流程,尽管支持BPM CoE,但它们是关于变革管理的执行的。因此,它将是BPM CoE中更具策略性和操作性的变革管理部分。这些变革也是通过BPM CoE中的各种任务和角色执行的。

正如已经指出的那样,虽然它们不同,但以下五个关键领域共同构成了BPM CoE变革管理以及BPM变革管理的主体:

(1)需求管理;

(2)规划与包括流程项目组合、项目集和项目管理的链接;

(3)价值和绩效管理;

(4)BPM治理;

(5)BPM根据变更程度(低、中、高)进行持续改进反馈循环。

超出上述范围的变革属于比BPM服务变革或BPM项目变革产生更广泛的影响。这些变革包括:

- 需要通过流程组合管理渠道引导的业务组织变革;
- 需要通过BPM CoE管理引导的结构变革;
- 需要通过业务变革管理小组进行的人员变革。

我们认识到,管理变革的有效性依赖于流程在变革管理过程的使用,而这意味着BPM CoE将不得不采取属于自己独特的措施,我们还认识到,这将是大多数组织的基本变革。这将意味着所有BPM更改都将被记录和评估,并且授权的变革在BPM CoE更改管理过程中享有优先级。这使得流程生命周期成为执行BPM CoE变革管理的五个关键领域的理想选择,可以实现任何阶段的变革,并为持续改进提供反馈循环支持(图2)。

将BPM变革管理集成到流程生命周期中包括对新流程和现有流程的变革。它增加了整个生命周期中更改的透明性、减少了未授权更改的数量、将任何影响和

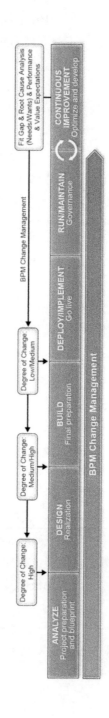

FIGURE 2

The process life cycle with BPM change management.[9]

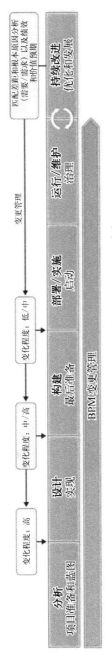

图2　BPM变革管理的流程生命周期[9]

throughout the life cycle and reduces the number of unauthorized changes, minimizing the severity of any impact and disruption. However, a recent study showed that even though we know we should do something we often do not,[11] even if life depends on it! The results of the study demonstrate that even when doctors tell heart patients they will die if they don't change their habits, only one in seven will be able to follow through successfully. Desire and motivation are not enough: even when it is literally a matter of life or death, the ability to change remains maddeningly elusive. Given that the status quo is so ingrained in our psyches, how can we change ourselves and our organizations? Whereas people and organizations know what they should do, they often do not know how to go about it.

Therefore, we will explore the following in detail:

- The life cycle phase
- The roles involved
- The strategic, tactical, and operational tasks
- The notations that are relevant for the subject (both BPMNotations, x-BPM-Notations, and other business notations)
- Typical enterprise standards involved

We will in Figure 3 illustrate the entire BPM Change Management Life Cycle complete with all the various roles, their tasks, and how the different modeling, engineering, and architecture disciplines interact and are used around each step in the life cycle.

Although the tasks are not that challenging in themselves, the challenge lies in the change control and management of the diverse roles in the various tasks, some that need to interlink and be governed by BPM CoE Portfolio Change Management, and some by BPM Change Management, relating to multiple enterprise standards to enable business and IT change.

BPM Change Management in the Analyze Phase

The first phase of the BPM Life Cycle is the phase in which the organization's processes are analyzed, captured, and defined, based on the business goals and specific process requirements (e.g. business needs and wants), as well as any interlinked business and process demands (Figure 4) Process goals and detailed process requirements are defined, and process choices are clarified through process blueprinting, and the initial process maps are populated with the identified processes. This phase includes Continuous Improvement through Change Management of the BPM Life Cycle, and the degree of changes made during this phase is considered high (Figure 2).

Tasks that have a link to strategic aspects:

1. Log and review change requests.
2. Identify change opportunities.
3. Assess change proposals (depending on scope, it links to appropriate processes)

Tasks that have a link to tactical aspects:

1. Refer to appropriate processes (depending on scope).
2. Raise Requests for Change (RFC) for changes flagged in scope.

破坏的严重性降到最低。然而,最近的一项研究表明,即使我们生活依赖于某些方面,但我们仍然做不到我们应该做的事情,例如,即使医生告诉心脏病患者,如果他们不改变习惯就会死亡,但事实上只有七分之一的人能够遵从医生的建议。

因此,我们将对以下内容进行详细探讨:

- 生命周期阶段;
- 所涉及的角色;
- 战略、战术和操作任务;
- 与主题相关的符号(BPM符号、x-BPM符号和其他业务符号);
- 涉及的典型企业标准。

我们将在图3中展示完整的BPM变革管理生命周期,包括所有不同的角色、它们的任务,以及不同的建模、工程和架构原则如何在生命周期的每个步骤中相互作用和使用。

虽然这些任务本身并不具有挑战性,但挑战在于各种任务中各种角色的变更控制和管理,因为一些需要通过BPM CoE项目组合变革管理进行链接和管理,另一些则需要与多个企业标准相关的BPM变革管理,才能实现业务和IT变革。

1. BPM变革管理的分析阶段

BPM生命周期的第一个阶段是根据业务目标和特定流程需求(如业务需要和需求)以及任何相互关联的业务和流程需求(图4)分析、捕获和定义组织流程的阶段。同时,定义流程目标和详细的流程需求,以及通过流程蓝图澄清流程选择,并使用已识别的流程初始流程图。这个阶段包括通过对BPM生命周期的变更管理进行持续改进,并且在这个阶段所做的变更程度被认为是很高的(图2)。

与战略有关的任务:

(1)记录和审查变更申请;

(2)确定改变的机会;

(3)评估变革提案(取决于范围,与适当的流程相关联)。

与战术方面有关联的任务:

(1)参考适当的过程(取决于范围);

(2)针对范围内标记的变动提出更改请求(requests for change,RFC);

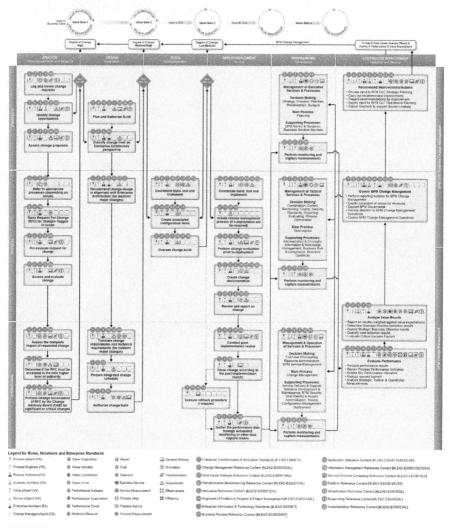

FIGURE 3

The BPM change management life cycle at a glance.[11]

3. Preevaluate request for change.
4. Assess and evaluate change (leads to assessment of the complete impact of requested change).

Tasks that have a link to operational aspects:

1. Assess the complete impact of requested change.
2. Determine if the RFC must be escalated to the next higher level of authority.
3. Perform Change Assessment of RFC by the Change Advisory Board (CAB) for significant or critical changes.

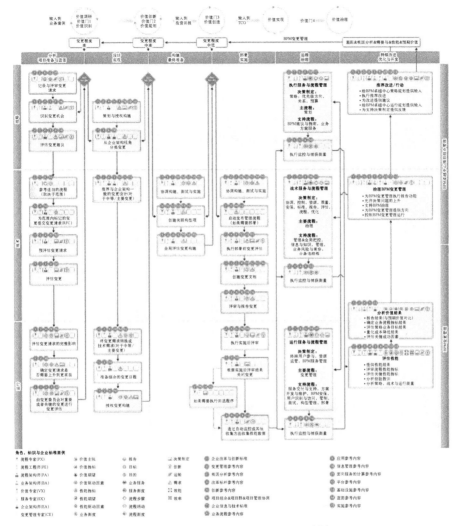

图3　BPM变革管理生命周期概览[11]

（3）预先评估变更请求；

（4）评估和评价变化（评估导致变更发生的全面影响）。

　与操作有关的任务：

（1）评估请求变更的全部影响；

（2）确定是否必须将RFC升级到下一个更高级别的权限；

（3）对于重大或关键的变更，由变革咨询委员会（Change Advisory Board，CAB）对RFC进行变更评估。

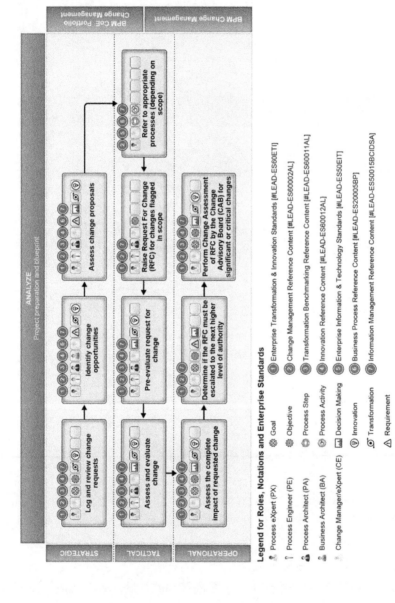

Legend for Roles, Notations and Enterprise Standards

Process eXpert (PX)
Process Engineer (PE)
Process Architect (PA)
Business Architect (BA)
Change Manager/eXpert (CE)

Goal
Objective
Process Step
Process Activity
Decision Making
Innovation
Transformation
Requirement

① Enterprise Transformation & Innovation Standards [#LEAD-ES60ETI]
② Change Management Reference Content [#LEAD-ES60002AL]
③ Transformation Benchmarking Reference Content [#LEAD-ES60011AL]
④ Innovation Reference Content [#LEAD-ES60012AL]
⑤ Enterprise Information & Technology Standards [#LEAD-ES50EIT]
⑥ Business Process Reference Content [#LEAD-ES200005BP]
⑦ Information Management Reference Content [#LEAD-ES500015BCIDSA]

(©LEADing Practice Business Process Reference Content [#LEAD-ES200005BP]

FIGURE 4

The BPM change management life cycle's analysis phase and how tasks, roles, notations, and standards could be applied.[12]

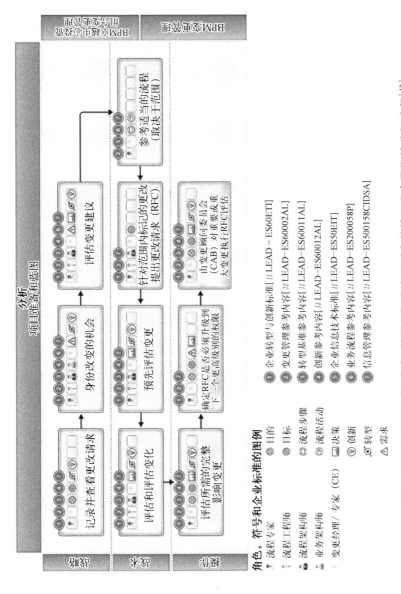

角色，符号和企业标准的图例

角色，符号
- ☝ 流程专家
- ⚙ 流程工程师
- ⚒ 流程架构师
- ⬚ 业务架构师
- ⬚ 变更经理／专家（CE）
- ⊛ 目的
- ⬒ 目标
- ⬡ 流程步骤
- ⬚ 流程活动
- ▦ 决策
- ◉ 创新
- ◢ 转型
- △ 需求

① 企业转型与创新标准[#LEAD＝ES60ETI]
② 变更管理参考内容[#LEAD＝ES60002AL]
③ 转型基准参考内容[#LEAD＝ES60011AL]
④ 创新参考内容[#LEAD＝ES60012AL]
⑤ 企业信息技术标准[#LEAD＝ES50EIT]
⑥ 业务流程参考内容[#LEAD＝ES200058P]
⑦ 信息管理参考内容[#LEAD＝ES50015SCIDSA]

图 4 BPM 如何将任务，角色，符号和标准应用到变革管理生命周期的分析阶段[12]

BPM Change Management in the Design Phase

The objective of this process is to take the information gathered in the preceding analyze phase and use it to develop a structured set of activities that *tell the BPM Change Management story for the BPM CoE* (Figure 5).

- Identify and design an enterprise BPM Change Management implantation plan (if required).
- Define and document all work flows or processes required by the stakeholders to perform change management within the BPM CoE; these will be the operations of change management (how to handle a process change, a project change, a service change, an RFC, etc.)
- Develop all supporting policies, guidelines, or best practices required to support the running of change management processes within BPM CoE for our

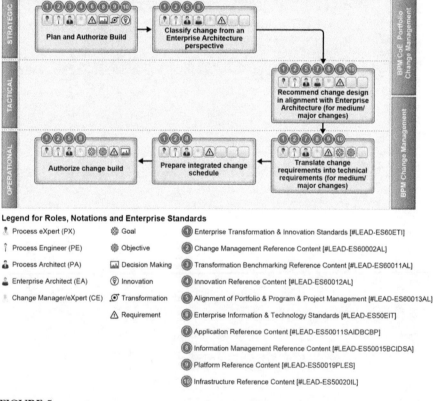

FIGURE 5

The BPM change management life cycle's design phase and how tasks, roles, notations, and standards could be applied.[13]

2. BPM变革管理的设计阶段

这个过程的目标是获取在前面的分析阶段收集到的信息,并使用它来设计一组结构化的行为,这些行为"诉说BPM CoE的BPM变革管理故事"(图5)。

- 识别和设计企业BPM变革管理实施计划(如果需要)。
- 定义并记录利益相关者在BPM CoE内执行变革管理所需的所有工作流程或步骤,这些将是变革管理的举措(如何处理流程变革、项目变革、服务变革、RFC等)。
- 为我们的客户开发在BPM CoE内运行变革管理流程所需的所有支持政

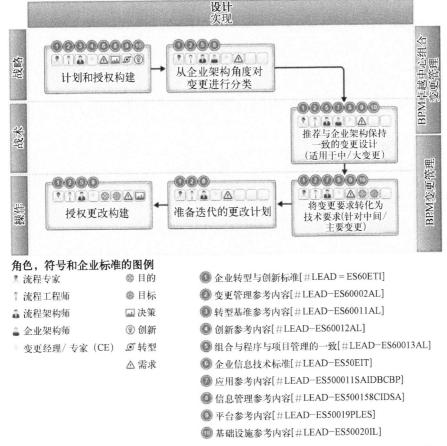

图5　BPM变革管理生命周期设计阶段以及如何应用任务、角色、符号和标准[13]

clients (governance structures, communication mechanisms, change plans, etc.)

Tasks that have a link to strategic aspects:

1. Plan and authorize build.
2. Classify change from an Enterprise Architecture perspective.

Tasks that have a link to tactical aspects:

1. Recommend change design in alignment with Enterprise Architecture (for medium/major changes).

Tasks that have a link to operational aspects:

2. Translate change requirements into technical requirements (for medium/major changes).
3. Prepare integrated change schedule.
4. Authorize change build.

BPM Change Management in the Build Phase

In the build phase (Figure 6) of the BPM Change Management Life Cycle, the tasks focus on carrying out the already-identified design solutions that have been developed during the previous design phase. The new processes and process structures, as well as the redesign of existing processes within the current process portfolio, should be constructed within a coordinated and isolated process testing and simulation environment, and primarily involves participants like process experts, engineers, and architects, as well as change management experts. Configuration foundations are likewise important to adapt to the environment around new processes, in particular, as well as a continuous governance aspect should be implemented to control and oversee the build phase from start to finish, and possibly enforce a construction rollback, if deemed necessary.

Tasks that have a link to tactical aspects:

1. Coordinate build, test, and implement (simulated production environment).
2. Create associated configuration items.
3. Oversee change build.

Tasks that have a link to operational aspects:

1. Execute rollback procedure if required.

BPM Change Management in the Deploy/Implement Phase

In the deployment and implementation phase (Figure 7), all reconfigured, reengineered, as well as new, processes are evaluated for go-live readiness. Change documentation will also need to be created to fully document and define all change aspects that have occurred to the new process portfolio and/or structures. A thorough review analysis will follow up release documentation, and the management will

策、指南或最佳实践(治理结构、沟通机制、变革计划等)。

与战略有关的任务:

(1)构建计划和授权;

(2)从企业架构的角度对变革进行分类。

与战术有关的任务:

推荐与企业架构一致的变革设计(对于中期/重大变革)。

与操作有关的任务:

(1)将变革需求转化为技术需求(针对中期/重大变革);

(2)准备综合变革计划;

(3)构建授权改变方案。

3. BPM 变革管理的构建阶段

在BPM变革管理生命周期的构建阶段(图6),任务的重点是执行在前一个设计阶段开发的已经确定的设计解决方案。新的流程和流程结构,以及对当前流程组合中现有流程的重新设计,均应在流程专家、工程师和架构师以及变革管理专家等参与者下的流程测试和模拟环境的协调和环境内构建。基础配置对于适应新流程周围的环境同样重要,特别是应该实施持续的治理来控制和监督构建阶段的从开始到结束,并且如果认为有必要可以构建回溯反馈机制。

与战术有关的任务:

(1)协调构建、测试和实施(模拟生产环境);

(2)创建关联的配置项目;

(3)监督构建变革。

与操作有关的任务:

如果需要,执行回溯过程。

4. BPM 变革管理的部署/实施阶段

在部署和实施阶段(图7),所有重新配置、重新设计以及新的流程都会针对上线前的准备情况进行评估。此外,还需要创建变革文档,以全面记录和定义新流程组合或结构发生的所有变革方面。完整的审核分析将形成记录文档,然后管理层作出上线和流程发布的最终决策。发布后不久,还需要实施后审查和报告,以跟进

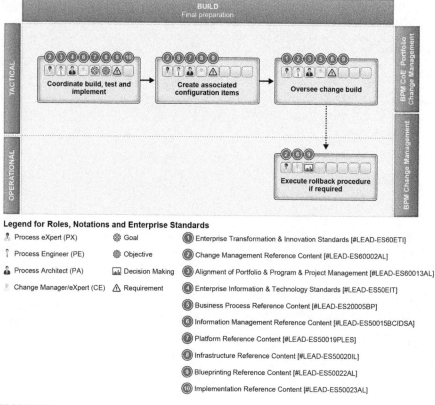

FIGURE 6

The BPM change management life cycle's build phase and how tasks, roles, notations, and standards could be applied.[14]

then carry out the final decision for go live and process release. Post-implementation review and reporting is also necessary shortly after release to follow up on successes and/or shortcomings of the live process environment.

Tasks that have a link to tactical aspects:

1. Coordinate build, test, and implement (go live).
2. Initiate release management process (if a deployment will be required).
3. Perform change evaluation prior to deployment.
4. Create change documentation.
5. Review and report on change.

Tasks that have a link to operational aspects:

1. Conduct postimplementation review.

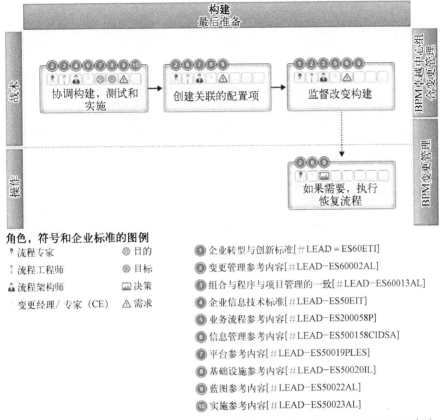

角色，符号和企业标准的图例

- 🕯 流程专家
- 🕯 流程工程师
- 👤 流程架构师
- 变更经理/专家（CE）

- ⊛ 日的
- ⊛ 目标
- 📊 决策
- ⚠ 需求

- ① 企业转型与创新标准[#LEAD = ES60ETI]
- ② 变更管理参考内容[#LEAD-ES60002AL]
- ③ 组合与程序与项目管理的一致[#LEAD-ES60013AL]
- ④ 企业信息技术标准[#LEAD-ES50EIT]
- ⑤ 业务流程参考内容[#LEAD-ES200058P]
- ⑥ 信息管理参考内容[#LEAD-ES500158CIDSA]
- ⑦ 平台参考内容[#LEAD-ES50019PLES]
- ⑧ 基础设施参考内容[#LEAD-ES50020IL]
- ⑨ 蓝图参考内容[#LEAD-ES50022AL]
- ⑩ 实施参考内容[#LEAD-ES50023AL]

图6　BPM变革管理生命周期的构建阶段以及如何应用任务、角色、符号和标准[14]

流程的成功或不足。

与战术有关的任务：

（1）构建协调、测试和实施（上线）；

（2）启动发布管理过程（如果需要部署）；

（3）在部署之前执行变革评估；

（4）创建变革文档；

（5）审查并报告实施前后的变化情况。

与操作有关的任务：

（1）进行实施后的审查；

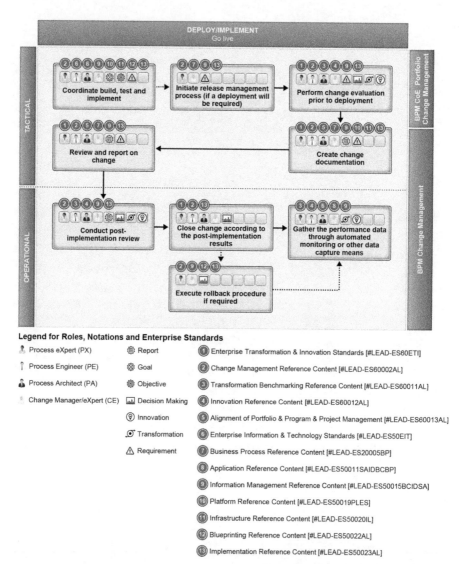

FIGURE 7

The BPM change management life cycle's deploy/implement phase and how tasks, roles, notations, and standards could be applied.[15]

2. Close change according to the postimplementation results (execute rollback procedure if required).

3. Gather the performance data through automated monitoring or other data capture means.

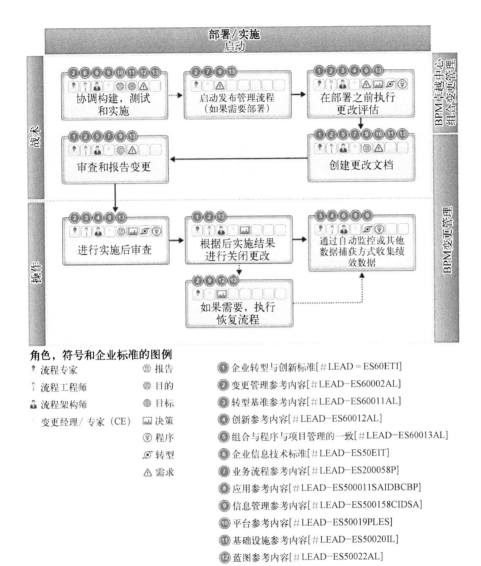

图7 BPM变革管理生命周期的部署/实施阶段以及如何应用任务、角色、符号和标准[15]

（2）根据实现结果停止更改（如果需要,执行回溯过程）;

（3）通过自动监控或其他数据捕获方式收集性能数据。

BPM Change Management in the Run/Maintain Phase

The objective of this phase (Figure 8) is to provide a set of processes that, when executed, become the working portfolio change management work flows for the BPM CoE. They are the change management requests that satisfy the client's change

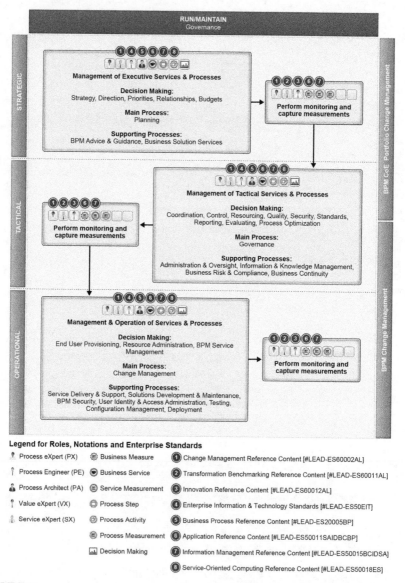

Legend for Roles, Notations and Enterprise Standards

Process eXpert (PX) · Business Measure · ① Change Management Reference Content [#LEAD-ES60002AL]

Process Engineer (PE) · Business Service · ② Transformation Benchmarking Reference Content [#LEAD-ES60011AL]

Process Architect (PA) · Service Measurement · ③ Innovation Reference Content [#LEAD-ES60012AL]

Value eXpert (VX) · Process Step · ④ Enterprise Information & Technology Standards [#LEAD-ES50EIT]

Service eXpert (SX) · Process Activity · ⑤ Business Process Reference Content [#LEAD-ES20005BP]

Process Measurement · ⑥ Application Reference Content [#LEAD-ES50011SAIDBCBP]

Decision Making · ⑦ Information Management Reference Content [#LEAD-ES50015BCIDSA]

⑧ Service-Oriented Computing Reference Content [#LEAD-ES50018ES]

FIGURE 8

The BPM change management life cycle's run/maintain phase and how tasks, roles, notations, and standards could be applied.[1,16]

5. BPM变革管理的运行/维护阶段

　　此阶段的目标(图8)是提供一组流程,这些流程在执行BPM CoE时将成为组合变革管理的工作流程。它们是满足客户端需求变革的变革管理申请。它们还应

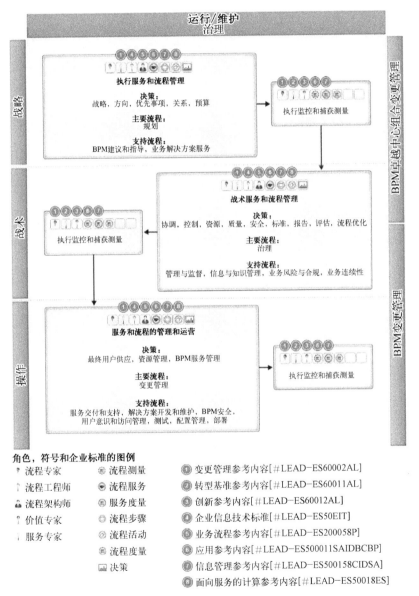

角色,符号和企业标准的图例

🕴 流程专家	🎯 流程测量	① 变更管理参考内容[#LEAD-ES60002AL]
🕴 流程工程师	⚙ 流程服务	② 转型基准参考内容[#LEAD-ES60011AL]
👤 流程架构师	⚙ 服务度量	③ 创新参考内容[#LEAD-ES60012AL]
🕴 价值专家	⚙ 流程步骤	④ 企业信息技术标准[#LEAD-ES50EIT]
⎮ 服务专家	⚙ 流程活动	⑤ 业务流程参考内容[#LEAD-ES200058P]
	⚙ 流程度量	⑥ 应用参考内容[#LEAD-ES500011SAIDBCBP]
	📊 决策	⑦ 信息管理参考内容[#LEAD-ES500158CIDSA]
		⑩ 面向服务的计算参考内容[#LEAD-ES50018ES]

图8　BPM变革管理生命周期的运行/维护阶段以及如何应用任务、角色、符号和标准[16]

needs. They should also include step-by-step instructions for performing procedural activities such as impact analysis or user profile assessments when required.

Tasks that have a link to strategic aspects:

1. *Management of Executive Services and Processes*
   a. Decision Making: Strategy, Direction, Priorities, Relationships, Budgets.
   b. Main Process: Planning.
   c. Supporting Processes: BPM Advice and Guidance, Business Solution Services.
2. Perform monitoring and capture measurements

Tasks that have a link to tactical aspects:

1. *Management of Tactical Services and Processes*
   a. Decision Making: Coordination, Control, Resourcing, Quality, Security, Standards, Reporting, Evaluating, and Process Optimization.
   b. Main Process: Governance.
   c. Supporting Processes: Administration and Oversight, Information and Knowledge Management, Business Risk and Compliance, and Business Continuity.
2. Perform monitoring and capture measurements

Tasks that have a link to operational aspects:

1. Management and Operation of Services and Processes
   a. Decision Making: End User Provisioning, Resource Administration, BPM Service Management.
   b. Main Process: Change Management.
   c. Supporting Processes: Service Delivery and Support, Solutions Development and Maintenance, BPM Security, User Identity and Access Administration, Testing, Configuration Management, Deployment.
2. Perform monitoring and capture measurements

BPM Change Management in the Continuous Improvement Phase

The objective of this phase (Figure 9) is to control, monitor, and evaluate the operational change management processes, while addressing escalated issues/concerns and feedback for improvement of the running processes and the BPM portfolio as a whole. They should also include step-by-step instructions for performing any required procedural activities.

Tasks that have a link to strategic aspects:

1. Recommend Improvements/Actions
   a. Provide input to BPM CoE Strategic Planning.
   b. Carry out recommended improvements.
   c. Present recommendations for improvement.
   d. Supply input for BPM CoE Operational Planning.
   e. Deliver feedback to support decision-making.

该包括在需要时执行程序性活动(如影响分析或用户概况评估)的分步指导。

与战略有关的任务如下。

(1)执行与服务和流程有关的管理。

a. 决策:战略、方向、优先事项、关系、预算。

b. 主要流程:计划。

c. 支持流程:BPM建议和指导、业务解决方案服务。

(2)绩效监控与收益评估。

与战术有关的任务如下。

(1)战术服务和流程的管理。

a. 决策:协调、控制、资源、质量、安全、标准、报告、评估和流程优化。

b. 主要流程:治理。

c. 支持流程:管理与监督、信息与知识管理、业务风险与合规、业务连续性。

(2)绩效监控与收益评估。

与操作有关的任务如下。

(1)服务和流程的管理和操作。

a. 决策:终端用户配置、资源管理、BPM服务管理。

b. 主要流程:变革管理。

c. 支持流程:服务交付与支持、解决方案开发与维护、BPM安全、用户标识与访问管理、测试、配置管理、部署。

(2)绩效监控与收益评估。

6. BPM变革管理的持续改进阶段

此阶段的目标(图9)是控制、监控和评估流程变革的结果,同时解决升级的问题/关注事项,并对整个运行流程和BPM组合进行改进。它们还应该包括执行任何必需的程序性活动的步骤说明。

与战略有关的任务如下。

改进/行动建议:

a. 为BPM CoE战略规划提供资源;

b. 执行改进建议;

c. 提出改进建议;

d. 为BPM CoE运营计划提供资源;

e. 提供反馈以支持决策。

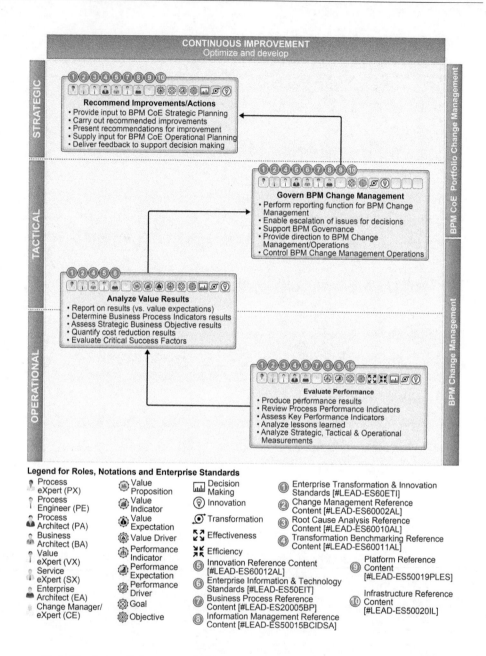

Tasks that have a link to tactical aspects:

1. Govern BPM Change Management
   a. Perform reporting function for BPM Change Management.
   b. Enable escalation of issues for decisions.
   c. Support BPM Governance.

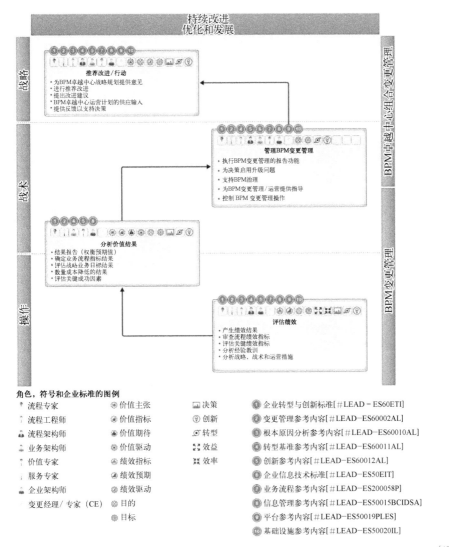

图9 BPM变革管理生命周期的持续改进阶段,以及如何应用任务、角色、符号和标准[17]

与战术有关的任务如下。

(1)管理BPM变革管理:

a. 为BPM变革管理的执行效果撰写报告;

b. 将问题升级为决策的依据;

c. 支持BPM治理;

      d. Provide direction to BPM Change Management Operations.

      e. Control BPM Change Management Operations.

2. Analyze Value Results

      a. Report on results (weighed against value expectations).

      b. Determine Business Process Indicators results.

      c. Assess Strategic Business Objective results.

      d. Quantify cost-reduction results.

      e. Evaluate Critical Success Factors.

Tasks that have a link to operational aspects:

1. Evaluate Performance

      a. Produce performance results

      b. Review Process Performance Indicators

      c. Assess Key Performance Indicators

      d. Analyze lessons learned

      e. Analyze Strategic, Tactical, and Operational Measurements

2. Analyze Value Results

      a. Report on results (weighed against value expectations).

      b. Determine Business Process Indicators results.

      c. Assess Strategic Business Objective results.

      d. Quantify cost-reduction results.

      e. Evaluate Critical Success Factors.

The key is to identify, document, and categorize the continuous improvement changes into the degree of change (see Figure 2) and thereby loop it into any of the process life-cycle phases:

- *Degree of Change: Low*
  Changes that can be achieved with low work amount are referred to by different names in the various BPM CoEs. Some of the names are, among others, fast changes, quick changes, or even standard changes. All of them have one thing in common; they have a low degree of change. Low degree changes are, for the most part, preauthorized as they have low risk, are relatively common, and follow a known procedure or work flow.
- *Degree of Change: Medium*
  Changes that can be achieved with medium work amount are for the most part referred to as normal change requests.
- *Degree of Change: High*
  Changes that include a high degree of change can also have different names. Some of the names are, among others, big changes, strategic changes, major incident changes, and emergency changes. All of them do not only have in common that they have a high degree of change, but also that they have a significant relationship to performance and value creation, and therefore must be implemented as soon as possible. This is not always the case, but we see a pattern in the high degree change requests.

d. 为BPM变革管理提供指南；

e. 管控BPM变革管理运行。

（2）分析价值结果：

a. 结果报告（与价值预期进行对比）；

b. 确定业务流程结果指标；

c. 评估SBO结果；

d. 量化成本降低结果；

e. 评估CSF。

链接到操作方面的任务如下。

（1）评估性能：

a. 产生性能结果；

b. 审查PPI；

c. 评估KPI；

d. 分析吸取的教训；

e. 分析战略、战术和作战措施。

（2）分析价值结果：

a. 结果报告（与价值预期进行对比）；

b. 确定业务流程结果指标；

c. 评估SBO结果；

d. 量化成本降低结果；

e. 评估CSF。

关键是对持续改进变革的程度进行识别、记录并分类（图2），从而将其在流程生命周期的各阶段内进行循环。

- 变革程度：较低程度的变革可以通过低工作量在BPM CoE中实现，这些变革是快速变化、快速更改，甚至是标准更改。它们都有一个共同点：它们的变化程度很低。在大多数情况下，低程度的变革大多是预先授权的，因为它们具有较低的风险，且比较常见，并且遵循已知的程序或工作流程。

- 变革程度：中等程度的变革可通过中等工作量实现，这部分也被认为是属于正常变革。

- 变革程度：按照变革的程度和类型，变更可分为不同的级别，包括重大变化、战略变化、重大事件变化和紧急变化。它们不仅具有高度变化的共同点，而且与绩效和价值创造有紧密的关系，因此必须尽快实施。这种情况并不总是如此，但我们在高度变更需求中看到了这个方面。

Perform Managerial Governance Activities Across all Phases

The objective of this process area is to perform the management and administrative activities to support the running of change management. They should also include step-by-step instructions for performing any required procedural activities.

CONCLUSION

The implementation of a BPM Change Management program demands a whole new way of working in an organization, and implies looking differently at one's organization. Many organizations underestimate this. Old, existing ways of working and managing/directing people must be changed. This fact alone begs for a clear change at the management level, but it also requires change at the lower organization levels. Examples of such changes may include[18]:

- Management and supervision will have to change: management and supervision should be done over the entire process (horizontal point of view) rather than on a hierarchical basis. Attention is to be given to the process instead of on keeping the department up and running.
- The process manager gets a much more prominent role in the organization.
- Cubicle thinking between departments will have to be eliminated.
- Roles and responsibilities will have to be defined better and are likely to change significantly compared to the situation before implementing BPM.
- Another type of management, with a different attitude toward the work floor and the execution in the organization's operations.

This new way of working should be accepted before working in a process-oriented manner can become successful. When organizations decide to implement process improvements and/or BPM, they must not only pay attention to the new possibilities and the factors that stimulate successful implementation, they must also be aware of the restrictions. These restrictions or barriers are often bound to the organization culture, to the comfort one obtains from holding a certain position, and to power and status. Management must deal with these barriers and actively deal with the factors that stimulate implementation as well (Figure 10).

Finally, clear and accurate communication is important for successful change management. This implies a need to build integrity and trust, which will have implications for the specific tactics that will be adopted in implementing the changes required.[19]

Essentially, an organization should focus on these core values:

- Ensure that the need for change is strategically driven
- Outline the BPM strategy together with its program of works
- Develop a change management approach, and devise a tool kit to be used with the implementation of the program of works around three areas of change:
  - Preparing for change

7. 在所有阶段执行管理治理活动

此目标是通过执行管理和管理活动支持变更管理的运行,此外,除了这些管理活动,还应包括执行任何活动的步骤说明。

5.3.6　结论

BPM变更管理流程的实现要求,必须改变旧的、现有的工作方式和管理/指导人员的方式。因此在组织中采用一种全新的工作方式,意味着以不同的视角看待组织,许多组织低估了这一点。所以需要对组织的管理级别进行明确的更改,这些变化的例子可包括如下方面[18]。

- 管理和监督将会改变:管理和监控应该在整个流程中进行(横向的观点),而不只是存在于各层级之上。与保持该部门的正常运作相比,应该特别注意这一点。
- 流程经理在组织中扮演更重要的角色。
- 必须消除部门间的思维障碍。
- 角色和职责可能会发生重大变化,因此相比于实施BPM之前,必须更好地定义角色和职责。
- 工作场所和组织运作中的执行采取另一种类型的管理态度。

这种新的工作方式应该在以流程为导向的工作方式成功之前被接受。当组织决定实施流程改进或BPM时,他们需要同时关注导致结果成功和失败的积极和消极因素,消极因素往往与组织文化、某一职务所拥有的权力和地位有关,因此,管理层必须应对这些障碍,并积极刺激能够促进实施的积极因素(图10)。

最后,清晰准确的沟通对于成功的变革管理至关重要,这意味着需要建立诚信和信任,这将对实施变革时所采用的具体策略产生影响[19]。

本质上,一个组织应该关注以下这些核心价值。

- 确保变革的需求是由战略驱动的。
- 概述BPM策略及其工作计划。
- 制定变更管理方法,并设计一套工具包,用于围绕三个变更领域实施工作计划:
  - 准备变革;

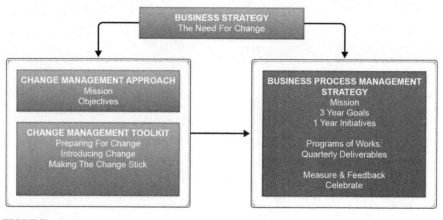

FIGURE 10

The BPM change management approach.[20]

- Introducing change
- Making the change stick

Many tactics can be selected from the tool kit for each area, and the actual tactics adopted will need to match the particular business, but if you have a framework from which to select, the likely success of your BPM Change Management project is increased.

End Notes

1. Kübler-Ross, E, On Death and Dying, Routledge, (1969) ISBN: 0-415-04015-9.
2. Kotter, J.P, "Leading change: why transformation efforts fail," *Harvard Business Review*, (May 1995).
3. Micheal Axelsen, "Business process management and change management analysing the human factor: people, change and governance, Applied Insight Pty Ltd, (2012).
4. Hammer, M., Champy, J., Reengineering the Corporation – A Manifest for Business, Revolution, Harperbusiness, (1994), p.272.
5. Jeston, J., and Nelis, J., Business process management: practical guidelines to successful implementations, Elsevier (2008).
6. Balmes, G., The role of organizational change management in BPM, business process management, organizational performance, BPM institute.org, (2011).
7. Harmon, P., Business Process Advisor, (May 10, 2011).
8. See note 3 above.
9. BPM Change Management Life Cycle Model, LEADing Practice Business Process Reference Content #LEAD-ES20005BP.
10. See note 5 above.
11. See note 9 above.
12. See note 9 above.
13. See note 9 above.

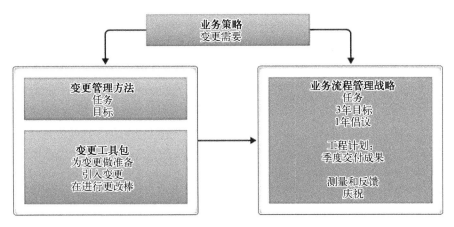

图10　BPM变更管理方法[20]

- 对变更进行导入；
- 让变更更加持久。

可以从每个领域的工具包中选择许多策略，并且所采用的实际策略将需要与特定业务相匹配，但是如果您从中选择一个整体框架，那么您的BPM变更管理项目的成功率可能会更高。

参考文献

［1］Kübler-Ross, E, On Death and Dying, Routledge, (1969) ISBN: 0-415-04015-9.

［2］Kotter, J.P, "Leading change: why transformation efforts fail," *Harvard Business Review*, (May 1995).

［3］Micheal Axelsen, "Business process management and change management analysing the human factor: people, change and governance, Applied Insight Pty Ltd, (2012).

［4］Hammer, M., Champy, J., Reengineering the Corporation — A Manifest for Business, Revolution, Harperbusiness, (1994), p.272.

［5］Jeston, J., and Nelis, J., Business process management: practical guidelines to successful implementations, Elsevier (2008).

［6］Balmes, G., The role of organizational change management in BPM, business process management, organizational performance, BPM institute.org, (2011).

［7］Harmon, P., Business Process Advisor, (May 10, 2011).

［8］See note 3 above.

［9］BPM Change Management Life Cycle Model, LEADing Practice Business Process Reference Content #LEAD-ES20005BP.

［10］See note 5 above.

［11］See note 9 above.

［12］See note 9 above.

［13］See note 9 above.

14. See note 9 above.
15. See note 9 above.
16. See note 9 above.
17. See note 9 above.
18. "The importance of change management when implementing BPM", Freek Hermkens.
19. Micheal Axelsen, Business process management and change management.
20. See note 9 above.

[14] See note 9 above.
[15] See note 9 above.
[16] See note 9 above.
[17] See note 9 above.
[18] "The importance of change management when implementing BPM", Freek Hermkens.
[19] Micheal Axelsen, Business process management and change management.
[20] See note 9 above.

Business Process Management Governance

Maria Hove, Gabriella von Rosing, Bob Storms

INTRODUCTION

Companies create value for customers and shareholders (value streams) via the effectiveness and efficiency of activities that flow across organization boundaries—often referred to as the firm's cross-functional business processes. Business process management (BPM) spans both business and technology and provides a layer of visibility and control over the processes. To optimize and sustain business process improvements, it is essential to overlay some form of governance that creates the right structures, metrics, roles, and responsibilities to measure, improve, and manage the performance of a firm's end-to-end business processes.[1] In this chapter we will therefore focus on the concepts of BPM governance from the angles of what it is, why it is needed, where it can be applied, and the benefits of applying it. We believe that the principles of BPM governance are essential to any organization. This chapter is intended for executives, project leaders, and BPM methodology experts, those responsible for governance, and process owners who are responsible for daily operations and are interested in BPM governance.

WHY IS BPM GOVERNANCE IMPORTANT?

Most business processes happen through serendipitous need. Action is needed, someone does something that works, and the organization accepts this as the way things should be done going forward. As such, many organizations do not fully understand how things happen nor have they considered alternative ways to improve their processes, especially in relation to their content and how it moves through the organization. To truly understand organizational processes and the impact they have on content or content with the process, you need to map the process and document the interaction with content.[2] BPM is a management practice that provides for governance of the business process environment toward the goal of improving agility and operational performance. BPM is a structured approach employing methods, policies, metrics, management practices, and software tools to manage and continuously optimize an organization's activities and processes.[3] This means that BPM goes to the root of organizational structures, methods, and operations, and it is important to ensure that the changes it initiates are the right ones.[4]

Besides the ability to govern the processes and provide governance methods, policies, and metrics to ensure BPM governance, BPM governance is required to link

5.4　BPM治理

Maria Hove, Gabriella von Rosing, Bob Storms

5.4.1　介绍

公司通过跨组织边界的活动的有效性和效率为客户和股东创造价值(价值流),这些活动通常被称为公司的"跨职能业务流程"。BPM扩展了业务和技术,并提供对流程的可视性和控制级别。为了优化和维持BPI,必须进行某种形式的治理,以创建正确的结构、指标、角色和职责,有助于评估、改进和管理公司端到端业务流程[1]。因此,在本节中,我们将从什么是BPM治理、为什么需要BPM治理、在哪里可以应用以及应用的好处等角度来关注BPM治理的概念。我们相信BPM治理的原则对于任何组织都是必不可少的。本节针对的对象包括:主管、项目领导和BPM方法理论专家、负责治理的人员以及负责日常运营并对BPM治理感兴趣的流程责任人。

5.4.2　为什么BPM治理很重要?

大多数业务流程都是通过偶然的需求实现的。行动是必要的,而且有人做了一些有用的事情,组织接受这一点作为今后应该做的事情的方式。因此,许多组织并不完全理解事情是如何发生的,也没有考虑改进其流程的其他方法,尤其是与内容以及如何在组织中移动相关的方法。要真正了解组织流程及其对流程内容的影响,您需要规划流程并记录与内容的相互关系[2]。BPM是一种管理实践,目标是提高敏捷性和操作性能,它为业务流程环境的治理提供支持。同时,BPM是一种结构化方法,它使用方法、策略、指标、管理实践和软件工具来管理和持续优化组织的活动和流程[3]。这意味着BPM是组织结构、方法和操作的根源,确保它发起的更改是正确的[4]。

除能够管理流程并提供治理方法、策略和评估以确保BPM治理之外,BPM治理还需要将日常流程治理与公司治理、价值治理、绩效治理和IT治理联系起来。

the daily process governance to corporate governance, value governance, performance governance, and information technology (IT) governance. As illustrated in Figure 1, BPM governance provides the connection to the other governance disciplines within an organization. It governs performance and conformance at an operational level, and at the strategic level focuses on value identification, planning, and creation. IT related activities of BPM governance focus cost reduction, whereas corporate aspects of BPM governance focus on activities to ensure revenue generation.

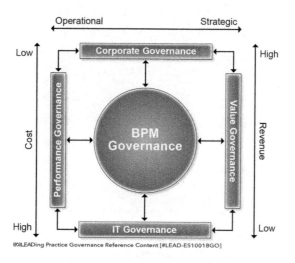

FIGURE 1

How business process management (BPM) governance connects to other governance disciplines, and the revenue and cost associated with them.[5]

WHAT IS BPM GOVERNANCE?

A business process is a continuous series of enterprise tasks undertaken for the purpose of creating output. Business processes enable the value chain of the enterprise and focus on the customer when the output is created. The purpose is to make the business process as significant as possible and to link it to multiple functions.[6] All companies have business processes, regardless of size or industry. When maintained and optimized, they will ensure competitiveness and survival in the marketplace.

Business processes have to be managed within an organization to enable and support a long-term business success. This means the continuous reassessment, realignment, and adaptation of business processes that enable strategic objectives to be implemented consistently and translated into everyday operational activity. It also means realignment and continuous adaptation of the related organizational and IT structures to meet the requirements of the market. Based on process analysis, it is possible to make the right decisions, significantly improve product and service quality, boost efficiency, and cut costs.[5]

如图1所示,BPM治理提供了与组织内其他治理规程的连接。它管理操作级别上的性能和一致性,在战略级别关注价值识别、规划和创建。BPM治理的IT相关活动侧重于降低成本,而BPM治理的企业方面则侧重于确保产生收入的活动。

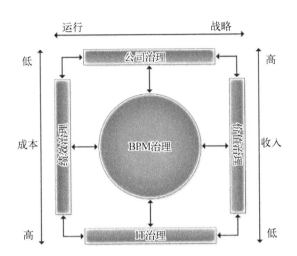

图1　BPM治理如何连接到其他治理规程,以及与之相关的收入和成本[5]

5.4.3　什么是BPM治理?

业务流程是为得到结果而执行的一系列连续的企业任务。业务流程支持企业的价值链,并在得到结果时关注客户,其目的是使业务流程尽可能重要,并将其链接到多个功能[6]。所有公司都有业务流程,无论规模大小或行业。当保持和优化时,它们将确保市场竞争力和生存。

业务流程必须在组织内进行管理,以支持长期的业务成功。这意味着必须对业务流程进行持续的重新评估、重新组合和调整,使战略目标能够转化为日常操作活动。它还意味着相关组织和IT结构的重新调整和持续适应,以满足市场的需求。基于过程分析,可以做出正确的决策,显著提高产品和服务质量、提高效率、降低成本[5]。

BPM is a management discipline that provides governance in a business environment, with the goal of improving agility and operational performance.[7] It is a structured approach employing methods, policies, metrics, management practices, and software tools to coordinate and continuously optimize an organization's activities and business processes. Its objective is to control and improve an organization's business through active, coordinated governance of all aspects of the specification, design, implementation, operation, measurement, analysis, and optimization of business processes to effectively and efficiently deliver business objectives.[5,8]

BPM CENTER OF EXCELLENCE AND GOVERNANCE

BPM governance supports the establishment of enterprise-wide processes, rules, methodologies, guidelines, tools, measures, and roles for BPM. For the most part, either BPM governance is found within the BPM center of excellence (CoE) or its concepts are applied within to oversee and manage projects throughout the entire BPM life cycle. Ensuring the operation of one's processes is consistent across the enterprise and reusable and efficient. The development and maintenance of policy, conventions, and standards for an enterprise-wide BPM approach are main tasks of the BPM CoE[5] and are executed through BPM governance (Figure 2).

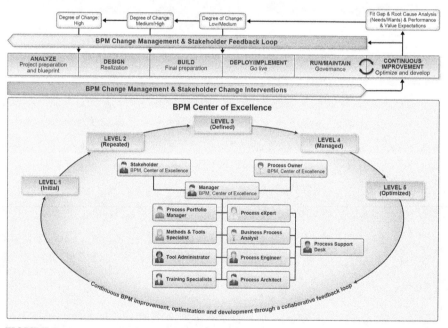

FIGURE 2

Business process management (BPM) governance applied throughout the BPM life cycle and done by the BPM CoE.[9]

BPM是一种在业务环境中提供治理的管理规程,其目标是提高敏捷性和操作性能[7]。它是一种结构化的方法,使用方法、策略、度量、管理实践和软件工具来协调和持续优化组织的活动和业务流程。其目标是通过对业务流程的规范、设计、实现、操作、测量、分析和优化的各个方面进行积极的、协调的治理来控制和改进组织的业务,从而有效地、高效地实现业务目标[5,8]。

5.4.4 BPM CoE 与治理

BPM治理为BPM建立企业范围的流程、规则、方法、指导方针、工具、度量和角色提供支持。在大多数情况下,BPM治理要么以BPM CoE为基础,要么将其概念应用于整个BPM生命周期中的项目监督和管理。确保流程的操作在整个企业中是一致的、可重用的和有效的,企业范围BPM方法的策略、约定及标准的开发和维护是BPM CoE[5]的主要任务,并通过BPM治理执行(图2)。

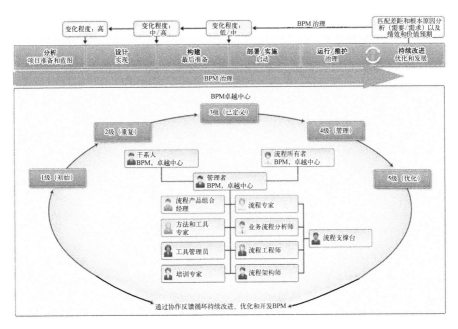

图2 由BPM CoE负责BPM治理应用于整个BPM生命周期[9]

Process governance consists of the set of guidelines and resources that an organization uses to facilitate collaboration and communication when it undertakes enterprise process initiatives, such as implementing a new contracts administration process or a new budgeting system. In this context, there are five basic steps for effective process governance[10]:

1. **Establish standards** for implementing new BPM projects. Examples of process standards include:
   a. Implementation methodology: Organizations should subscribe to a specific set of process implementation methodologies
   b. Process modeling notation: When modeling processes, it is important to standardize how different activities and events should be graphically represented
   c. Development platform: Implementing enterprise-wide process initiatives require organizations to standardize on a set of development tools
   d. Integration protocols: Enterprise-wide processes typically require tight integration with back-end data and other enterprise systems. It is important to decide early on which integration standards will be used to connect processes with other internal and external systems
2. **Prioritize BPM projects** so that you work on the most achievable ones first.
   a. Level of complexity: How complex is the proposed process? At the beginning of your process initiative it is important to establish quick wins. If possible, avoid implementing complex processes first
   b. Process reach and impact: How many people will the process impact and how much pain is caused by the current process? Initially, priority should be given to processes that represent the greatest impact to a small group of users
   c. Executive support: Is the process currently a hot topic in the executive boardroom? It is important to decide how priority will be given to processes that are widely supported by the executive team
   d. Subject matter expertise: How well documented is the process on paper and in carbon life forms? Priority should be given to processes that are well documented. You should also consider accessibility to subject matter experts
   e. Process selection guidelines should list and weight each criterion. This will create some level of transparency regarding how processes are selected for implementation. It is important to gain consensus throughout the organization on the slate of process selection criteria
3. **Clearly define the roles** and responsibilities of everyone involved in the BPM project. At a minimum, responsibilities should be defined for the following roles:
   a. Executive sponsor: As outlined in the previous section, the executive sponsor provides high-level visibility for the process initiative
   b. Process steward: The process steward provides guidance and direction for a particular process. This is typically the business unit manager or director who realizes a direct benefit or pain as a result of the process

流程治理由一组指导方针和资源组成,组织在执行企业流程计划(如实施新的合同管理流程或新的预算系统)时,使用这些指导方针和资源来促进协作和沟通。在这种情况下,有效流程管理有五个基本步骤[10]。

(1)为实施新的BPM项目建立标准,流程标准的例子包括如下几点。

a. 实现方法:组织应采用一套具体的流程实现方法。

b. 流程建模符号:在流程建模时,将不同活动和事件以标准化、图形化表示方式表示。

c. 开发平台:实现企业范围的流程活动需要组织对一组开发工具进行标准化。

d. 统一协议:企业范围的流程通常需要与后端数据和其他企业系统紧密集成。尽早决定将使用哪些集成标准将流程与其他内部和外部系统连接起来是很重要的。

(2)确定BPM项目的优先级,以便您首先处理最可实现的项目。

a. 复杂程度:所提议的流程有多复杂? 在您的流程初始阶段,快速取得成效是十分重要的。如果可能,首先要避免实现复杂的流程。

b. 流程实现和影响:流程将影响多少人,当前流程造成多少痛苦? 最初,应优先考虑对一小部分用户产生最大影响的流程。

c. 高管支持:该流程目前是否是高管关心的? 决定如何优先处理由执行团队广泛支持的流程是很重要的。

d. 专业知识:纸上的流程有多好的记录? 应优先考虑记录良好的过程。您还应该考虑到主题专家的可访问性。

e. 为每个流程确立和选择操作指南:操作指南应列出并权衡每个标准,这将为如何选择实施过程创造一定程度的透明度,在整个组织内就过程选择标准的制定达成一致意见是很重要的。

(3)明确定义BPM项目中涉及的每个人的角色和职责。至少应该为以下角色定义职责。

a. 执行发起人:如第(2)点所述,执行发起人为流程活动提供高级可见性。

b. 流程专员:流程专员为特定流程提供指导和指引。这通常是业务单元经理或主管,他们通过流程实现了直接的好处或坏处。

    c. Process manager: The process manager is the person charged with leading the implementation of a particular business process. This individual is held accountable for the success or failure of deploying the business process

    d. Functional lead: This individual is responsible for leading process analysis and requirements gathering. Functional leads oversee process discovery sessions with end users and managers, in addition to modeling the process and business rules

    e. Technical lead: The technical lead oversees implementation of technical components of the process. This includes system installation and configuration, application and forms development, and back-end integration

4. **Put someone in charge** with authority to enforce BPM governance rules

    a. Executive sponsorship is the single most important ingredient required for successful process governance. Without executive sponsorship, most enterprise-wide process initiatives lack a decisive voice capable of resolving process-related conflicts that arise during implementation. Within some organizations, the executive sponsor is a full-time vice president or manager who is accountable to the chief executive officer. However, in most cases the executive sponsor has a part-time role in addition to his or her regular full-time job duties. With either scenario, the executive sponsor must be empowered to enforce agreed-upon governance rules and should have budget authority for process initiatives

5. **Establish a BPM CoE** to ensure that Steps 1–4 are followed on every initiative. These CoEs serve as internal practices that support deployment of enterprise-wide business processes. BPM CoEs are usually chartered to accomplish the following objectives:

    a. Prioritize and implement processes: The CoE works with the executive sponsor and business managers to identify and prioritize process projects. Process selection guidelines should be developed to help the BPM CoE rank the processes to be implemented. After processes have been prioritized, the CoE can focus on its primary objective: developing and deploying processes

    b. Maintain the process portfolios: This consists of maintaining knowledge and documentation captured for each process. This knowledge is often contained in process requirements documents, training manuals, and project plans. The CoE is tasked with maintaining these artifacts in a physical library or within a virtual knowledge base

    c. Establish process practices: Upon completion of each deployment, the BPM CoE conducts post-project reviews and identifies lessons learned. These reviews are used to establish leading, or at the least, best practices that can be applied to future process implementations

    d. Evaluate process performance: Working with the process steward and executive sponsor, the CoE periodically evaluates the effectiveness of deployed processes. Process effectiveness is evaluated based on key performance metrics established before process deployment

c. 流程经理：流程经理是负责领导特定业务流程实现的人员。这个人对部署业务流程的成功或失败负责。

d. 职能领导：负责领导过程分析和需求收集。功能领导除了对流程和业务规则建模，还负责监督与最终用户和管理人员的流程发现会话。

e. 技术领导：技术领导监督过程中技术组件的实施。这包括系统安装和配置、应用程序和表单开发以及后端集成。

（4）让专人负责实施BPM治理规则。

高管支持是成功的流程治理所需要的最重要的因素。如果没有执行人员的支持，大多数企业内的流程活动都缺乏解决在实现过程中出现的流程冲突的能力。在一些组织中，执行发起人是对CEO负责的全职副总裁或经理。然而，在大多数情况下，执行发起人除了他或她的全职工作职责，还有一个兼职的角色。对于任何一种场景，执行发起人都必须被授权来执行商定的治理规则，并且应该拥有流程计划的预算权限。

（5）建立BPM CoE，以确保每个计划都遵循步骤（1）~（4）。这些CoE作为支持企业范围业务流程部署的内部实践，BPM CoE通常用于实现以下目标。

a. 流程的优先级和实现：CoE与执行发起人和业务经理一起确定流程项目的优先级。应该开发并使用流程选择指南以帮助BPM CoE对要实现的流程进行优先级排序。在确定流程的优先级之后，CoE可以将重点放在其主要目标上：开发和部署流程。

b. 维护流程组合：这包括维护每个流程的知识和文档。这些知识通常包含在流程需求文档、培训手册和项目计划中，CoE的任务是在物理库或虚拟知识库中维护这些制品。

c. 建立流程实践：在每次部署完成后，BPM CoE进行项目后审查，并确定经验教训。这些评审用于建立可应用于未来流程实现的领先或至少是最佳实践。

d. 评估流程绩效：与流程管理员和执行发起人合作，通过CoE定期评估已部署流程的有效性。根据流程部署前建立的KPI评估流程有效性。

BPM CoEs are growing in popularity as many organizations begin to expand departmental BPM initiatives to encompass the enterprise. CoE are most appropriate for organizations looking to deploy three or more processes that will need to interact with multiple departments. Working closely with the executive sponsor, the BPM CoE should be assigned responsibility for defining and enforcing process governance rules. Once process governance rules have been established, these rules should be institutionalized and automated by the CoE.

HOW DOES BPM GOVERNANCE WORK?

In practice, BPM governance works on two levels: program and project.

At the program level, whereas executive management sets the strategic direction for the BPM initiative, the governance body is responsible for its measurement and enforcement, ensuring alignment using approved frameworks and tools, and managing the road map and selection of projects to meet strategic goals. At the project level, BPM governance directs the project management of solution delivery to work within an engineering and project framework.[4]

BPM governance directs both the conformance and performance aspects in alignment within enterprise governance.

Performance includes:

- Maximizing economies of scale across coexisting processes and value streams
- Efficiency gains (processing time, costs, and throughput capability)
- Customer satisfaction/customer experience
- BPM program/project management (benefits realization and value management)
- Effective and efficient use of resources (e.g., human, financial, assets)

Conformance includes:

- Process compliance (standardization)
- Compliance with enterprise policies and external regulatory standards
- Risk management and control
- Integration of BPM governance across the organization (coexistence with other governances)
- BPM guiding principles and standards
- Alignment of BPM governance with business architecture (e.g., processes mapped to value streams; processes linked to business competencies)

This is accomplished within a continuous improvement life cycle methodology, ensuring BPM governance is relevant, effective, and always in alignment with the strategic objects of the organization.[5] It includes the governance of business process improvement using BPM, as well as governance of the implementation and adoption of BPM in the organization. The life cycle can be described as follows:

随着许多组织开始将部门BPM活动扩展到整个企业,BPM CoE越来越受欢迎。CoE最适合部署在三个或多个流程的组织中需要与多个部门交互的场景。通过与执行发起人紧密合作,BPM CoE应该被分配定义和实施流程治理规则的职责。流程治理规则一旦建立,CoE就应该对这些规则进行制度化和自动化。

5.4.5 如何开展BPM治理工作?

在实践中,BPM治理在两个级别上工作:程序和项目。

在项目层面,虽然管理层为执行BPM计划设定了战略方向,但负责其测量和执行的是治理机构,治理机构确保使用批准的框架和工具保持一致,并管理路线图和项目选择以实现战略目标。在项目级别方面,BPM治理可以指导项目管理解决方案在工程和项目框架内工作和交付[4]。

BPM治理指导企业治理中的一致性和绩效方面。

绩效方面包括:

- 在共存的流程和价值流中最大化实现规模经济;
- 效率提高(处理时间、成本和吞吐量能力);
- 客户满意度/客户体验;
- BPM计划/项目管理(利益实现和价值管理);
- 有效和高效地利用资源(如人力、金融、资产)。

一致性包括:

- 流程合规性(标准化);
- 遵守企业政策和外部监管标准;
- 风险管理和控制;
- 跨组织整合BPM治理(与其他治理共存);
- BPM指导原则和标准;
- 将BPM治理与业务架构协调一致(如映射到价值流的流程、与业务能力关联的流程)。

这是在持续改进生命周期方法中完成的,确保BPM治理是相关的、有效的,并且始终与组织的战略目标保持一致[5]。它包括使用BPM治理对业务流程改进,以及对组织中BPM治理的实现和使用。生命周期可以描述如下。

Analyze:

- Analyze as-is state of BPM CoE governance (e.g., methodology, organizational structure) and maturity of business processes
- Identify BPM CoE governance objectives and requirements
- Assess alignment of business processes to strategic business objectives and critical success factors of organization
- Identify present pain points, bottlenecks, and weakness clusters
- Analyze the structure and efficiency of current business processes (e.g., process architecture, process flows, times, and costs)
- Assess current state of process accountabilities and compliance with standards

Design:

- Define to-be state of BPM governance CoE (e.g., framework/model, structure)
- Assess BPM CoE resource capacity
- Define process control objectives and practices
- Define targets to support strategic business objectives and critical success factors of organization
- Define enterprise-wide process architecture
- Define BPM standards and policies
- Define measurements and monitoring methods (benefits realization and value management)

Build:

- Develop accountability framework (e.g., Responsible, Accountable, Consulted, Informed (RACI))
- Develop BPM policies and standards (modeling approach and best practices)
- Select process improvement and re-engineering projects (e.g., Six Sigma), grouping projects into BPM programs
- Create to-be state of selected processes (e.g., optimization, agility)
- Allocate BPM CoE governance resources
- Create BPM CoE governance structures

Deploy/implement:

- Implement BPM CoE governance structures
- Direct process improvement projects and programs, including automation opportunities (e.g., using process reference models)
- Implement BPM standards, policies, and procedures
- Deploy BPM CoE change deployment management, including communications and training

Run/maintain:

- Monitor and govern compliance and accountability (audit)
- Monitor effectiveness of BPM CoE governance structures

分析：

- 分析BPM CoE治理的现状（如方法、组织结构）和业务流程的成熟度；
- 确定BPM CoE治理目标和需求；
- 评估业务流程与SBO和组织CSF的一致性；
- 确定当前的痛点、瓶颈和弱点集群；
- 分析当前业务流程的结构和效率（如流程架构、流程流、时间和成本）；
- 评估流程的当前状态和符合标准的状况。

设计：

- 定义BPM治理CoE的未来状态（如框架/模型、结构）；
- 评估BPM CoE资源能力；
- 定义过程控制目标和实践；
- 定义目标以支持SBO和组织的CSF；
- 定义企业范围的流程体系结构；
- 定义BPM标准和策略；
- 定义度量和监视方法（利益实现和价值管理）。

构建：

- 制定问责制的RACI框架［如责任（responsible）、负责（accountable）、咨询（consulted）、知情（informed）］；
- 制定BPM政策和标准（建模方法和最佳实践）；
- 选择流程改进和重新设计项目（如六西格玛），将项目分组为BPM项目；
- 创建选定流程的状态（如优化、敏捷性）；
- 分配BPM CoE治理资源；
- 创建BPM CoE治理结构。

部署/实现：

- 实现BPM CoE治理结构；
- 指导过程改进项目和程序，包括自动化机会（如使用流程参考模型）；
- 实现BPM标准、策略和流程；
- 部署BPM CoE变更部署管理，包括通信和培训。

运行/维护：

- 监督和管理合规和问责制（审计）；
- 监控BPM CoE治理结构的有效性；

- Monitor process effectiveness and efficiency (e.g., resource use, costs, processing times) through qualitative and quantitative measures against targets (key performance indicators, Service Level Agreements (SLAs))
- Monitor success of process improvement projects (benefits realization)

Continuous improvement:

- Identify processes with potential for redesign, improvement, and greater productivity
- Identify opportunities to streamline value streams
- Identify/eliminate process redundancy
- Identify opportunities to improve accountability and compliance
- Identify opportunities to improve BPM standards, policies, procedures, and enterprise-wide process architecture
- Identify opportunities to optimize BPM CoE governance structures
- Identify opportunities to optimize resource use
- Modify targets to support changes in strategic business objectives and critical success factors of organization

BPM GOVERNANCE AND INCIDENT MANAGEMENT

In the run and maintain phase of the BPM governance and continuous improvement life cycle (Figure 3), incident management is an important service that supports process governance and control. Incident management should be used for handling

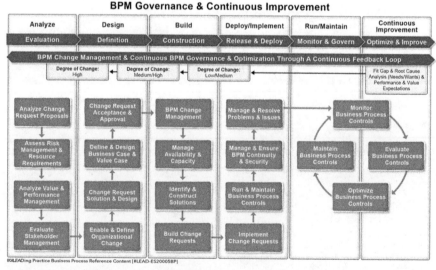

FIGURE 3

The Business process management (BPM) governance and continuous improvement model.[11]

- 通过针对目标(KPI、SLA)的定性和定量措施,监控流程有效性和效率(如资源使用、成本、处理时间);
- 监控流程改进项目的成功(利益实现)。

持续改进:
- 识别具有重新设计、改进和提高生产力潜力的流程;
- 识别机会,使价值流流线化;
- 识别/消除流程冗余;
- 确定改善问责制和合规的机会;
- 确定改进BPM标准、策略、流程和企业范围流程体系结构的机会;
- 确定优化BPM CoE治理结构的机会;
- 确定优化资源使用的机会;
- 修改目标以支持SBO和组织CSF的变更。

5.4.6 BPM治理和事件管理

在BPM治理和持续改进生命周期的运行和维护阶段(图3),事件管理是支持流程治理和控制的重要服务。事件管理应用于处理所有与人工/劳动密集型和自

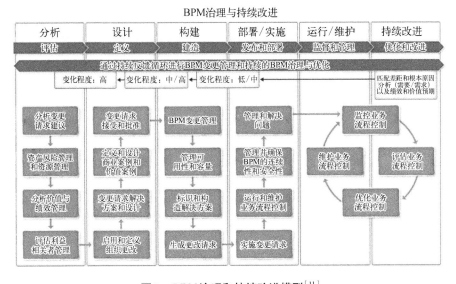

图3 BPM治理和持续改进模型[11]

all incidents related to both manual/labor-intensive and automated processes. This includes requests for BPM projects (process improvement/redesign), process failures, degradation of services (value streams), and process questions and concerns reported by users or technical staff, or automatically detected and reported by event monitoring tools. In the context of BPM governance the primary goal of the incident management process is to restore normal operations as quickly as possible and minimize the adverse impact on services. BPM governance also uses the information created by incident management to monitor and direct performance and conformance governance aspects. As illustrated in Figure 4, there are there seven main phases of incident management,[12] all of which are interlinked with the BPM CoE, BPM roles, BPM life cycle, process templates, and BPM change management.

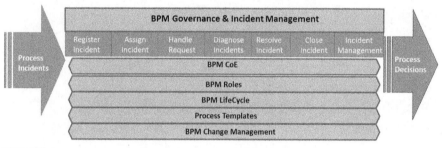

FIGURE 4

Incident management phases are interlinked with various business process management (BPM) concepts.

BPM PORTFOLIO MANAGEMENT AND GOVERNANCE

Process portfolio management has a pivotal role in successful BPM (Figure 5) for a number of reasons. It provides an approach or mind set that is essential in directing limited resources in terms of funds, people, etc., into the processes with the highest demand for an increased process orientation. In the true sense of a balanced portfolio, process portfolio management can be used to diversify the BPM governance activities, leading to parallel projects in different stages of the business process life cycle. In summary, process portfolio management marks the difference between the isolated and uncoordinated improvement and management of a single process and the holistic process-based management of an organization.[13]

In the BPM portfolio management cycle, the following general questions should be asked for every item in your organization's process portfolio[14]:

- Does this process align with the organization's strategic objectives and goals?
- Based on defined key performance indicators and probable risks, does this process initiative meet the requirements for portfolio inclusion?

动化流程相关的事件。这包括对BPM项目(流程改进/重新设计)的申请、流程失败、服务降级(价值流),以及由用户或技术人员报告的流程问题和关注点,或由事件监控工具自动检测和报告的流程问题和关注点。在BPM治理背景中,事件管理流程的主要目标是尽快恢复正常操作,并将对服务的负面影响降到最低。BPM治理还使用事件管理创建的信息来监视和指导性能和一致性的治理方面。如图4所示,事件管理有7个主要阶段[12],其中都与BPM CoE、BPM角色、BPM生命周期、流程模板和BPM变更管理相关联。

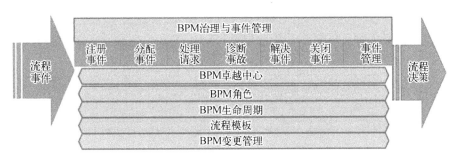

图4　事件管理阶段与各种BPM概念相互关联

5.4.7　BPM项目组合管理和治理

由于许多原因,流程组合管理在成功的BPM中扮演着关键角色(图5)。它提供了一种方法或思维模式,这种方法或思维模式对于将资金、人员等方面的有限资源引导到对增加的流程导向有最高需求的流程中至关重要。在平衡的投资组合的真正意义上,流程投资组合管理可用于使BPM治理活动多样化,从而在业务流程生命周期的不同阶段产生并行项目。总之,流程项目组合管理标志着单个过程的孤立和不协调的改进和管理与组织的基于过程的整体管理之间的区别[13]。

在BPM项目组合管理周期中,您的组织流程项目组合中的每一个项目都应该被问到以下一般性问题[14]:

- 这个流程是否与组织的战略目标一致?
- 基于定义的KPI和可能的风险,这个流程主动性是否满足投资组合包含的要求?

FIGURE 5

Process portfolio management in the context of BPM governance activities.[15]

- Does it still meet the requirements for portfolio inclusion compared with other competing process portfolio items also in consideration?

Many organizations do not have a portfolio management process. To create a process where none exists, the right steps must be taken.

1. Define a process portfolio management structure (organizational chart)
2. Define a process portfolio management plan
3. Define key performance indicators
4. Define process performance indicators

The following are essential steps or activities for process portfolio management, although more process mature organizations may introduce additional steps into the process:

1. Identify process portfolio items
2. Define process portfolio items
3. Evaluate process portfolio items
4. Select your process portfolio
5. Reassess process portfolio
6. Approve process portfolio
7. Transition to process initiative
8. Communicate. Track and report
9. Accommodate and adjust for changes

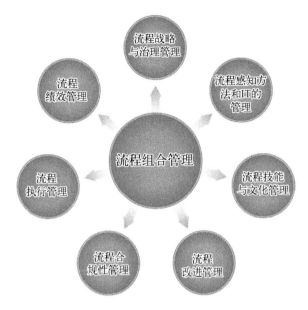

图5　BPM治理活动背景下的流程组合管理[15]

- 与同样在考虑中的其他相互竞争的流程项目组合相比,它是否仍然满足项目组合包含的要求?

许多组织没有项目组合管理流程。要在不存在流程的地方创建流程,必须采取正确的步骤。

(1)定义流程项目组合管理结构(组织图);

(2)定义流程项目组合管理计划;

(3)定义主要表现指标;

(4)定义流程性能指标。

以下是流程组合管理的基本步骤或活动,尽管流程成熟的组织可能会在流程中引入如下其他步骤。

(1)确定流程组合项。

(2)定义流程组合项。

(3)评估流程组合项。

(4)选择流程组合。

(5)重新评估流程组合。

(6)批准流程组合。

(7)向流程的主动性过渡。

(8)沟通。跟踪和报告。

(9)适应和调整变化。

Although many organizations have significantly matured in their understanding of the opportunities and constraints of BPM, many lack well-defined accountability for their entire portfolio of business processes. In the same way that a manager in charge of strategic marketing is not the product manager, there will be a clear distinction between the role of the centralized process portfolio manager and the duties of decentralized process owners. In terms of BPM governance, a process portfolio manager can be seen as an additional role within the corporate BPM team. The process portfolio manager will be measured by his or her ongoing awareness of the entire range of an organization's business processes and the capability of allocating BPM resources to the most promising processes. The more advanced the organization is in its BPM maturity—i.e., the more it moves from simple overview architectures to a widely populated model repository, even with information about actual process executions—the more critical it will be to shift the initial focus on individual processes to a view that manages the entire landscape of organizational business processes.[16]

LESSONS LEARNED

The organization-wide adoption of BPM typically goes through multiple stages,[17] which can be described as follows:

1. Awareness of the benefits and methodologies of BPM
2. Desire to adopt BPM based on a business-related driver (e.g., mergers, IT system implementations, new business segments, cost reduction activities)
3. Set up, execute, and monitor individual BPM projects
4. Convert from multiple BPM projects to a governing and more centralized BPM program
5. Productize BPM through a BPM CoE

To reach this goal of a holistic BPM approach, the same elements have to be in place that are necessary to run every process efficiently[5]:

- Definition and understanding of the BPM process
- Clear objectives regarding the outcome and benefits of the BPM process
- A BPM organization and a BPM CoE with appropriate knowledge, roles, and responsibilities
- A BPM methodology, BPM standards, and BPM service offerings
- A mature BPM technology and tools that optimally support an efficient BPM approach across the organization.

There are five main responsibilities of a centralized and enterprise-wide BPM CoE[5]:

- BPM leadership
- Regulatory framework
- Project support
- Training and communication
- Process controlling

尽管许多组织在理解 BPM 的机会和限制方面已经非常成熟,但许多组织对其整个业务流程组合缺乏明确的问责制。正如负责战略推广的经理不是产品经理一样,集中化流程组合经理的角色和分散流程所有人的职责之间也将有明显的区别。就 BPM 治理而言,流程组合经理可以被视为企业 BPM 团队中的一个额外角色。流程组合经理将通过他或她对组织的整个业务流程范围的持续认识以及将 BPM 资源分配给最有希望的流程的能力来度量。组织的 BPM 成熟度越高,它从简单的概述架构转移到广泛填充的模型存储库(即使包含关于实际流程执行的信息)的程度越高,就越需要将最初的关注点从单个流程转移到管理整个组织业务流程的视图[16]。

5.4.8　经验教训

整个组织对 BPM 的采用通常经历多个阶段[17],可以描述如下:

(1)了解 BPM 的好处和方法;

(2)希望采用基于业务相关驱动程序的 BPM(如合并、IT 系统实现、新业务部门、降低成本活动);

(3)设置、执行和监视各个 BPM 项目;

(4)从多个 BPM 项目转换为一个治理和更集中的 BPM 程序;

(5)通过 BPM CoE 产品化 BPM。

为了实现全面 BPM 方法的这一目标,必须具备有效运行每个流程所必需的相同元素[5]:

- BPM 流程的定义和理解;
- 关于 BPM 流程的结果和好处的明确目标;
- 拥有适当知识、角色和职责的 BPM 组织和 BPM CoE;
- BPM 方法、BPM 标准和 BPM 服务产品;
- 成熟的 BPM 技术和工具,最优地支持整个组织的高效 BPM 方法。

集中式和企业级 BPM CoE[5]有 5 个主要职责:

- BPM 的领导;
- 监管框架;
- 项目支持;
- 培训与沟通;
- 过程控制。

Rosemann proposed analyzing and managing the BPM services portfolio offered by the BPM CoE based on the two dimensions of demand and capability.[9] Demand reflects the current organizational needs and appetite for a specific BPM service. The capabilities describe the readiness of the BPM CoE to provide a certain service. This dimension reflects the accumulated knowledge, skills, and experience of the BPM CoE, as well as the technological capacities to successfully deliver.[5]

To implement a BPM CoE within an organization to support a BPM governance approach, specific service offerings for stakeholders should be defined, along with internal roles and responsibilities. A service offering is a combination of methodology, tools, and communication activities that together address a strategic BPM target field of the organization. The organization's BPM target fields should be analyzed and prioritized first to identify the necessary BPM service offerings. Every target field (e.g., strategic decision support, Information Technology Infrastructure Library (ITIL) implementation and review, IT system implementation, cost reduction initiatives, or introduction of new products) has to be identified and described.[5]

The greatest value from BPM is found in automating enterprise-wide value streams, and this is where good governance becomes critical. Unfortunately, a number of challenges commonly stand in the way of this kind of cross-functional, enterprise-wide governance, including the following[4]:

- Organizational structures are typically organized by functional silos
- Related to the organizational concern, there is typically a resistance to change and even to cross-functional cooperation
- Formal governance frameworks for this scope are rare
- There is inadequate infrastructure and tools support

It is necessary to understand the scope of BPM governance in relation to other governance concerns and practices to avoid duplication or conflict, but also where appropriate to integrate them for greatest continuity.[4]

BENEFITS AND VALUE OF BPM GOVERNANCE

A Gartner research note stated that 80% of enterprise companies conducting BPM projects will experience a return on investment (ROI) greater than 15%. The survey looked at responses from 20 companies that had completed 154 BPM projects, and 95% of the companies experienced more than a 90% success rate among their BPM projects. All successful projects had an ROI greater than 10%, Gartner found. Seventy-eight percent of the respondents had ROI rates greater than 15%.[18] Organizations with an identified BPM CoE and governance in place can achieve a five times greater ROI over those with no CoE or dedicated process team.[19] Similarly, those with a dedicated business process team in place reported nearly twice the ROI of those without any dedicated team in place.[5,11]

BPM governance is an essential program overlay activity that enforces standards, engineering practices, organization structures, roles, and responsibilities, to measure, manage, and improve the effectiveness of BPM itself. It ensures that your BPM

Rosemann提出基于需求和能力两个维度分析和管理BPM CoE提供的BPM服务组合[9]。需求反映了当前组织对特定BPM服务的需求和喜好。这些功能描述了BPM CoE提供特定服务的准备情况。这个维度反映了BPM CoE所积累的知识、技能和经验，以及成功交付的技术能力[5]。

要在组织内实现BPM CoE以支持BPM治理方法，应该定义面向利益相关者的特定服务供给以及内部角色和职责。服务供给是方法、工具和通信活动的组合，它们共同处理组织的战略性BPM目标领域。应该首先分析组织的BPM目标字段并确定其优先级，以确定必要的BPM服务提供。每个目标领域（如战略决策支持、ITIL的实施和审查、IT系统的实施、降低成本的举措或新产品的引进）都必须被识别和描述[5]。

BPM最大的价值在于将企业范围的价值流自动化，而这正是良好治理变得至关重要的地方。不幸的是，这种跨功能的、企业范围的治理通常面临许多挑战，包括以下方面[4]：

- 组织结构通常由功能孤岛组成；
- 与组织的关注点相关，通常存在对变更甚至跨职能合作的抵制；
- 用于此范围的正式治理框架非常少见；
- 基础设施和工具支持不足。

我们有必要了解BPM治理的范围与其他治理关注点和实践的关系，以避免重复或冲突，并在适当的情况下集成它们以实现最大的连续性[4]。

5.4.9　BPM治理的好处和价值

Gartner的一份研究报告指出，执行BPM项目的80%的企业将体验到超过15%的ROI。该调查查看了完成154个BPM项目的20家公司的回复，其中95%的公司的BPM项目成功率超过90%。Gartner发现，所有成功的项目的ROI都超过10%。78%的受访者的ROI超过15%[18]。拥有确定的BPM CoE和治理的组织可以比没有CoE或专门流程团队的组织获得5倍的ROI[19]。类似地，那些拥有专门的业务流程团队的企业报告的ROI几乎是没有专门团队的企业的两倍[5,11]。

BPM治理是一个基本的程序覆盖活动，它强制执行标准、工程实践、组织结构、角色和职责，以度量、管理和改进BPM本身的有效性。它确保您的BPM活动

initiative is aligned with your corporate strategies and objectives and that BPM delivers measurable value.[4] BPM governance will help the organization to find and reinforce its control structure and thereby develop its agility and organizational effectiveness.

CONCLUSIONS

In this chapter we have focused on BPM governance and how it is a top priority for executives and BPM CoE. We covered what BPM governance is, why it is important, and how and where it can or should be applied. Furthermore, we detailed typical phases and tasks of BPM governance and illustrated lessons learned in terms of what works well and what does not. We ended with the benefits and value of BPM governance, which is critical for monitoring enterprise-wide and automated value streams.

End Notes

1. BPM Governance, Andrew Spanal, http://www.bpminstitute.org/resources/articles/bpm-governance.

2. *How to Unclog Your Business by Automating Content-Intensive Process* (AIIM Training) http://www.trindocs.com/Portals/3/HowtoUnclogYourBusiness.pdf.

3. Gartner report, *Business Process Management: Preparing for the Process-Managed Organization* (2005).

4. Oracle Practitioner Guide—A Framework for BPM Governance, Release 3.0, E24090-03 (August 2011).

5. LEADing Practice Governance Reference Content #LEAD-ES10018GO.

6. Hammer and Champy, *Reengineering the Corporation: A Manifesto for Business Revolution* (New York: HarperBusiness, 1993).

7. Melenovsky M. J., *Business Process Management as a Discipline*, Gartner Research, ID Number: G00139856 (August 1, 2006).

8. Brabaender E. and Davis R., *ARIS Design Platform—Getting started with BPM* (Springer: London, 2007).

9. LEADing Practice Business Process Reference Content #LEAD-ES20005BP.

10. Clay Richardson, "Process Governance Best Practices: Building a BPM Center of Excellence," (2006).

11. LEADing Practice Governance Reference Content #LEAD-ES10018GO.

12. Lisa Callihan, "Incident Management," from http://www.mais.umich.edu/methodology/service-management/incident-management.html.

13. Michael Rosemann, "Process Portfolio Management" (2006).

14. *Essential Steps to Building a Profitable Portfolio Process*, See more at: http://epmlive.com/portfolio-management/essential-steps-to-building-a-profitable-portfolio-process/#sthash.yxwwdMs3.dpuf.

15. LEADing Practice Portfolio Management Reference Content #LEAD-ES10019AL.

16. See the note 13 above.

17. Rosemann M., *ARIS TV—Episode 6—What are your BPM Services?* (2008) http://www.youtube.com/watch?v=LQ1ZqUq9q-k&feature=channel_page.

18. Dubie D., BPM and ROI (2004) NetworkWorld.com http://www.networkworld.com/weblogs/management/005640.html accessed June 03, 2008.

19. Nathaniel Palmer, "Introduction: Workflow and BPM in 2007: Business Process Standards See a New Global Imperative," *2007 BPM & Workflow Handbook*.

与您的企业战略和目标保持一致，并且BPM交付可度量的价值[4]。BPM治理将帮助组织找到并加强其控制结构，从而开发其敏捷性和组织有效性。

5.4.10　结论

在本节中，我们重点讨论了BPM治理以及它如何成为高管和BPM CoE的最高优先级。我们讨论了BPM治理是什么、为何它如此重要，以及如何及应该在何处应用它。此外，我们详细描述了BPM治理的典型阶段和任务，并说明了从哪些工作良好，哪些工作不佳方面获得的经验教训。我们以BPM治理的好处和价值为结束，BPM治理对于监控企业级和自动化的价值流非常重要。

---------- 参考文献 ----------

[1] BPM Governance, Andrew Spanal, http://www.bpminstitute.org/resources/articles/bpm-governance.
[2] *How to Unclog Your Business by Automating Content-Intensive Process* (AIIM Training) http://www.trindocs.com/Portals/3/HowtoUnclogYourBusiness.pdf.
[3] Gartner report, *Business Process Management: Preparing for the Process-Managed Organization* (2005).
[4] Oracle Practitioner Guide—A Framework for BPM Governance, Release 3.0, E24090-03 (August 2011).
[5] LEADing Practice Governance Reference Content #LEAD-ES10018GO.
[6] Hammer and Champy, *Reengineering the Corporation: A Manifesto for Business Revolution* (New York: Harper Business, 1993).
[7] Melenovsky M. J., *Business Process Management as a Discipline*, Gartner Research, ID Number: G00139856 (August 1, 2006).
[8] Brabaender E. and Davis R., *ARIS Design Platform—Getting started with BPM* (Springer: London, 2007).
[9] LEADing Practice Business Process Reference Content #LEAD-ES20005BP.
[10] Clay Richardson, "Process Governance Best Practices: Building a BPM Center of Excellence," (2006).
[11] LEADing Practice Governance Reference Content #LEAD-ES10018GO.
[12] Lisa Callihan, "Incident Management," from http://www.mais.umich.edu/methodology/service-management/incident-management.html.
[13] Michael Rosemann, "Process Portfolio Management" (2006).
[14] *Essential Steps to Building a Profitable Portfolio Process*, See more at: http://epmlive.com/portfolio-management/essential-steps-to-building-a-profitable-portfolio-process/#sthash.yxwwdMs3.dpuf.
[15] LEADing Practice Portfolio Management Reference Content #LEAD-ES10019AL.
[16] See the note 13 above.
[17] Rosemann M., *ARIS TV—Episode 6—What are your BPM Services?* (2008) http://www.youtube.com/watch?v=LQ1ZqUq9q-k&feature=channel_page.
[18] Dubie D., BPM and ROI (2004) Network World.com http://www.networkworld.com/weblogs/management/005640.html accessed June 03, 2008.
[19] Nathaniel Palmer, "Introduction: Workflow and BPM in 2007: Business Process Standards See a New Global Imperative," *2007 BPM & Workflow Handbook*.

Business Process Portfolio Management

Mark von Rosing, Hendrik Bohn, Gabriel von Scheel, Richard Conzo, Maria Hove

FROM BUSINESS PROCESS MANAGEMENT TO BUSINESS PROCESS PORTFOLIO MANAGEMENT

In today's society, work environment and customers' expectations change on a daily basis. Consequently, it is crucial for the modern enterprise to find a way to adapt itself to new requirements.[1]

After more than 25 years of existence, business process management (BPM) has become the de facto standard in eliminating historically grown and chaotic business procedures by organizing them in a structured, transparent, and standardized way.[2] The growth in acceptance of BPM has been supported by a wide range of tools such as BPM suites, which are entering the plateau of productivity on the Gartner Hype Cycle.[3] This, in turn, will increase the widespread adoption of BPM and lift the BPM maturity in organizations. However, with increasing maturity, organizations face the next challenge.[4] How can they manage the entire set of processes, improve them as a whole, or decide on which processes their limited resources should be deployed?

Organizations apply various process re-engineering, process innovation, optimization, and management techniques to improve their business processes. The impetus for these improvements can range from a specific business problem to a management directive or a need to reduce costs and improve efficiency. Organizations typically undertake periodic process improvements that are focused on specific business processes and may or may not align with the business strategy. Too often, once a project is completed, management attention goes elsewhere and things revert to the way they were. The value realization potential is not used enough, resulting in high cost and low value for the organization. Often the anticipated benefits are not realized or even audited to see whether the goals were reached. Similarly, many pitfalls can appear when process improvement is attempted one process at a time. In these cases, it is difficult to tell which processes contribute the most to achieving the business objectives, which process is the critical process to improve, or which processes are interdependent and therefore influence each other.

To realize all of the benefits of sometimes disparate BPM efforts, there needs to be an ongoing, organization-wide effort to assess and measure the results and continue to use the successful implementations. This organizational wide effort needs to be governed and controlled through a BPM Center of Excellence (CoE) already described elsewhere. Emphasis should be on what needs to be done rather

The Complete Business Process Handbook. http://dx.doi.org/10.1016/B978-0-12-799959-3.00030-6

5.5   BPPM

Mark von Rosing, Hendrik Bohn, Gabriel von Scheel, Richard Conzo, Maria Hove

5.5.1   从BPM到BPPM

在当今社会,工作环境和顾客的期望每天都在变化。因此,对于现代企业来说,找到一种方法来适应新的需求是至关重要的[1]。

经过超过25年的发展,BPM通过以结构化、透明和标准化的方式组织业务流程[2],已经成为消除历史上增长和混乱的业务流程的标准。在BPM套件等很多工具的支持下,BPM被广泛接受,它们在Gartner的宣传中进入了快速发展阶段[3]。然而,随着成熟度的提高,组织将面临下一个挑战[4]:他们如何管理整个流程集,将其作为一个整体进行改进?或者决定应该部署哪些流程?

组织应用各种流程再造、流程创新、优化和管理技术来改进其业务流程。这些改进的动力来自具体的业务问题、管理指令或降低成本和提高效率的需要。组织通常进行定期的流程改进,这些改进集中于特定的业务流程,并且可能与业务策略保持一致,也可能不一致。通常,一旦项目完成,管理人员的注意力就会转移到其他地方,事情就会恢复到原来的样子,因此,价值实现潜力不足,导致组织成本高、价值低。通常预期的好处没有实现,甚至没有进行评估以确定是否达到了目标。类似地,当一次尝试一个流程改进时,可能会出现许多缺陷。在这些情况下,很难判断哪些流程对实现业务目标的贡献最大,哪些流程是需要改进的关键流程,或者哪些流程是相互依赖、相互影响的。

为了充分实现BPM的所有好处,需要持续的、组织范围的工作来评估和度量结果,并持续推动其成功。这种组织范围的工作需要通过BPM CoE进行治理和控制,该中心已经在其他地方进行了描述。重点应该放在需要做什么,而不是在哪里

than the location of where it is done, such as a department or area of the company. The processes need to be managed end-to-end to optimize the different silos in an organization where processes normally exist. There is a good chance that in most organizations no one person exists who manages the process end-to-end or even understands the process in that context.

Value chain and value network concepts are approaches to provide a more high-level view on value generation and improvement, linking sequential activities to business strategies and outcomes.[5] However, they often do not allow drilling down or breaking down to of specific business processes where the value is created, and thus lack support for process and inter-process relationships.[6] Business process portfolio management (BPPM) is the solution as well as the concept that should be applied, as described in subsequent sections.

COMMON PITFALLS WHEN IMPLEMENTING BPPM

Most organizations, both private and public sector, experience a number of pitfalls when implementing and sustaining successful BPPM, as shown in Figure 1.

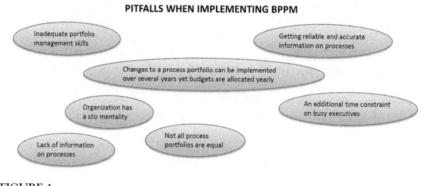

PITFALLS WHEN IMPLEMENTING BPPM

- Inadequate portfolio management skills
- Getting reliable and accurate information on processes
- Changes to a process portfolio can be implemented over several years yet budgets are allocated yearly
- Organization has a silo mentality
- An additional time constraint on busy executives
- Lack of information on processes
- Not all process portfolios are equal

FIGURE 1

Common pitfalls when implementing business process portfolio management (BPPM).

Not All Process Portfolios are Equal

BPPM ought to take place wherever investment decisions are being made; however, the views of the portfolio and the questions that need to be answered will depend on whose perspective affects the decisions. For example, a portfolio of enterprise processes will need much more rigorous evaluation from a number of perspectives than, for example, a portfolio of behind-the-scenes information technology (IT) processes. An enterprise's alignment to corporate strategy may be a score used to determine value at the board level, whereas alignment to Information System (IS)/Information Technology (IT) strategy will be of primary interest to the chief information officer.

做,如公司的一个部门或区域。需要对流程进行端到端的管理,以优化组织中通常存在的流程障碍。然而在大多数组织中,很有可能没有一个人管理端到端流程,甚至不理解该体系中的流程。

价值链和价值网络的概念是提供关于价值生成和改进更高级可视化的方法,它们将活动顺序、业务策略和结果联系起来。然而,它们通常不允许向下钻取或分解到创建价值的特定业务流程,因此缺乏对流程和流程间关系的支持,BPPM正适合该问题的解决方案,后面章节将有所述。

5.5.2 实现BPPM时常见的陷阱

包括私有和公共部门在内的大多数组织,在实现和维持成功的BPPM时都会遇到许多陷阱,如图1所示。

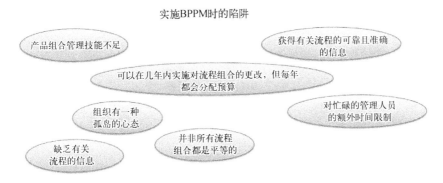

图1　实现BPPM时常见的陷阱

1. 并不是所有的流程组合都是平等的

BPPM应该发生在作出决定组合的任何地方。然而,组合管理的观点和需要回答的问题将取决于谁的观点影响决策。例如,企业流程组合需要从多个角度进行更严格的评估,如幕后IT的组合。企业经营与企业战略的一致性可用于对董事会层面的评价,而与信息系统(IS)/IT战略的一致性将是CIO的主要任务。

Changes to a Process Portfolio Can be Implemented Over Several Years, Yet Budgets are Allocated Yearly

A change initiative to a process portfolio is often a multi-year initiative that is not always aligned to the financial allocation of budget to individual process changes within the portfolio. Furthermore, in most organizations budgets are allocated yearly, which makes the continuum of the portfolio challenging considering that resources might be lowered at key stages based on budget allocation and reallocation. The most challenging periods are normally over the financial year end when existing budgets must be balanced and new budgets have yet to be fully allocated.

Organization Has a Silo Mentality

Silo mentality is an attitude found in some organizations that occurs when several departments or groups do not want to share information or knowledge with other individuals in the same company. A silo mentality reduces efficiency and can be a contributing factor to a failing corporate culture. This occurrence is detrimental in BPPM as processes devalue one another. Inefficiency is exaggerated owing to resources and knowledge being withheld.

Lack of Information on Processes

To prioritize the work successfully on a large portfolio of processes and allocate limited resources adequately, it is important to gather the same information consistently for all involved processes. As mentioned earlier, this information might consist of the risk of failure, client exposure, frequency of use, and cost to run. If desired information is not available for all processes of a portfolio, objective priority calls are difficult to make and remain subjective.

Getting Reliable and Accurate Information on Processes

BPPM also requires a place to store all of the data for processes, preferably a central source that is regarded as being the single source of these data and under change control. Having a single source for reporting also enables the elimination of the double counting that may otherwise arise if each process is permitted to use different sources of data and measures for its reports.

Inadequate Portfolio Management Skills

Managing at the process level is no longer sufficient for organizations. Increasingly, a higher-level perspective is required to ensure that entire ecosystem of processes deliver desired services and products as BPPM allows an organization to take a holistic view of a group (or groups) of its processes to improve return on investment (ROI) and strategic alignment. Process owners or key managers need to be trained to become qualified process portfolio managers. They need to combine strategic alignment and sound financial and risk management to drive large portfolios. The combination of all three elements is often missing and created pain points in all three mentioned areas for executives and organizations looking to improve the ROI successfully in their portfolios. Therefore,

2. 对流程组合的变更可以在数年内实现，但是预算是每年分配的

流程组合的变更计划通常是一个多年的计划，它并不总是与流程组合中单个流程变更的预算相一致。此外，在大多数组织中，预算是按年分配的，考虑到资源可能在预算分配和重新分配的关键阶段减少，这使得流程组合的连续性具有挑战性。最具挑战性的时期通常是在财政年度终了时，新的预算尚未充分分配，那时必须平衡现有预算。

3. 组织有一种孤岛心态

孤岛心态是一些组织中常见的一种态度，即当多个部门或团体不想与同一公司的其他个人共享信息或知识时，就会出现这种态度。孤岛心态会降低效率，并可能是导致企业文化失败的一个因素。由于资源和知识被截留，效率低下且被夸大，这种情况在 BPPM 中是有害的。

4. 缺少流程相关信息

为了在大型流程组合中成功地确定工作的优先级并充分分配有限的资源，重要的是为所有涉及的流程一致地收集相同的信息。如前所述，此信息可能包括失败风险、客户弊端、使用频率和运行成本。如果一个流程组合的所有过程都不能获得所需的信息，那么就很难进行客观的优先级评定，并且仍然是主观的。

5. 获取可靠和准确的流程信息

BPPM 还需要一个地方来存储流程的所有数据，最好是一个被视为这些数据的单一来源并受变更控制的中心源。如果允许每个流程使用不同的数据源和评价标准，那么使用单一的数据源可以避免重复计数。

6. 项目组合管理技能不足

对于组织而言，在流程级别进行管理已经不够了，因为 BPPM 允许组织对其一个（或多个）流程组进行全面的查看，以提高 ROI 和战略一致性，所以越来越多的人需要更高层次的视角来确保整个流程生态系统交付所需的服务和产品。为此，流程责任人或关键经理需要接受培训，以成为合格的流程组合经理。它们需要结成战略联盟、健全的财务和风险管理来推动大型组合。所有这三个主体的组合常常是缺失的，并且在所有这三个提到的领域中为那些希望在他们的组合中成功地提高 ROI 的管理人员和组织创建了痛点。因此，不论在一个领域理论上可获得的

irrespective of the benefits theoretically obtainable in one area, a focus in one of these areas will not deliver value without the combination of all three elements of strategic alignment and sound financial and risk management. Furthermore, a focus on the outcome, not the individual performance of processes, is required when dealing with BPPM, a fact with which many previous process owners experience difficulties.

Additional Time Constraint on Busy Executives

Improving the way an enterprise creates and manages value is a change program in itself that will need its own business case. Even with senior sponsorship and a strong appetite for positive change, once executives realize that the effort involved in doing things properly will require time and effort on their part, there is often nervousness, uncertainty, and sometimes push-back. Executives may also struggle with the new data and processes involved and will need guidance, coaching, and support.

The intention is to save these busy executives much more in terms of resources than might otherwise be wasted by enabling them to allocate scarce resources more wisely across the process portfolio, with an acceptable level of risk and more in line with their target investment mix. Gathering lots of data and then putting the data in front of those who make decisions is not necessarily the right answer. Few executives are detail focused and very few would wish to be. They will simply see this data dump as an additional demand on their already overstretched time, or worse, may feel unable to make a decision for fear of making the wrong one. What are called for are the services of impartial experts who do not take the place of the decision makers, but who can offer insight, guidance, and recommendations to decision makers. These services should be offered both proactively and on-demand, so that they can answer questions, including: Are we doing the right things? Are we getting the benefits? The home of such a service might well be called a value management office, process portfolio office, or center of excellence. Whatever the title, its role is essentially that of trusted advisor or secretariat to those who make the investment decisions.

ESTABLISHING BPPM

BPPM borrows its concept from project portfolio management (PPM), which has gone through a similar evolution. The key concepts of PPM can be easily applied: From project to process, from program to cross-functional processes, and from project portfolio to process portfolio.[7] BPPM flips the vertical BPM process horizontally to manage vertical silo organizational processes. The BPPM process will maximize organizational performance as a whole, identify and reduce duplication, and manage process interdependencies.

This section will start with a comparison of PPM, BPPM, and BPM to set the context, which is followed by a discussion of required portfolio information to effectively manage a business process portfolio. To support organizations moving to BPPM, the important steps and considerations in creating a BPPM Office are outlined afterward. The section closes with an overview of the BPPM life cycle.

利益如何,如果没有战略协调的所有三个要素以及健全的财务和风险管理的结合,在这些领域中集中注意一个领域是不会产生价值的。此外,在处理BPPM时,需要关注结果,而不是流程的单个性能,这是以前许多的流程责任人遇到的困难。

7. 对忙碌的管理人员的额外时间限制

改进企业创造和管理价值的方式本身就是一个变革计划,它需要自己的经验。即使有高管的支持和积极变革的强烈欲望,一旦高管们意识到正确做事需要投入时间和精力,他们往往会感到紧张、不确定,有时还会退缩。高管们也可能在新数据和流程中面临不知所措,所以也需要指导、培训和支持。

这样做的目的是让这些忙碌的高管能够更明智地在整个流程投入组合中分配稀缺的资源,并在可接受的风险水平上更符合他们的目标,从而为他们节省资源。很少有高管愿意关注细节,因此尽管能够收集大量的数据,但把这些数据放在决策者面前不一定是正确的答案。他们只会把这个数据积累看作是对他们已经超负荷工作外的额外需求,或者更糟的是,他们可能会因为害怕做出错误的决定而不愿意做出决定。高管们所需要的是专家提供的客观服务,专家的作用不是取代决策者,而是能够向决策者提供洞察力、指导和建议。这些服务应该是主动和按需提供的,这样他们就可以回答问题,包括:我们做的是正确的事情吗? 我们得到好处了吗? 这样的服务的总部可以称为"价值管理办公室、流程组合办公室或CoE"。无论其名称是什么,其作用本质上是为那些作出管理决策的人提供受信任的顾问或秘书处。

5.5.3 建立BPPM

BPPM借鉴了项目组合管理(PPM)的概念,PPM经历了类似的演变。PPM的关键概念可以很容易地应用:从项目到流程,从项目到跨职能流程,以及从项目组合到流程组合。BPPM流程将最大限度地提高整个组织的绩效,识别和减少重复,并管理流程的相互依赖性。

本节将首先比较PPM、BPPM和BPM以设置相互关系,然后讨论有效管理业务流程组合所需的组合信息。为了支持组织迁移到BPPM,接下来将概述创建BPPM办公室的重要步骤和注意事项。本节结尾是对BPPM生命周期的概述。

COMPARISON OF PPM, BPPM, AND BPM

PPM deals with how to undertake the right projects at the right time, whereas process portfolio management addresses the right processes at the right time and process management focuses on ;performing processes right. Compared with BPM, BPPM focuses on the selection, prioritization, and monitoring of a portfolio of processes to optimize enterprise strategic outcomes, whereas BPM focuses on the individual delivery of a service or product. A comparison is presented in Table 1.

Table 1 *Comparison Between Project Portfolio Management (PPM), Business Process Portfolio Management (BPPM), and Business Process Management (BPM)*

	Project Portfolios (PPM) "Doing the Right Projects"	Process Portfolios (BPPM) "Doing the Right Processes"	Processes (BPM) "Doing the Process Right"
Scope	Project portfolios have horizontal scope that aligns with the strategic framework, objectives, goals, and priorities of the organization.	Process portfolios have horizontal scope that aligns with the strategic framework, objectives, goals, and priorities of the organization.	Processes produce a specific service or product.
Change	Project portfolio managers continuously monitor change in the broader internal and external environments.	Process portfolio managers continuously monitor change in the broader internal and external environments.	Process owners manage change on processes and keep change controlled.
Planning	Project portfolio managers create and maintain processes and communication relative to the aggregate portfolio of projects.	Process portfolio managers create and maintain planning and communication relative to the aggregate portfolio of processes.	Process owners manage detailed plans throughout the process life cycle.
Management	Project portfolio managers manage or coordinate portfolio interdependencies, communications, and benefits in the aggregate portfolio of projects.	Process portfolio managers manage or coordinate portfolio interdependencies, communications, and benefits in the aggregate portfolio of processes.	Process owners manage the performance of a process.
Benefits	Success is measured in terms of the aggregate investment performance, stakeholder satisfaction, and benefit realization of the aggregate portfolio of projects.	Success is measured in terms of the aggregate performance, stakeholder satisfaction, and benefit realization of the aggregate portfolio of processes.	Success is measured by service and product quality, timeliness, budget compliance, and degree of stakeholder satisfaction.
Monitoring	Project portfolio managers monitor strategic changes and aggregate benefits, resources, interdependencies, performance results, and risks of the portfolio of projects.	Process portfolio managers monitor strategic changes and aggregate benefits, resources, interdependencies, performance results, and risks of the portfolio of processes.	Process owners monitor and control the performance of processes to deliver products or services.

5.5.4　PPM、BPPM和BPM的比较

PPM处理如何在正确的时间进行正确的项目,而BPPM在正确的时间处理正确的过程,BPM关注于正确地执行过程。与BPM相比,BPPM侧重于流程组合的选择、优先级和监控,以优化企业战略结果,而BPM侧重于服务或产品的单独交付。比较结果如表1所示。

表1　PPM、BPPM和BPM之间的比较

	PPM "做正确的项目"	BPPM "做正确的过程"	BPM "正确处理流程"
范围	项目组合具有与组织的战略框架、目标、目标和优先级一致的水平范围	流程投资组合具有与组织的战略框架、目的、目标和优先级一致的水平范围	流程产生特定的服务或产品
改变	项目组合经理持续监控更广泛的内部和外部环境中的变化	流程组合经理持续监控更广泛的内部和外部环境中的变化	流程责任人负责流程上的更改并保持对更改的控制
规划	项目组合经理创建并维护相对于项目组合的聚合的过程和沟通	流程组合经理创建并维护相对于过程的总体投资组合的计划和沟通	流程责任人在整个流程生命周期中管理详细的计划
管理	项目组合经理负责总体组合中各项目关系的管理、协调和利益分配	流程组合经理负责总体组合中各项目关系的管理、协调和利益分配	流程责任人管理流程的运行效率
好处	成功是根据项目组合的总体投资绩效、利益相关者满意度和效益实现来衡量的	成功是根据流程的总体组合的总体绩效、利益相关者满意度和利益实现来度量的	成功由服务和产品质量、及时性、预算遵从性和利益相关者满意度来衡量
监控	项目组合经理监控战略变更,并聚合项目组合的利益、资源、相互依赖、性能结果和风险	流程组合经理监控战略变化,并聚合流程投资组合的利益、资源、相互依赖、性能结果和风险	流程责任人监控流程的性能,以交付产品或服务

As can be seen, the conjunction of PPM and BPPM can be found in change initiatives. Major process improvements are usually run as projects which are planned, executed, and governed through PPM.

CREATING A BPPM COMPETENCY

Once an organization has decided to implement BPPM, the BPPM competency must be planned and its mandate defined; then it can be integrated into the organization. A structured approach to implement BPPM is presented in Figure 2, adapted to BPPM from the portfolio reference content.[8]

ESTABLISHING A BPPM COMPETENCY

Analyze	Design	Build	Implement / Deploy	Run / Maintain	Continuous Improvement
Identify change	Define change	Structure change	Execute change	Monitor change	Enhance change
Define Portfolio Management Objectives	Design Portfolio Structures	Define Charter & Governance			Prioritize
Identify Portfolio Management Requirements	Plan Portfolio Structures	Create Portfolio Structures	Implement & Deploy Portfolio Structures	Monitor	Enhance
Assess Portfolio Management Readiness					Categorize
Communication					

FIGURE 2

Implementing and sustaining a business process portfolio management (BPPM) competency.

The implementation of BPPM is framed by the need for change management, which is highlighted through the ongoing deliverable of communication.

Guiding Principles

The BPPM principles represent the foundations upon which effective portfolio management is built. They provide the organizational environment in which portfolio management practices can operate effectively. These principles apply equally within any portfolio of processes and process improvement projects, whether the individual investment is occurring in the business, application, or technology layers.

Guiding principles are:

• Business process portfolios are aligned to the strategic business objectives as well as the critical success factors of an organization.

可以看到,PPM和BPPM的结合可以在变更计划中找到。主要的流程改进通常作为计划、执行和通过PPM治理的项目来运行。

5.5.5　创建BPPM的能力

一旦组织决定实施BPPM,就必须制订BPPM的能力并定义其任务,然后可以将其集成到组织中。为说明BPPM的适应过程,图2中给出了一种实现BPPM的结构化方法[8]。

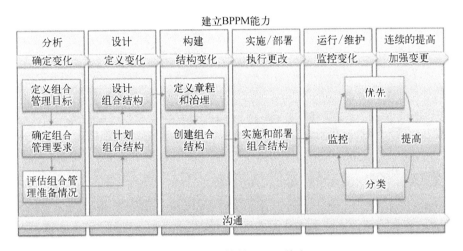

图2　实现和维护BPPM能力

BPPM的实现是由变更管理的需求引起,而变更管理的需求通过持续沟通而得到强化。

1.指导原则

BPPM原则代表了构建有效的项目组合管理的基础。它们提供了项目组合管理实践能够有效运作的组织环境。这些原则同样适用于流程和流程改进项目的任何组合中,无论个人投资是发生在业务、应用程序还是技术层中。

指导原则是:

- 业务流程组合与SBO以及组织的CSF保持一致;

- Interdependencies are managed at a business continuum level across portfolios and assessed against business transformation, delivery, and stakeholder commitments.
- Portfolio governance and decision making is clearly defined and integrated across the organization's corporate governance to proactively balance resource capacity against organizational performance.
- Active stakeholder engagement is in place by integrating and coordinating stakeholder requirements within a portfolio where they are recognized as key contributors to the delivery of organizational outcomes.

Analysis Phase

The purpose of this phase is to identify the need for BPPM within an organization. It sets out the basic requirements in terms of what objectives the portfolio management within the organization will fulfill. It details and decomposes the specific requirements needed to meet the expectations and objectives of BPPM and then assesses the portfolio management readiness across the organization.

The section on alignment considerations when implementing BPPM will provide an overview of the different factors to which a BPPM initiative needs to align.

Design Phase

The purpose of this phase is to plan and design the BPPM structures within the organization. The plan and design activities, roles, and deliverables ensure that portfolio management structures are ready for the build phase. Design phase activities should account for variation factor types and feedback communications. Both activities will consider uncertainty events and near real-time performance feedback for process effectiveness. The section on business process hierarchy will introduce a hierarchy concept to provide structure for the business processes in an organization to facilitate prioritization. To compare different process improvement initiatives, certain information is required for the business process portfolio, which is described in the section on BPPM information, measurements, and reporting.

Build Phase

The purpose of this phase is to take the designs created in the preceding phases and build the portfolio management structures and governance within the organization.

Deploy/Implement Phase

The deploy phase executes the change into existence within the organization. It launches BPPM as a framework into the organization whereby process improvement programs and projects fall under the governance of a BPPM office (BP-PMO).

- 在跨项目组合的业务连续级别上管理相互依赖关系,并根据业务转换、交付和涉众承诺进行评估;
- 项目组合治理和决策在组织的公司治理中得到了清晰的定义和集成,以主动地平衡资源能力和组织绩效;
- 积极的参与是通过集成和协调项目组合中的相关方需求来实现的,这些需求被认为是组织成果交付的关键贡献者。

2. 分析阶段

这个阶段的目的是确定组织中对BPPM的需求。它根据组织内的项目组合管理将要实现的目标列出了基本的需求。它详细描述并分解满足BPPM的期望和目标所需的具体需求,然后评估整个组织的项目组合管理准备情况。

在实现BPPM时,关于对齐考虑事项的部分将概述BPPM计划需要对齐的不同因素。

3. 设计阶段

这个阶段的目的是计划和设计组织中的BPPM结构。计划和设计活动、角色和可交付成果确保项目组合管理结构为构建阶段做好准备。设计阶段的活动应该考虑变化因素类型和反馈沟通。这两个活动都将考虑不确定性事件和过程有效性的实时性能反馈。关于业务流程层次结构的部分将引入层次结构概念,为组织中的业务流程提供结构,以促进优先级划分。为了比较不同的流程改进计划,业务流程组合需要特定的信息,这在5.5.9小节BPPM信息、测量和报告中进行了描述。

4. 构建阶段

此阶段的目的是采用在前面阶段中创建的设计,在组织中构建项目组合管理的结构和治理内容。

5. 部署和实施阶段

部署阶段执行组织中存在的更改。它将BPPM作为一个框架启动到组织中,流程改进计划和项目在BPPM办公室(BP-PMO)的治理之下完成。

Run/Maintain Phase

The run/maintain phase takes the BPPM life cycle into the business as usual/operational space. BPPM now requires ongoing monitoring, re-prioritization via ranking, and categorization after enhancements have taken place in the continuous improvement phase. The run phase and continuous improvement phase are linked together through the prioritization and categorization activities that lead to the ongoing improvements in monitoring from the continual enhancement of portfolio management.

Continuous Improvement Phase

The continuous improvement phase objective is to look for ongoing enhancements to the BPPM life cycle and method so that these can be used to improve portfolio management over time. The continuous improvement phase and run phase are linked together through the prioritization and categorization activities that lead to prioritized improvements entering the continuous improvement phase; these improvements are then enhanced further where possible before being categorized and applied to the organization's BPPM.

Once an organization has successfully implemented BPPM, the overall BPPM life cycle can be applied in effectively executing, sustaining, and continuously improving their portfolio management. The section on BPPM life cycle deals with the effective daily portfolio management operation.

ALIGNMENT CONSIDERATIONS WHEN IMPLEMENTING BPPM

Alignment to an organization's strategy is the cornerstone of creating a BPPM competency. Clark and Cameron described four alignment factors that need to be taken into consideration when establishing a BPPM competency (as shown in Figure 3): strategic, "processual" (process-centric), social, and technical alignment.[9]

BPPM STRATEGIC ALIGNMENT FACTORS

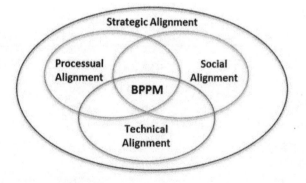

FIGURE 3

Strategic alignment factors when establishing a business process portfolio management (BPPM) competency.

6.运行/维护阶段

运行/维护阶段将BPPM生命周期带入业务运行过程。BPPM现在需要持续监控,通过排名重新确定优先级,并在持续改进阶段、改进之后进行分类。运行阶段和持续改进阶段通过优先级和分类活动联系在一起,这些活动保障从项目组合管理的持续强化中获得改进。

7.持续改进阶段

持续改进阶段的目标是根据BPPM生命周期寻找持续改进的方法,以便随着时间的推移可以使用它们来改进项目组合管理。持续改进阶段和运行阶段通过优先级和分类活动联系在一起,这些活动确保优先级的改进进入持续改进阶段,在分类并应用到组织的BPPM之前,这些改进将在可能的环节得到进一步的增强。

一旦一个组织成功地实现了BPPM,整个BPPM生命周期就可以应用于有效地执行、维持和持续地改进他们的项目组合管理。关于BPPM生命周期的部分涉及有效的日常项目组合管理操作。

5.5.6　实现BPPM时的对齐注意事项

与组织的策略保持一致是创建BPPM能力的基石。Clark和Cameron描述了在建立BPPM能力时需要考虑的四个对齐因素(如图3所示):战略、流程(以流程为中心)、社会和技术对齐[9]。

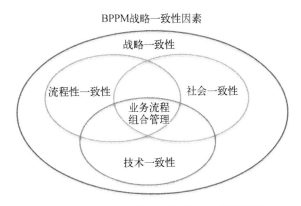

图3　建立BPPM能力时的战略一致性因素

Strategic Alignment

The strategic alignment factors describe when, where, and how BPPM will be applied in the organization and how it will be leveraged for strategic advantage. Vision, scope, and distinctive competencies need to be described here. According to Clark and Cameron, potential outcomes of BPPM might be operational transparency, dynamic executive dashboards tracking process performance, auditing capabilities, ease of business process configuration, and highlighting processes delivering competitive advantage.

Social Alignment

Social alignment factors look at aspects such as the facilitation of collaboration, knowledge sharing, and convergence in perceptions through BPPM as well as the degree of inclusion in decision making. The underlying premise is that process management is highly social and collaborative in many aspects. Clark and Cameron also highlighted that governance is a key social factor.

Processual Alignment

Processual alignment factors deal with methods performed to execute BPPM. According to Clark and Cameron, these include decomposition strategies (top-down, bottom-up, or a hybrid), taxonomies, process classifications, definitions, documentation, and performance metrics, knowledge management, and dealing with process repositories.

Technical Alignment

Technical alignment factors describe the technology used to achieve the desired outcomes of a BPPM strategy incorporating social and processual factors. Technology should be an enabler and not an inhibiter of a successful rollout of BPPM. Clark and Cameron also talked about the facilitation of meeting the top-down approach for BPPM with the bottom-up approach often used in service-oriented architectures deployments that should form part of the BPPM strategy.

BPPM LIFE CYCLE

The BPPM life cycle consists of a series of phases spanning from the opportunity management phase to the realization of an opportunity, right through to its transition into the appropriate operational environment; it concludes with continuous improvement, as shown in Figure 4, which has been adapted from the portfolio reference content.[10] This figure also highlights the value identification, planning, creation, realization, and governance flow across the portfolio management life cycle. It is obvious that the highest value potential can be realized during the earliest phase of the life cycle as well as in the continuous improvement phase.

1. 战略一致性

战略一致性因素描述了何时、何地与如何在组织中应用BPPM，以及如何利用BPPM获得战略优势。这里需要描述远景、范围和独特的能力。Clark 和 Cameron 认为，实施BPPM的产出包括：运营透明度、通过实时的高管仪表盘实现对流程性能的跟踪、评审功能、业务流程配置，并突出显示流程的竞争优势。

2. 社交联盟一致性

社交联盟因素着眼于协作的便利性、知识共享、感知融合以及决策的包容性等方面。基本前提是流程管理在许多方面具有高度的社会性和协作性。Clark 和 Cameron还强调，治理是一个关键的社会因素。

3. 流程一致性

流程对齐因素与BPPM执行方法有关。根据Clark 和 Cameron的说法，这些因素包括：分解策略（自顶向下、自底向上或混合的）、分类、流程分类、定义、文档和性能指标、知识管理，以及管理流程仓库。

4. 技术一致性

技术一致性因素体现在实现结合社会和过程因素的BPPM策略预期结果的技术。技术应该是成功推出BPPM的推动者而不是抑制者。Clark 和 Cameron还谈到了使用自下而上的方法实现BPPM自上而下的方法的便利性，这种方法通常用于应构成BPPM策略一部分的SOA部署中。

5.5.7 BPPM生命周期

BPPM生命周期由一系列阶段组成，从机会管理阶段到机会的实现，一直到机会过渡到适当的操作环境，它以持续改进结束，如图4所示，这是从项目组合参考内容改编而来的[10]。图4还强调了跨项目组合管理生命周期的价值识别、计划、创建、实现和治理流。很明显，最大的价值潜力可以在生命周期的早期实现，也可以在持续改进阶段实现。

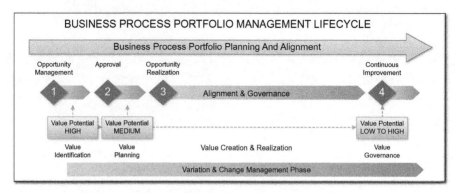

FIGURE 4

Business process portfolio management (BPPM) life cycle.

Each phase in the portfolio management life cycle is described below.

Business Process Portfolio Planning and Alignment Phase

The purpose of this phase is to define the portfolio and ensure the alignment of portfolio planning with organizational strategic goals, priorities, and direction. The portfolio needs to be approved by executive stakeholders.

Business process portfolio planning and alignment phase defining documents include the portfolio road map, portfolio baseline, executive change management strategy, and executive portfolio communication road map.

Opportunity Management Phase

The purpose of this phase is to rank required process improvement projects within the portfolio by assessing the project proposals for portfolio improvements against a defined set of criteria to generate objective ranking (identify, elect, prioritize, and rank) in relation to the organizational strategic outcomes provided by the executive stakeholders.

The opportunity management phase defining document is the project proposal that, at a minimum, describes required process changes by clearly articulating current as-is state, end-state vision, funding, high-level business requirements, benefits, risks, and stakeholders.

The outcome of this phase is to have a prioritized list of project proposals within the portfolio that have preliminary approval in principle with validation that the business unit or organization has the capacity and competency to carry out proposed projects.

Variation/Change Management Phase

Successful portfolio management techniques must consider the effects of variation both internal and external to the business across the life cycle of the portfolio. Programs and projects can start at any time and in most cases do not all start together, so there must be

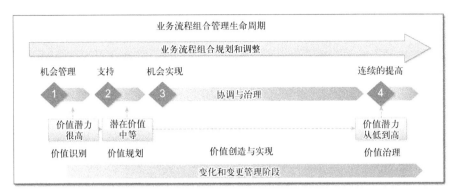

图4　BPPM生命周期

下面描述项目组合管理生命周期中的每个阶段。

1. 业务流程组合规划和校准阶段

此阶段的目的是定义项目组合,并确保项目组合规划与组织的战略目标、优先级和方向保持一致。投资组合需要得到执行利益相关者的批准。

业务流程组合规划和校准阶段定义文档包括组合路线图、组合基线、执行变更管理策略和执行组合沟通路线图。

2. 机会管理阶段

本阶段的目的是通过根据一组确定的标准评估项目组合改进建议,对组合中所需的流程改进项目进行排序,以生成与执行目标提供的组织战略成果相关的目标排序(识别、选择、优先排序和排名)。

业务机会管理阶段定义文档是项目建议,它通过清楚地阐明当前状态、最终状态远景、资金、高级业务需求、利益、风险和利益相关者来描述所需的流程更改。

本阶段的结果是,在投资组合中有一个优先的项目建议列表,原则上,这些建议经过初步批准,并确认业务部门或组织有能力执行建议的项目。

3. 偏差/变更管理阶段

成功的项目组合管理技术必须考虑跨项目组合生命周期的业务内部和外部变化的影响。程序和项目可以在任何时候启动,事实上在大多数情况下,它们不是一起启动的,因此必须有一个阶段来考虑新项目或新程序强加的更改。这就是偏差/

a phase that considers the changes imposed by the new project or new program. This is the variation/change management phase. Internal variations may be defined as missing processes, system requirements, or missing business objectives. External variation may be defined as missing legal/regulatory impacts or missing weather/environmental factors. The portfolio must have the capability to adjust to variations with which it is presented.

The BPPM process could use additional language to support change created by variation. Four variation factor types have been defined that have been found to be most problematic when managing multiple programs within a portfolio. In general, concern must be expressed about exposing what we do not know. The BPPM process should be designed to expose unknown deficiencies, gaps, or duplication as quickly as possible. A feedback loop is required within the process to react to the change and challenge the process to ultimately apply corrective action for each variation factor type. The effect on the portfolio and its related programs is indicated in Figure 5, with portfolio risk increasing as the scope clarity decreases.

Variation Factor Types:

- Variation Type 1: Known project with known "unknowns" (an approved and planned project that knows there are things which it does not know)
  - Documented project with documented requirements or processes to address the known "unknowns"

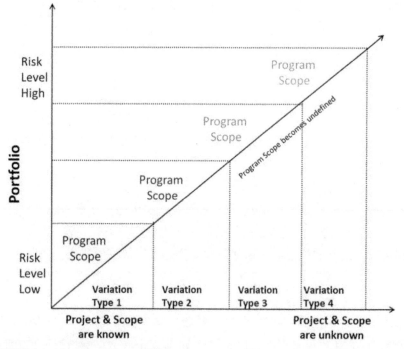

FIGURE 5

Change risk and variation factor type.

变更管理阶段。内部变化可以定义为缺少流程、系统需求或缺少业务目标。外部变化可定义为法律/法规影响缺失或天气/环境因素缺失。流程组合必须有能力适应它所呈现的变化。

　　BPPM流程可以使用额外的语言来支持由变更创建的变化。已经定义了四种变更因素类型,它们被发现在管理一个投资组合中的多个项目时问题最大。一般来说,必须对揭露我们所不知道的情况表示关切。BPPM过程应该被设计成尽可能快地暴露未知的缺陷、差距或重复。流程中需要一个反馈循环来对变更作出反应,并要求流程最终对每个变更因素类型应用纠正措施。项目组合及其相关程序的影响如图5所示,随着范围清晰度的降低,组合风险会增加。

　　变更因素类型如下。

　　变更类型1：已知项目中存在未知(一个已批准和计划中的项目,包含一些位置因素)。

- 可通过已有的文件,根据需求或流程来解决已知的未知。

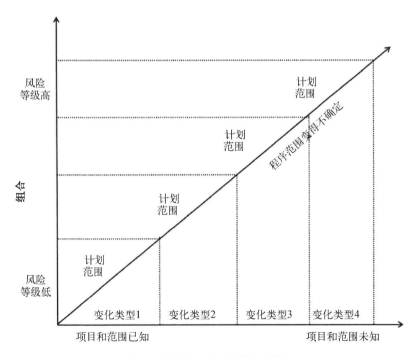

图5　变更风险和变更因素类型

- Impacts: Minimal, the team has a known project and is tailoring requirements and processes to the program to address the known "unknowns." Because the project is known the team may have time to apply corrective action
- Variation Type 2: Known project with unknown "unknowns" (an approved and planned project that is unaware that there are things which it does not know)
  - Documented project with no documented requirements or processes to address the unknown "unknowns"
  - Impacts: Minor, the team has a known project and is not tailoring requirements and process to the program to address the unknown "unknowns." Because the project is known the team may have time to apply corrective action
- Variation Type 3: Unknown project with known "unknowns" (an unapproved or unplanned project that knows there are things it does not know)
  - Undocumented project with documented requirements or processes to address the known "unknowns"
  - Impacts: Major; the team has an unknown project that could have a significant impact on the program. However, the team is tailoring requirements and processes to address the known "unknowns." Because the project is unknown and has documented known "unknowns," the team may not have time to apply corrective action.
- Variation Type 4: Unknown project with unknown "unknowns" (an unapproved or unplanned project that that is unaware that there are things which it does not know)
  - Undocumented project with no documented requirements or processes to address the unknown "unknowns"
  - Impacts: Critical because the team has an unknown project that could have a significant impact on the program. However, the team is not tailoring requirements and processes to address the unknown "unknowns." Because the project is unknown and has unknown "unknowns," the team will not have time to apply corrective action.

Approval Phase

The purpose of this phase is to formally authorize the business process portfolio improvements. The portfolio improvement projects are presented to executive stakeholders for approval by either the business process portfolio manager or the BP-PMO. The executive stakeholders evaluate the portfolio improvements at an enterprise level and then grant authorization for the improvement projects to proceed based on organizational strategic objectives and corporate capacity.

Once the approval is obtained, the process portfolio is rebalanced. The BP-PMO updates the portfolio delivery strategy within the portfolio road map document to provide an overview of the sub-portfolios, projects, costs, interdependencies, risks, stakeholders, and benefits along with a time frame used within the alignment and governance portfolio phase for effective portfolio monitoring and control.

Opportunity Realization Phase

The purpose of this phase is to provide alignment, oversight, and direction for the effective and timely management of the various process improvement

- 影响：最小。团队有一个已知的项目，正在根据项目调整需求和流程，以解决已知的未知。因为项目已知，团队可能有时间采取纠正措施。

变更类型2：未知已知项目存在未知（未意识到已批准和计划项目存在其不知道的事物的）。

- 文件化项目，没有文件化的要求或流程来解决未知的未知。
- 影响：轻微。团队有一个已知的项目，并且没有根据项目调整需求和流程，以解决未知的未知。因为项目已知，团队可能有时间采取纠正措施。
- 变化类型3：已知未知项目中的未知（一个未经批准或计划外的项目，知道它不知道的事情）。
- 无文件记录的项目，有文件化的需求或流程，以解决已知的未知。
- 影响：重大。团队有一个未知项目，可能对整体项目产生重大影响。但是，团队正在调整需求和流程以解决已知的未知。因为项目未知，并且记录了已知的未知，所以团队可能没有时间应用纠正措施。
- 变化类型4：未知未知项目中的未知（未经批准或未计划的项目，不知道有不知道的事情）。
- 无文件记录的项目，没有文件化的要求或流程来解决未知的未知。
- 影响：至关重要。因为团队有一个未知项目，它可能对项目产生重大影响。然而，团队并没有为解决未知的未知而定制需求和流程。因为项目未知，并且有未知的未知，所以团队将没有时间应用纠正措施。

4. 审批阶段

此阶段的目的是正式批准对业务流程组合进行改进。项目组合改进项目被提交给执行人员，由业务流程项目组合经理或BP-PMO批准。执行人员在企业级别上评估组合改进，然后根据组织战略目标和公司能力授权改进项目进行。

一旦获得批准，流程组合将重新平衡。BP-PMO在项目组合路线图文档中更新项目组合交付策略，以提供子项目组合、项目、成本、相互依赖、风险、涉众和利益的概述，以及用于有效的项目组合监视和控制的校准和治理项目组合阶段中使用的时间框架。

5. 机会实现阶段

此阶段的目的是为项目组合中的各种流程改进计划及项目的有效和及时管理提供一致性、监督和方向。机会实现阶段将管理层的注意力集中在项目组合级别，

initiatives and projects within the portfolio. The opportunity realization phase focuses executive attention at the portfolio level to primarily address the following considerations:

- Continual evaluation of performance and alignment of portfolio improvement projects against the portfolio road map approval of changes to project and portfolio baselines
- Alignment with other change initiatives (e.g., pure technology improvements with no business process impact) in the organization that might be managed by a separate project management office as well as prioritization across the different portfolios if necessary

This phase also serves to balance and optimize adjoining portfolios with respect to:

- Portfolio improvement project benefits and interdependencies within and between the adjoining portfolios
- Interdependencies between processes that are affected by improvement projects within and between adjoining portfolios
- Limited resource capacity/capability versus substantial demand
- Changes in business strategy/business opportunities
- Managing the portfolio to a predetermined risk profile

Figure 6 shows the relationship between the BPPM life cycle and the project management life cycle for portfolio improvement projects during the opportunity realization phase.[11]

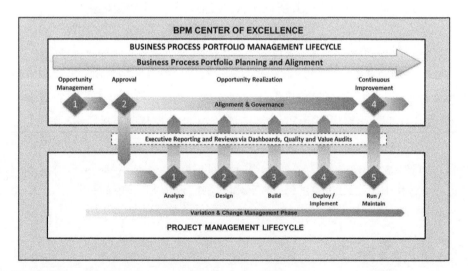

FIGURE 6

Relationship between BPPM life cycle and project management life cycle of process portfolio improvement projects.

主要解决以下问题：

- 根据项目组合路线图批准对项目和项目组合基线的变更，持续评估项目组合改进项目的性能和一致性；
- 与组织中可能由单独的项目管理办公室管理的其他变更活动（如没有业务流程影响的纯技术改进）保持一致，并在必要时跨不同的流程组合确定优先级。

这个阶段还可以平衡和优化相邻的组合，包括：

- 组合改进项目收益及相邻组合内部和之间的相互依赖关系；
- 受改进项目影响的过程之间以及相邻项目组合之间的相互依赖关系；
- 有限的资源容量/能力与大量的需求；
- 商业战略/商业机会的变化；
- 按照预先确定的风险配置管理组合。

图6显示了在机会实现阶段，项目组合改进项目的BPPM生命周期和项目管理生命周期之间的关系[11]。

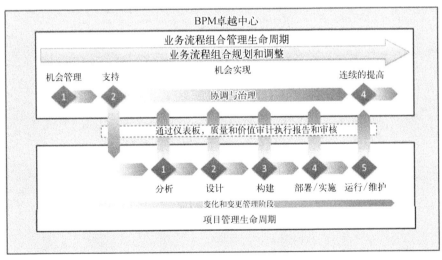

图6 流程组合改进项目的BPPM生命周期与项目管理生命周期的关系

Feedback Loop Communication

Successful portfolio management techniques must also consider the effects of communication. A concern in many organizations is about the language used in the documentation that addresses portfolio plans or process requirements approval with senior management. Pushing communication up to senior management and then back down to working management is time-consuming and loaded with risk. How do organizations improve that process so that approval or requirements, especially when impacted by variation and change management, are addressed faster? How do organizations compensate the loop to enable faster response time? Empowerment of middle management ranks that have access to or reside with the BPM CoE to make decisions that do not require executive level approval is an idea or provides a vehicle for frontline communications to flow directly back to senior management. Essentially the organization should create a distributed architecture of empowered middle managers who are part of the BPM CoE and who have the authority to approve tactical and operational changes and program and project management direction and provide a direct line of feedback to the executive team. This builds both the BPPM maturity and strengthens the BPM CoE in its governance.

Executive-level approval and decisions do not always work, for the following reasons:

1. Information is filtered and condensed on executive-level reporting
   a. Executives typically like to see one to three PowerPoint slides
   b. Many times critical information is missing
   c. Complex issue cannot always be quantified in three PowerPoint slides.
2. Executives are primarily concerned with money and budgets
   a. Many executives are too disconnected from the operations/technical issues to make the connection on how this impacts business
   b. Many executives do not have a long-term vision
   c. Many executives are too concerned about short-term gains and putting out immediate daily fires
3. Critical decisions are funneled and bottlenecked for a few select people to review
   a. Vertical communication up to senior management is time-consuming and slow
   b. Often there is no real value added in the process other than rubber stamping the request and moving forward with the project
   c. The decisions is often already been made, so why bother with the exercise?

Feedback Loop

The feedback loop requires the empowerment of a middle manager who has direct dotted line access to senior management. The existing manager, director, executive director, vice president (VP), and senior vice president (SVP) reporting structure stays in place; however, the empowered middle manager provides real team feedback to the VP/SVP team. This new channel will forward unfiltered and near real-time feedback to the senior management team. This new feedback channel will also keep

6. 循环反馈与沟通

成功的项目组合管理技术还必须考虑沟通的影响。许多组织中的一个关注点是文档中使用的语言,对语言处理与项目组合计划或与高级管理人员一起的过程需求批准有重要关系。将沟通提升到高层管理,然后再回到工作管理,既耗时又充满风险。组织如何改进这个过程,使批准或需求,特别是受到变更和变更管理的影响时,更快地得到处理? 组织如何补偿循环以使响应时间更快? 授权能够访问BPM CoE 或与 BPM CoE 一起做出不需要执行级别批准的决策的中层管理人员,是一种想法,或为前线通信直接流回高级管理人员提供了一种工具。从本质上说,组织应该创建一个分布式体系结构,由被授权的中层管理人员组成,他们是 BPM CoE 的一部分,有权批准战术和操作更改以及规划和项目管理方向,并向执行团队提供直接的反馈。这既构建了 BPPM 成熟度,又加强了 BPM CoE 的治理。

执行层的批准和决策并不总是有效,原因如下。

(1)在执行级报告上对信息进行过滤和压缩。

a. 高管们通常喜欢看一到三张幻灯片;

b. 很多时候,关键信息缺失;

c. 复杂的问题不可能总是用三张 PowerPoint 幻灯片来量化。

(2)高管们主要关心的是资金和预算。

a. 许多执行人员与业务/技术问题的联系过于脱节,无法就这些问题如何影响业务进行联系;

b. 许多高管没有长远的眼光;

c. 许多高管过于关注短期收益,并过度关注每日大火。

(3)关键的决策被汇集在一起,让少数经过挑选的人来审核。

a. 直达高层的垂直沟通费时且缓慢;

b. 往往在过程中除了盖章的请求和项目的推进没有真正的附加价值;

c. 决策通常已经做出了,为什么还要做这个操作呢?

7. 反馈环

反馈环要求授权给一个直接接触高级管理层的中层经理。现有的经理、董事、执行董事、副总裁(VP)、高级副总裁(SVP)汇报结构不变,然而,被授权的中层经理向 VP/SVP 团队提供真实的团队反馈。这个新渠道将向高级管理团队提供未经过滤的、近乎实时的反馈。这个新的反馈渠道还将检查现有的报告结构和管理团

the existing reporting structure and management team in check and ensure that both reports and feedback loops are in phase and in agreement.

An example of this is presented graphically in Figure 7.

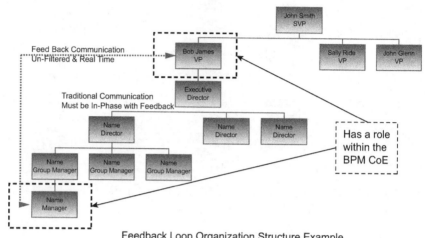

Feedback Loop Organization Structure Example

FIGURE 7

Example of feedback loop.

Continuous Improvement Phase

The purpose of this phase is to evaluate whether the portfolio benefits and contribution to the organization have been effectively realized. This is a consolidation and review of each of the individual projects' outcomes. Furthermore, remaining pain points and new opportunities for portfolio improvements are identified during this phase.

The portfolio continuous improvement phase defining documents are the portfolio benefits assessment and portfolio health assessment.

A portfolio health assessment can be gathered from the health assessment of individual business processes. Process health assessments should focus on costs versus value to the organization.[12]

BUSINESS PROCESS HIERARCHY

One of the tasks performed during the design phase of establishing a BPPM is to create a business process hierarchy, organize existing business processes into the hierarchy, and ideally store the information in a central repository.

Figure 8 presents a way to organize business process into a hierarchy as adapted from the business process reference content.[13] Business process areas and business process groups provide means to categorize existing business process hierarchies including business processes, steps, and activities. Business process areas and business process groups are not business processes.

队,确保报告和反馈循环处于一致的阶段。

图7以图形形式给出了一个示例。

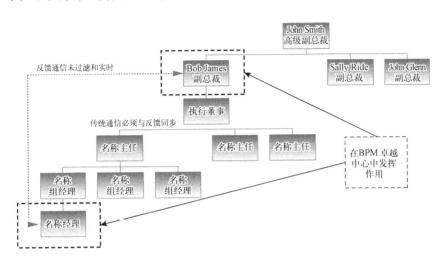

图7   反馈环示例

8. 持续改进阶段

这个阶段的目的是评估项目组合的效益和对组织的贡献是否已经有效地实现。这是对每个单独项目成果的整合和审查。此外,在此阶段还确定了项目组合改进的剩余痛点和新机会。

项目组合持续改进阶段定义文档是项目组合收益评估和项目组合健康评估的依据。

可以从单个业务流程的健康评估中收集投资组合健康评估。流程健康评估应该关注组织的成本与价值。

5.5.8   业务流程的层次结构

在建立BPPM的设计阶段执行的任务之一是创建业务流程层次结构,将现有的业务流程组织到层次结构中,并在理想情况下将信息存储在中央存储库中。

图8展示了一种将业务流程组织成层次结构的方法,该层次结构根据业务流程引用内容进行了调整[13]。业务流程领域和业务流程组提供了对现有业务流程层次结构(包括业务流程、步骤和活动)进行分类的方法,但业务流程领域和业务流程组不是业务流程。

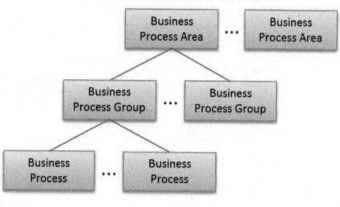

FIGURE 8

Example of a business process hierarchy.

Business Process Area

A business process area consists of business process groups with the same business goal, thus spanning the organization end-to-end or even across to business partners if they are involved in fulfilling that business goal. A business process portfolio should cover at least one business process area. Portfolio prioritization and improvements are also performed at this level.

Business Process Group

A business process group encapsulates logically related first-level business processes that are executed to realize a defined, measurable business outcome for a particular internal or external customer. Process improvements performed concurrently inside a business process group or cutting across different business process groups should be grouped into a program of projects to manage their interdependencies.

Business Process

The business process layer defines the first-level business processes that are organized into a flow to achieve a defined business outcome. Process improvement projects usually operate at this level.

BPPM INFORMATION, MEASUREMENTS, AND REPORTING

Once the requirements for a BPPM and a business process hierarchy are established, it is important to define the information that needs to be tracked for the business process areas, groups, and business processes for measurement and reporting.

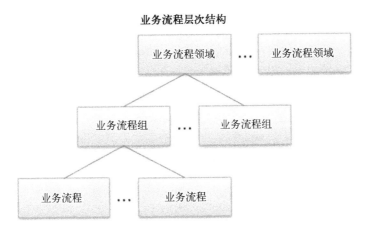

图8　业务流程层次结构的示例

1. 业务流程领域

业务流程领域由具有相同业务目标的业务流程组组成,因此跨组织端到端,甚至跨业务合作伙伴(如果他们参与实现该业务目标的话)。业务流程组合应该至少覆盖一个业务流程领域。项目组合的优先级和改进也在这个级别上执行。

2. 业务流程组

业务流程组封装逻辑上相关的第一级业务流程,这些业务流程被执行以实现特定内部或外部客户定义的、可度量的业务结果。在业务流程组内并发执行的流程改进或跨不同业务流程组执行的流程改进应该分组到项目的一个程序中,以管理其相互依赖关系。

3. 业务流程

业务流程层定义了第一级业务流程,这些业务流程被组织到流中以实现已定义的业务结果。流程改进项目通常在这个级别上运行。

5.5.9　BPPM信息、测量和报告

一旦建立了对BPPM和业务流程层次结构的需求,就必须为业务流程区域、组和用于度量及报告的业务流程定义需要跟踪的信息。

BPPM Information for Measurement and Reporting

The required information depends on the strategy behind the establishment of BPPM. The primary objective of BPPM is to increase business value for an organization, whereas contemporary BPM focuses on driving effectiveness and efficiency through optimizing operating models. With that in mind, gathered information around processes must include indicators for business value creation and improvement.

Recommended information to be gathered includes:

- **General information**: Name, goal, and description
- **Value**: Value to the organization as well as value classification (e.g., which processes contribute to competitive advantage)
- **Strategic alignment**: Alignment to strategic goals and objectives, and customer satisfaction
- **Resources and stakeholders**: Stakeholders and resource requirements
- **Interrelationships**: Dependencies and interdependencies
- **Financials**: Costs and financial benefit
- **Risk**: Probability of failure multiplied by impact. Also, customer exposure and customer impact should be tracked, which in relation to frequency provide a holistic overview of risk (Rosemann, 2006)
- **Process metrics**: Volume, cycle time, elapsed time including wait time and nonproductive time, frequency, exceptions, defects, and rework
- **Process classification**: Process types such as core, support, and governing (Bilodeau)
- **Operational and change costs/benefit**: Past process changes including costs and tracked benefits compared with expectations

Figure 9 provides an overview of how important the different types of information are for the different levels of the process hierarchy. For example, gathering strategic and value information is more important at business process area level than at the business process level for portfolio improvements.

IMPORTANCE OF INFORMATION ACROSS THE PROCESS HIERARCHY

FIGURE 9

Importance of described types of information for different levels of the process hierarchy.

1. 用于测量和报告的BPPM信息

所需信息取决于建立BPPM背后的战略。BPPM的主要目标是为组织增加业务价值，而现代的BPM则通过优化操作模型来提高效率和效果。有鉴于此，围绕流程收集的信息必须包括业务价值创造和改进的指标。

建议收集的信息包括以下方面。

- 一般信息：名称、目标和描述。
- 价值：对组织的价值以及价值分类（例如，哪些过程有助于竞争优势）。
- 战略联盟：与战略目标和客户满意度保持一致。
- 资源和利益相关者：利益相关者和资源需求。
- 相互关系：依赖和相互依赖。
- 财务：成本和财务收益。
- 风险：失败的概率乘以影响。此外，还应该跟踪客户接触和客户影响，从而根据频率提供风险的整体概述（Rosemann，2006）。
- 流程度量：数量、周期时间、运行时间，包括等待时间和非生产时间、频率、异常、缺陷和返工。
- 流程分类：流程类型，如核心、支持和治理（Bilodeau）。
- 运营和变更成本/收益：过去的流程变更，包括与预期相比的成本和跟踪的收益。

图9概述了不同类型的信息对于流程层次结构的不同级别的重要性。例如，对于项目组合的改进，收集战略和价值信息在业务过程域级别比在业务过程级别更重要。

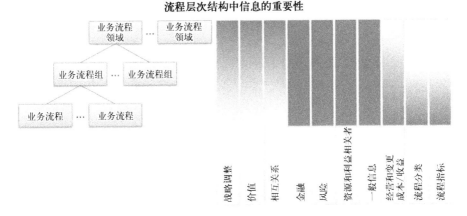

图9　描述的信息类型对于流程层次结构的不同级别的重要性

BPPM Measurements and Reporting

There are two levels where measurement and reporting should take place: at the portfolio level and at the process improvement project level.

At the portfolio level, the following measurements and reporting information could be beneficial beside some of the information described above:

- **Total number of processes**: A large process portfolio might be an indicator for increasing duplication of efforts
- **Percentage of processes per process owner**: Indicates accountability spread
- **Cost/benefit of processes**: Fosters decisions around improvement prioritization and driving effectiveness and efficiency
- **Percentage of processes above the desired maturity level and average maturity level**: Indicator of where to deploy resources and investments
- **Number of resources per process**: Indicator for possible composition or decomposition of processes

At the process improvement project level, the usual project measurement and reporting information can be used, such as:[14]

- Alignment to strategic goals
- Cost and return on investment
- Schedule and resourcing
- Scope/productivity
- Project cycle time (time to completion)
- Post project review and customer satisfaction
- Risk management

SUMMARY OF ESTABLISHING BPPM

This section provides an overview of the different considerations when establishing a BPPM in an organization. It discussed the creation of a BPPM competency, alignment issues, the BPPM life cycle, the process hierarchy for portfolio management, and required portfolio information for measurement and reporting. The next sections provides an overview of the lessons learned when implementing BPPM.

LESSONS LEARNED FROM IMPLEMENTING BPPM

This section presents an overview of considerations and lessons learned when implementing BPPM.

RIGHT TIME TO IMPLEMENT BPPM

The right time to implement BPPM and the extent depend on the BPM maturity of an organization. Rosemann distinguished among three phases along the maturity path.[15] The first is the process-unaware organization, in which BPPM can be used to provide an initial process structure with governing strategic (core) and support

2. BPPM 度量和报告

度量和报告应该在两个层次进行：组合层次和流程改进项目层次。

在组合层面，除了上述一些信息，以下度量和报告信息可能是有益的。

- 流程的总数量：大型流程组合可能是工作重复增加的指标。
- 每个流程所有人的流程百分比：表示责任分散。
- 过程的成本/效益：促进围绕改进优先级和驱动有效性及效率的决策。
- 超过期望成熟度级别和平均成熟度级别的流程百分比：指示在何处部署资源和投资。
- 每个流程的资源数量：流程可能的组成或分解的指示器。

在流程改进项目级别，可以使用通常的项目度量和报告信息，例如[14]：

- 与战略目标保持一致；
- 成本和投资回报；
- 计划和资源；
- 范围/生产力；
- 项目周期时间（完成时间）；
- 项目评审和客户满意度；
- 风险管理。

5.5.10 建立BPPM的总结

本节概述了在组织中建立BPPM时的不同注意事项。它讨论了BPPM能力的创建、对齐问题、BPPM生命周期、项目组合管理的过程层次结构，以及度量和报告所需的项目组合信息。5.5.11节将概述在实现BPPM时获得的经验教训。

5.5.11 从实施BPPM中获得的经验教训

本节概述在实现BPPM时需要考虑的事项和吸取的教训。

5.5.12 实施BPPM的正确时机

实施BPPM的正确时间和程度取决于组织的BPM成熟度。Rosemann区分了成熟道路上的三个阶段[15]。第一阶段是不了解流程的组织，在这个组织中，BPPM可以用来提供一个具有控制策略（核心）和支持过程的初始过程结构。第二阶段

processes. The process-aware organization already has a good understanding of its important processes, has established sufficient and consistent modeling guidelines, and has an integrated model repository. However, the models are usually underused. In this phase, BPPM can substantially contribute to the organization's next leap in BPM. The focus should be on cross-processes, interdependencies, and prioritization according to the strategy and the risk of the processes. The third is the process-mature organization with an established BPPM that is continuously tracking the process performance, providing appropriate process information at all levels of the organization and established sophisticated audit trails.

The BPM maturity model Figure 10 indicates the current state of organizations and their awareness of BPM process maturity. Most organizations surveyed exist at Level 2 maturity. An example of a BPM maturity path is also indicated with examples of what BPM deliverables can be expected along the journey. The BPM maturity model further indicates that as maturity is increased there is a proactive rather than reactive approach to processes management, and that industry performance improves for organizations with higher BPM maturity.

EFFECT OF LIMITED OR NO IMPLEMENTED BPPM IN THE LONG RUN

A limited BPPM or no BPPM can have severe effects on the operation and results of an organization, especially in the long term. This risk increases exponentially with the number of processes performed by an organization. Figure 11 shows an overview of direct effects of the lack of BPPM in an organization and their short-term and long-term results.

The sad reality is that many of these causes can be found in most organizations. Internal or external services and products are designed without a need and related processes are established to deliver them. Whereas external services and products fail when the customer does not accept them, unnecessary internal services are less obvious. The common approach to establishing a governance gate to facilitate the use of new internal services often leads to unnecessary or at least inefficient processes, binding resources that are required elsewhere.

CONCLUSIONS

BPPM constitutes the next logical step in implementing successful BPM. It provides an answer to organizations facing challenges with an ever-growing process portfolio with increasing duplication, with deploying their limited resources to the right process improvement initiatives at the right time, and with having an end-to-end view of how their processes create the desired business values. Organizations reaching higher BPM maturity will especially benefit from establishing BPPM.

This work discussed different aspects of implementing BPPM, such as the establishment of the BPPM competency, including areas to which it needs to align, the

是具有流程意识的组织。流程感知组织已经很好地理解了它的重要流程，建立了充分的和一致的建模指南，并拥有一个集成的模型存储库。然而，这些模型通常没有得到充分利用。在这个阶段，BPPM可以为组织在BPM中的下一个飞跃做出重大贡献。重点应该放在跨流程、相互依赖及根据策略和流程风险的优先级上。第三阶段是流程成熟的组织，它具有一个已建立的BPPM，该BPPM不断地跟踪流程性能，在组织的所有级别上提供适当的流程信息，并建立复杂的审计跟踪。

BPM成熟度模型（图10）指出组织的当前状态以及它们对BPM流程成熟度的认识。大多数被调查的组织都处于第2级成熟度。BPM成熟度路径的一个示例也通过BPM可交付成果在整个过程中的预期示例进行了说明。BPM成熟度模型进一步表明，随着成熟度的增加，流程管理采用了一种主动而非被动的方法，对于BPM成熟度较高的组织，行业性能将得到改善。

5.5.13　从长远来看，有限或没有实施BPPM

有限的BPPM或没有BPPM会对组织的运作和结果产生严重的影响，特别是从长期来看。这种风险随组织执行的流程数量呈指数级增加。图11概述了组织中缺乏BPPM的直接影响及其短期和长期结果。

可悲的现实是，这些原因中的许多可以在大多数组织中找到。内部或外部服务和产品的设计没有需求，并建立相关流程来交付它们。当客户不接受外部服务和产品时，外部服务和产品就会失效，而不必要的内部服务就不那么明显了。为方便使用新的内部服务而建立治理门的常用方法通常会导致不必要或至少效率低下的流程，从而将其他地方所需的资源绑定起来。

5.5.14　结论

BPPM构成了成功实现BPM的下一个逻辑步骤。它为面临挑战的组织提供了一种解决方案，即不断增长的流程组合、不断增加的重复、在适当的时间将有限的资源部署到适当的流程改进计划中，以及端到端地了解流程如何创建所需的业务价值。达到更高的BPM成熟度的组织将特别受益于建立BPPM。

这项工作讨论了实施BPPM的不同方面，如建立BPPM能力，包括它需要调整的领域、BPPM生命周期的运行、所需的业务流程层次结构以及在组合和流程改进

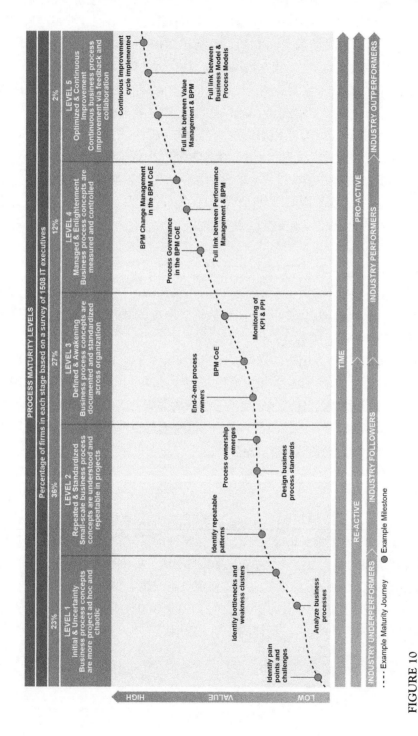

FIGURE 10

Business process management maturity levels.[16]

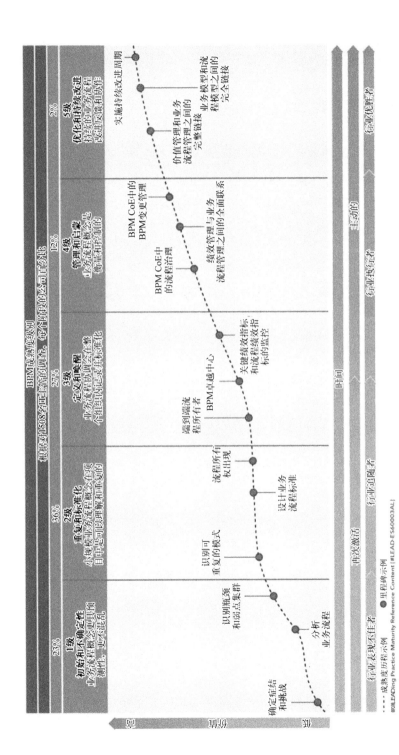

图 10 BPM 成熟度级别[16]

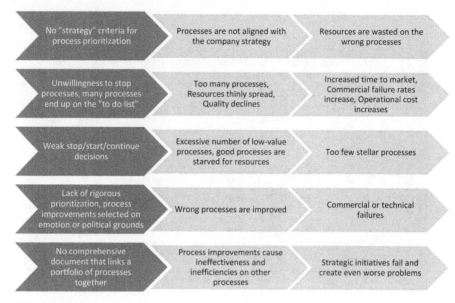

FIGURE 11

Causes of no or limited BPPM with short-term and long-term effects.

running of the BPPM life cycle, the required business process hierarchy, and portfolio information required for measurement and reporting at the portfolio and process improvement level. Furthermore, the timing of BPPM implementations depending on the organizations' BPM maturity was discussed, as well as common pitfalls and lessons learned when implementing BPPM.

Although, portfolio management in other areas such as project and PPM are well established, BPPM is relatively new. It is expected that successful implementations of BPPM in organizations will increase and that BPPM will be joined with PPM in the future.

End Notes

1. Darmani A. and Hanafizadeh P., "Business Process Portfolio Selection in Re-engineering Projects," *Business Process Management Journal* 19, no. 6 (2013): 892–916.
2. Scheer A-W., Feld T., and Caspers R., "BPM: New Architecture Driven by Business Process Planning and Control (BPPC). IM Journal for Information Management and Consulting," *Special Print* (2012): 1–8.
3. Robertson B., *Hype Cycle for Business Process Management*, 2013 (Gartner Inc., 2013).
4. Rosemann M., (2006). "Process Portfolio Management," Retrieved May 20, 2014, from BPTrends: http://www.bptrends.com/publicationfiles/04-06-ART-ProcessPortfolio Management-Rosemann1.pdf.
5. Allee V., *A Value Network Approach for Modeling and Measuring Intangibles*. Presented at Transparent Enterprise (Madrid, Spain, 2002).

图11　短期和长期没有或有限影响BPPM的原因

级别进行度量和报告所需的组合信息。此外,还讨论了基于组织的BPM成熟度的BPPM实施时机,以及在实施BPPM时所遇到的常见陷阱和经验教训。

　　尽管项目和PPM等其他领域的项目组合管理已经很好地建立了,但是BPPM较新。期望组织中BPPM的成功实现会增加,并且未来BPPM将与PPM结合。

参考文献

[1] Darmani A. and Hanafizadeh P., "Business Process Portfolio Selection in Re-engineering Projects," *Business Process Management Journal* 19, no. 6 (2013): 892−916.

[2] Scheer A-W., Feld T., and Caspers R., "BPM: New Architecture Driven by Business Process Planning and Control (BPPC). IM Journal for Information Management and Consulting," *Special Print* (2012): 1−8.

[3] Robertson B., *Hype Cycle for Business Process Management*, 2013 (Gartner Inc., 2013).

[4] Rosemann M., (2006). "Process Portfolio Management," Retrieved May 20, 2014, from BPTrends: http://www.bptrends.com/publicationfiles/04-06-ART-ProcessPortfolioManagement-Rosemann1.pdf.

[5] Allee V., *A Value Network Approach for Modeling and Measuring Intangibles*. Presented at Transparent Enterprise (Madrid, Spain, 2002).

6. Clark S. and Cameron B., "Business Process Portfolio Management: A Strategic Alignment Perspective," in *Business Enterprise, Process, and Technology Management: Models and Applications*, ed. V. Shankararaman, J.L. Zhao, and J.K. LeeHershey (Pennsylvania, USA: IGI Global, 2012), 18–31.

7. Bilodeau, N. Déjà Vu! From Project to Process Portfolio Management. Retrieved May 24, 2014, from BPMInstitute.org: http://www.bpminstitute.org/resources/articles/d%C3%A9j%C3%A0-vu-project-process-portfolio-management.

8. Taken from the Portfolio Management Reference Content LEAD-ES10019AL.

9. See note above 6.

10. Taken from the Portfolio Management Reference Content LEAD-ES10019AL.

11. Taken from the Portfolio Management Reference Content LEAD-ES10019AL.

12. See the note 1 above.

13. Taken from the Business Process Reference Content LEAD-ES20005BP.

14. Darmani A. and Hanafizadeh P., "Business process portfolio selection in re-engineering projects," *Business Process Management Journal* 19, no. 6 (2013): 892–916.

15. See note above 4.

16. LEADing Practice Maturity Reference Content #LEAD-ES60003AL.

[6] Clark S. and Cameron B., "Business Process Portfolio Management: A Strategic Alignment Perspective," in *Business Enterprise, Process, and Technology Management: Models and Applications,* ed. V. Shankararaman, J.L. Zhao, and J.K. LeeHershey (Pennsylvania, USA: IGI Global, 2012), 18−31.

[7] Bilodeau, N. Déjà Vu! From Project to Process Portfolio Management. Retrieved May 24, 2014, from BPMInstitute.org: http://www.bpminstitute.org/resources/articles/d%C3%A9j%C3%A0-vu-project-process-portfolio-management.

[8] Taken from the Portfolio Management Reference Content LEAD-ES10019AL.

[9] See note above 6.

[10] Taken from the Portfolio Management Reference Content LEAD-ES10019AL.

[11] Taken from the Portfolio Management Reference Content LEAD-ES10019AL.

[12] See the note 1 above.

[13] Taken from the Business Process Reference Content LEAD-ES20005BP.

[14] Darmani A. and Hanafizadeh P., "Business process portfolio selection in re-engineering projects," *Business Process Management Journal* 19, no. 6 (2013): 892−916.

[15] See note above 4.

[16] LEADing Practice Maturity Reference Content #LEAD-ES60003AL.

Real-Time Learning: Business Process Guidance at the Point of Need

Nils Faltin, Mark von Rosing, August-Wilhelm Scheer

INTRODUCTION

Management guru Peter Drucker coined the term "knowledge worker" in his 1969 book, *The Age of Discontinuity*.[1] Although knowledge workers were differentiated from manual workers at that time, Drucker concluded that new industries would primarily employ knowledge workers and that in the information-based economy the role of the knowledge worker would be at the heart of all organizations. When we accelerate and fast-forward to today, the terms "knowledge worker" and "manual worker" are no longer mutually exclusive.[2] People loading product onto rail cars certainly work with their hands, but they may also contribute knowledge to the business. Toffler[3] observed that typical knowledge workers in the age of knowledge economy must have some automated system at their disposal to create, process, and enhance their own knowledge. In some cases, he argued, they would also need to manage the knowledge of their co-workers; so although knowledge workers engage in "peer-to-peer" knowledge sharing across organizational and company boundaries, forming networks of expertise around their activities,[4] they are not currently enabled enough for their activities they execute. It is not only knowledge workers who would benefit from having more and better, readily available information about their work tasks. Employees working in business processes with complex applications need more than training to attain the needed competence level. Experience shows that formal training measures will only build the foundation of what users need to be able to use the new software to its full extent.

Just 48 h after being trained, learners will recall only 30% of the learned knowledge.[5] But could the knowledge workers and learners not just look up the needed knowledge when they need it? Employees usually receive training documents, business process models, and presentation slides, and have information in wikis and blogs at their disposal. However, to look up information the user has to interrupt work and concentrate on looking for a solution to the problem at hand. The same is true for other sources employees consult to fill their knowledge gap, whether it is searching the Internet or online forums or asking colleagues (see Figure 1). Current research provides evidence that knowledge workers spend at least 38% of their time searching for information.[6] Because of all of this, users can be overwhelmed by the

The Complete Business Process Handbook. http://dx.doi.org/10.1016/B978-0-12-799959-3.00031-8

5.6 实时学习：需求点的BPG

Nils Faltin, Mark von Rosing, August-Wilhelm Scheer

5.6.1 介绍

管理学大师 P. Drucker 在 1969 年出版的《不连续的时代》(*The Age of Disconnection*)一书中提出了知识工作者一词[1]。尽管当时知识工作者与体力劳动者有所不同，但 P. Drucker 得出的结论是：新行业将主要雇佣知识工作者，而在信息化经济中，知识工作者的角色将是所有组织的核心。当我们加速并快速前进到今天，术语"知识工人"和"体力工人"不再是互斥的[2]。人将产品装载到轨道车上当然是用手工作的，但他们也可能为企业贡献知识。Toffler[3]指出，在知识经济时代，典型的知识工作者必须有一些自动化的系统来创建、处理和增强他们自己的知识。他认为：在某些情况下，他们还需要管理同事的知识；因此，尽管知识工作者在组织和公司范围内进行"点对点"的知识共享，形成围绕其活动的专业知识网络[4]，但他们目前没有足够的能力执行其活动。不仅仅是知识型员工，他们还可以从掌握更多更好、随时可用的工作任务信息中获益。在具有复杂应用程序的业务流程中工作的员工需要的不仅仅是培训以达到所需的能力水平。经验表明，正式的培训措施只能为用户充分使用新软件奠定基础。

经验表明，学习者在学习 48 小时后仅能回忆起所学知识的 30%[5]。但是，知识工作者和学习者是否可以在需要的时候不只是查找所需的知识呢？员工通常会收到培训文档、业务流程模型和演示幻灯片，并在 wiki 和博客中随时掌握信息。但是，要查找信息，无论是在互联网上搜索、在线论坛，还是询问同事（图 1），用户必须中断工作，集中精力寻找手边问题的解决方案。员工为填补知识空白而咨询的其他资源也是如此。目前的研究表明，知识型员工至少有 38% 的时间是在搜索信息[6]。正因为如此，用户可能会被可用信息的数量所淹没，在找到与他们相关的

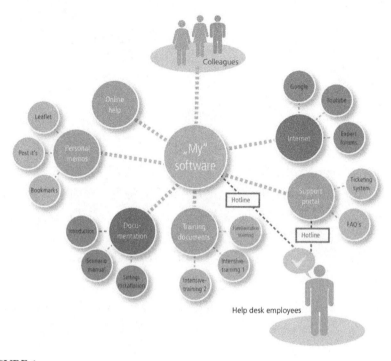

FIGURE 1

Knowledge workers have to find relevant information within a wealth of information sources.

amount of available information and lose much time until they find what is relevant for them. If users resort to trial and error, they may need even more time. As a result, productivity drops, usage errors sneak in, and acceptance of the newly introduced software system is at risk. As a consequence we see a whole new productivity, efficiency, performance management, and even effectiveness challenge for the modern organization. Thus, although there is a high level of standardization and integration enabled by automation, the potential for knowledge workers and other employees to call upon and use the embedded rules, guidelines, and knowledge is limited, hindering the organization's ability to innovate as well as harming its potential for growth.[7]

To give an example of the possible impact on an organization, such problems hit a large German clothing brand in 2006. When they replaced their outdated collection of enterprise resource planning (ERP) systems serving 2400 employees at five company sites with one new system, management expected more flexible, effective, and modern planning of clothing production. Because there are only a few delivery dates per year to dealers, production and logistics have to run reliably. However, the opposite happened. Employees were not able to use the new system correctly and made errors entering data. The wrong type and amount of

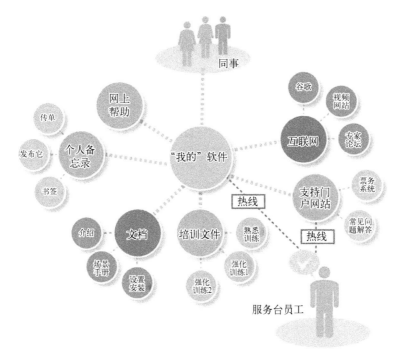

图1　知识工作者必须在丰富的信息源中找到相关的信息

信息之前会浪费很多时间。如果用户尝试错误,他们可能需要更多的时间。结果,生产力下降,存在使用错误风险,以及接受新引入软件系统存在障碍。因此,我们看到了一个全新的生产力、效率、绩效管理,甚至对现代组织的有效性的挑战。因此,尽管自动化实现了高度的标准化和集成,但是知识型员工和其他员工调用和使用嵌入的规则、指导方针和知识的潜力是有限的,这不仅阻碍了组织的创新能力,而且损害了其增长潜力[7]。

　　为了举例说明这些问题对一个组织可能产生的影响,举一个一家大型德国服装品牌在2006年遭遇的类似问题的例子。当他们用一个新系统取代过时的ERP系统时,因为经销商每年只有几个交货日期,所以生产和物流必须可靠运行,管理层期望服装生产的规划更加灵活、有效和现代化。然而,相反的事情发生了:员工无法正确使用新系统,在输入数据时出错,错误的面料类型和数量被送到工厂,导

fabric was delivered to the factories, which led to interruptions in the production of clothes. Insufficient supply to dealers was one reason the company later became insolvent.

All of this shows how important it is not only to train employees before a software rollout, but also to support them directly after going live and beyond. Learning and development (L&D) teams in the human resource departments become increasingly aware of this need and are looking for solutions. A benchmark by Towards Maturity published in 2013 highlighted that L&D teams are looking to technology to help them roll out new systems and processes faster.[8] Although 81% of these teams want technology to help them implement new information technology systems, only 28% of organizations already use such technology. What is troubling is that many organizations' value creation suffers because of lack of collaboration and reuse of existing knowledge[9] by not having the ability to provide the right information when the user needs it.

Because of this complexity and these challenges, today's workers and users want knowledge and skill elements that are concise and fit their need. Most expect the information to be available in a timely manner and easy to access, i.e., the time to consume should be as short as possible, requiring the content to be in small units with specific narrow and relevant topics, i.e., pragmatic and simple usage. These are often referred to as knowledge nuggets.

What seems to be most relevant in terms of these requirements is information that is discrete, concomitant, and directly relevant to the situation. It can be provided by knowledge situated or integrated into the activities, therefore being captured and managed within a process. Because this is a whole new way of continuous improvement of the learning organization, the following sections will explain how such technology support for learning new processes and applications can be made available to process workers.

REAL-TIME LEARNING TO CLOSE THE KNOWLEDGE GAP

Bite-Sized Learning Units

Real-time learning means providing the right information exactly when the user needs it. Information is presented in small, self-contained units to support building up knowledge quickly. This concept is called microlearning and enables learning with small learning units in the work context, adapted to the user. Users can obtain learning content in the midst of the work process, fitting to their context and the problem at hand. Later, if a user is confronted with the same problem again, the content can easily be obtained and consumed again, until it has been learned well.

Microlearning content is designed to enable multiple classes of workers in such a way that it can be easily understood and learned in a short time. Because it helps to solve a current problem, it enables operational excellence, improves efficiency,

致服装生产中断。最后对经销商的供应不足是该公司后来破产的原因之一。

　　所有这些都表明,不仅在软件发布之前对员工进行培训,而且在软件发布之后对他们提供支持是多么重要。人力资源部门的学习和发展(learning and development,L&D)团队越来越意识到这一需求,并正在寻找解决方案。2013年发表的《迈向成熟》(*Towards Maturity*)报告强调:学习和发展团队正在寻求技术帮助他们更快地推出新系统和流程[8]。尽管81%的团队希望技术能够帮助他们实现新的IT系统,但是只有28%的组织已经在使用这种技术[9]。令人不安的是,许多组织的价值创造受到损害,因为缺乏协作和现有知识的重用,无法在用户需要的时候提供正确的信息。

　　由于这种复杂性和挑战,今天的工人和用户希望知识和技能元素简明扼要,符合他们的需要。大多数人希望信息能够及时、容易获得,即消费的时间应尽可能短,要求内容碎片化并集中在相关的主题,即实用性强、使用简单。这些通常被称为"知识块"。

　　就这些需求而言,最相关的似乎是离散的、伴随的、与情况直接相关的信息。它可以由位于活动中的或集成到活动中的知识提供,因此可以在流程中捕获和管理。因为这是一种全新的持续改进学习型组织的方式,下面的部分将解释如何向流程工作者提供这种学习新流程和应用程序的技术支持。

5.6.2　实时学习,缩小知识差距

碎片化的学习单元。

　　实时学习意味着用户在需要的时候提供正确的信息,为快速构建知识体系,这些信息以碎片化的单元表示,这一概念被称为微学习,它能够在工作环境中使用适合用户的块状单元进行学习。用户可以在工作的过程中获得学习内容,适合自己的环境和手头的问题。之后,如果用户再次遇到同样的问题,内容可以很容易地再次获取和使用,直到学习完成为止。

　　微学习内容的设计目的是让多类工作者能够在短时间内轻松理解和学习。因为它有助于解决当前的问题,所以可以实现卓越的操作、提高效率,并有助于保持

and helps to keep cost low. Furthermore, it enables self-learning, gives a feeling of success, and strengthens motivation for further work and learning. In a sense, it empowers and therefore is a new perspective on learning processes in mediated environments. Microlearning is especially common in the area of e-learning, where it caters to different learning styles and media preferences. Examples are short text explanations or video sequences, test questions, pictures, screen shots, and Web-based trainings. Also, apps, quizzes, and learning games that are commonly used on mobile devices (such as smartphones and tablets) can be regarded as a kind of microlearning content. As Theo Hug pointed out in *Micro Learning and Narration*,[10] no matter whether learning refers to the process of building up and organizing knowledge, to the change of behavior, attitudes, values, mental abilities, cognitive structures, emotional reactions, action patterns, or societal dimensions, in all cases we have the possibility of considering micro aspects of the various views on more or less persisting changes and sustainable alterations of performance.

ELECTRONIC PERFORMANCE SUPPORT: DELIVERING KNOWLEDGE AT THE POINT OF NEED

An *electronic performance support system* (EPSS) is technology for implementing the idea of real-time learning. It supplies users with small context-related learning units directly at the workplace and increases users' productivity and effectiveness. Usually an EPSS is defined as "a computer-based system that improves worker productivity by providing on the job access to integrated information, advice and learning experiences."[11] It helps reduce work process complexity and processing time, providing exactly the information a user really needs and user decision support for solving specific problems.[12]

BUSINESS PROCESS GUIDANCE

Business process guidance (BPG) takes performance support to the next level. Instead of just supporting users working with a single software application, BPG shows them an overview of all steps (in the business process in which they work) and guides them step by step through the process across several applications. Both software-based and manual work steps can be supported. BPG leaves a degree of freedom to the user regarding how to execute the process. This is in contrast to *workflow management systems* that strictly enforce each step of a process. Another difference is that BPG works well even if the applications cannot be controlled from a central system, whereas workflow management systems need tight technical integration with the applications to be able to start each application and data entry screen automatically.

低成本。此外,它通过自我学习给人一种成功的感觉,并增强进一步工作和学习的动力。在某种意义上,它赋予了在中介环境中学习过程的能力,因此是一种新的视角。微型学习在电子学习领域尤其普遍,如简短的文字说明或视频、测试问题、图片、屏幕录制和基于web的培训,迎合了不同的学习风格和媒体偏好。此外,在移动设备(如智能手机和平板电脑)上普遍使用的应用程序、测验和游戏化学习也可以被视为一种微学习内容。Theo Hug(较早界定微型学习概念的学者之一)在《微型学习与叙事》中指出[10]:无论学习是指建立和组织知识的过程,还是行为、态度、价值观、心理能力、认知结构、情绪反应、行为模式或社会维度的变化,在所有情况下,我们都有可能考虑各种观点的微观方面,或多或少地持续变化以及可持续的性能改变。

5.6.3   在线支持:在需要时提供知识

电子绩效支持系统(EPSS)是一种实现实时学习思想的技术。它直接在工作场所为用户提供与环境相关的小型学习单元,提高用户的生产力和效率。EPSS通常被定义为"一种基于计算机的系统,它通过在工作中提供综合信息、建议和学习经验来提高工人的生产力[11]。"它有助于减少工作过程的复杂性和处理时间,为用户提供准确的信息,并为解决特定的问题为用户提供决策支持[12]。

5.6.4   BPG

BPG支持方式将绩效提升到一个更高的级别。BPG不只是支持用户使用单个软件应用程序,而是基于软件和手工的工作步骤,向他们展示所有步骤的概览(在他们工作的业务流程中),并在多个应用程序之间一步一步地指导他们完成流程。对于如何执行流程,BPG给用户留下了一定的自由度。这与严格执行流程的每个步骤的工作流管理系统形成了对比。另一个区别是,即使应用程序不能从中央系统进行控制,BPG也能很好地工作,而工作流管理系统需要与应用程序紧密地技术集成,以便能够自动启动每个应用程序和数据输入屏幕。

COMPONENTS OF A BPG SYSTEM

A *BPG system* will be used by content authors and process workers (end users). Each user group needs a different user interface to interact with the system.

Authors create support information for each role and each step in a business process. Support information can be new written text but it also links to existing content such as user manuals, Web-based trainings, user guidelines, business process diagrams, and any other media available on the intranet or Internet. Other important media to support the use of software applications are screen shots and screen recordings. Authors connect support information to user interface elements of the supported application, such as application windows, menus, forms, and data entry fields. This will later allow the BPG system to display the relevant support information automatically when the user reaches a certain process step with the respective application window.

For each business process and role, content authors can create a sequence of user interaction steps. This will enable the BPG system to present the business process as a whole to the user and then guide the user along the process steps.

Process workers use software that detects the application, application window, and process step the user is in. Only information relevant to the current application context and business process step is shown. Such a BPG system is called *context sensitive*. In addition, users can manually search for a process, to be guided along its steps. This will help users to complete the work task. Users can then rate how helpful the content is to their need, ask questions, or suggest improvements to the content. This feedback will be forwarded to the content authors so they can answer questions and improve the support information.

BPG IN PRACTICAL USE

BPG can be applied to all kinds of business processes and application systems. Among the most common areas are support for data entry, multi-application processes, and simplifying communication with the support desk.[13]

SUPPORTING ENTRY OF CORRECT DATA

Large and medium-sized companies rely on ERP systems. They are needed to administer products, customers, orders, employees, and projects and to manage complex production, service delivery processes, and supply chains. Although many data entry forms seem self-explanatory, users can have difficulty figuring out what exactly needs to be entered in a certain input field. Consider the example of adding a new corporate customer into an application: To what industry sector does the organization belong? What is the industry sector code? In what format will the user have to enter a tax number? Do telephone numbers have to be entered with international prefix codes?

5.6.5　BPG 系统的组件

内容作者和流程工作者(最终用户)是使用BPG系统的主要用户。每个用户组需要一个不同的UI来与系统交互。

作者为业务流程中的每个角色和每个步骤创建支持信息,支持信息可以是新的书面文本,但也可以链接到如用户手册、基于web的培训、用户指南、业务流程图等内容,以及内部网或Internet上可用的任何其他媒体。支持软件应用程序使用的其他重要媒体是屏幕录制和屏幕录音。作者将支持信息连接到受支持应用程序的UI元素,如应用程序窗口、菜单、表单和数据输入字段。这将允许BPG系统在用户使用各自的应用程序窗口到达某个流程步骤时自动显示相关的支持信息。

对于每个业务流程和角色,内容作者可以创建一系列用户交互步骤。这将使BPG系统能够将业务流程作为一个整体呈现给用户,然后沿着流程步骤指导用户。

流程工作人员使用软件来检测用户所在的应用程序、应用程序窗口和流程步骤。只显示与当前应用程序上下文和业务流程步骤相关的信息。这样的BPG系统称为步骤感知系统。此外,用户可以手动搜索流程,并按照流程的步骤进行指导,这将帮助用户完成工作任务。然后,用户可以根据内容对他们需求的帮助程度进行评级、提出问题,或者对内容提出改进建议。这些反馈将被转发给内容作者,以便他们回答问题并改进支持信息。

5.6.6　实际应用中的BPG

BPG可以应用于各种业务流程和应用系统。最常见的领域之一是支持数据输入、多应用程序流程以及简化与支持后台的通信[13]。

5.6.7　支持正确数据的输入

大中型企业依赖ERP系统。他们需要管理产品、客户、订单、员工和项目,并管理复杂的生产、服务交付流程和供应链。尽管许多数据输入表单似乎是不言自明的,但用户可能难以确定在某个输入字段中究竟需要输入什么。考虑向应用程序中添加新公司客户的示例:组织属于哪个行业部门? 什么是行业部门代码? 用户必须以什么格式输入税号? 电话号码必须输入国际前缀代码吗?

A BPG system stores organization specific knowledge and provides it to users while they work with the data entry application. The system detects which processing step the user is in and displays information that is relevant to this context.

SUPPORTING MULTIPLE APPLICATIONS

Process guidance needs to be supported across applications. This allows the business process and its parts to be described in a common platform, structure, and layout, which makes the support media much easier to understand. In any application in which they work, users can get the needed support and thus can work at full efficiency.

As an example, a salesperson creates a quote using several applications. He looks up some customer base data in, for example, the Customer Relationship Management (CRM) system. In Excel he calculates a price offer. He transfers the offer to a Word template, sends the document with Outlook to the customer, and changes the offer status in the CRM system. Classic help systems are installed together with their application and can only support this one application. Therefore, these help systems do not provide real process support. A real BPG system can provide support across applications, available at every work step. It brings business process descriptions for all applications of an organization in a common format to the employees' workplace.

ENHANCING COMMUNICATION WITH THE SUPPORT DESK

When a new application is deployed, users frequently contact the support desk, which can create a high workload on the side of the support desk. User requests can be triggered by software errors, but in most cases they are caused by user errors caused by employees' inexperience with the application or missing training on certain parts of it. Users often find it hard to explain their problem to the support desk with the required level of detail. Many cycles of the support desk asking questions to the user about the problem can occur, which is annoying to both sides and increases the time needed to solve the problem. A BPG system should allow users to send their automatically determined work context (process and process step worked on, application and screen used, and screen shot with data) together with a short problem statement to the support desk. This should be available to users while working with the application: for example, with special keys or buttons added by the BGP system to the user interface. This saves them the effort of switching to an external ticketing system and manually describing the context in detail. In a similar vein, users can comment on the support information provided by the BPG system (with their context being transmitted automatically) to ask for additional information or suggest improvements to the existing information. Both users and the support desk may benefit from such a BPG system, because users find it much easier to ask for support and send improvement requests. On the other hand, the support desk saves many clarification requests and can continuously improve

BPG 系统存储特定于组织的知识,并在用户使用数据输入应用程序时将其提供给用户。系统检测用户所处的处理步骤,并显示与此上下文相关的信息。

5.6.8　支持多个应用程序

流程指导需要跨应用程序的支持。这允许在一个通用的平台、结构和布局中描述业务流程及其部分,这使得支持内容更容易理解。在他们工作的任何应用程序中,用户都可以获得所需的支持,从而可以充分高效地工作。

例如,销售人员使用多个应用程序创建报价。他在CRM系统中查找一些客户基础数据。在Excel中,他计算出一个报价,并将报价转移到Word模板,将带有Outlook的文档发送给客户,并在CRM系统中更改报价状态。老版本的帮助系统是与其应用程序一起安装的,并且只能支持这一个应用程序。因此,这些帮助系统不能提供真正的流程支持。一个真正的BPG系统可以提供跨应用程序的支持,在每个工作步骤都可用。它将组织的所有应用程序的业务流程描述以一种通用格式带到员工的工作场所。

5.6.9　加强与技术支持部门的沟通

部署新应用程序时,用户经常与服务部门联系,这可能会在服务部门一侧创建高工作负载。用户请求可以由软件错误触发,但在大多数情况下,它们是由员工对应用程序缺乏经验或缺少对应用程序某些部分的培训而导致的理解和操作错误引起的。用户常常发现很难向服务部门解释他们的问题,服务部门可能会多次向用户询问有关问题,这对双方来说都很烦人,并且增加了解决问题所需的时间。BPG系统允许用户将自动确定的工作内容(处理的过程和过程步骤、使用的应用程序和屏幕,以及带有数据的屏幕快照)连同简短的问题描述发送到服务部门。此外,用户在使用应用程序时应该可以使用诸如将BPG系统的特殊键或按钮添加到UI的方法,这为他们节省了切换到其他系统的时间。类似地,用户可以评论BPG系统提供的支持信息(其评论内容将自动传输),以要求提供更多信息或对现有信息提出改进建议。用户和服务部门都可以从这样的BPG系统中受益,因为用户发出请求支持比发送改进请求要容易得多。另一方面,服务部门可以节省大量的请求,

the support information stored in the BPG system based on user suggestions. This, in turn, will give future users the answers they need so they will not need to ask the support desk.

INTRODUCING BPG IN AN ORGANIZATION

Creating a Repository of Microlearning Content

A BPG system offers bespoke microlearning and is therefore not pre-filled for any one application. The learning material must be sourced and entered into the BPG system before it is made available to users. Documentation, handbooks, project groups, specialist departmental knowledge, compliance-relevant information, work instructions, and organization-specific business process know-how from process repositories and databases are all relevant content sources (see Figure 2).

Taking into account the way people learn, decisions must be made on the form of content most applicable to support a specific function or process; these could range from short texts, images and screenshots, videos, and documentation to interactive online learning modules.

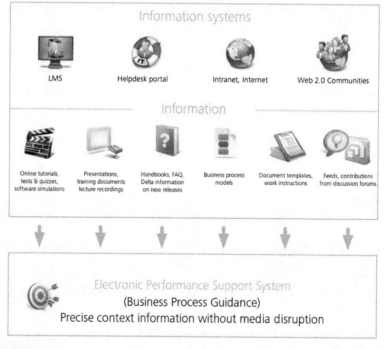

FIGURE 2

Learning content sources for Business Process Guidance.

并可以根据用户的建议不断改进存储在BPG系统中的支持信息。反过来,这将为未来的用户提供他们需要的答案,这样他们就不需要向技术服务部门询问了。

5.6.10　在组织中引入BPG

创建微型学习内容的存储库。

BPG系统提供定制的微学习,因此不需要任何应用程序预先填写。在向用户提供学习材料之前,必须提供来源并输入BPG系统。文档、手册、项目组、专业部门知识、与法规遵循相关的信息、工作指导以及来自流程存储库和数据库的特定于组织的业务流程知识都是相关的内容来源(图2)。

考虑到人们学习的方式,决策必须以最适用于支持特定功能或流程的内容形式作出,这些内容可以是短文本、图像和屏幕截图、视频和文档,也可以是交互式在线学习模块。

图2　BPG的学习内容来源

MAJOR STEPS TO CREATE THE REPOSITORY

Based on these principles, a typical approach to create a relevant, helpful, and up-to-date repository of BPG learning content is to:

1. Define the business process, applications, and functionality to be supported by the BPG system
2. Identify the individual learning groups (according to department, function, etc.)
3. Select and gather learning content (which topics need to be covered and at what level of detail). Focus should be on the most value-adding processes and those where users have most problems in interacting with the applications
4. Upload the content to the BPG system
5. Regularly analyze requests to support desk and improvement suggestions from users and update the learning content accordingly

CONCLUSIONS AND OUTLOOK

As shown, risks of introducing applications to an organization do not stem primarily from the existing or newly introduced technology. They stem from the quality of the software-supported processes and the ability of employees to use the technology correctly.

The same BPG systems support the implementation of new applications and the update of business processes that go with it. They will reduce the effort required for training users up front and the support desk efforts needed in the introduction phase. Users will learn to use new software applications more rapidly and thus become more efficient in their process work. They will also gain a better overview of the overall process and what role different applications have in it. This is an important part in ensuring that business processes are performed as designed.

It is expected that real-time learning through BPG will grow in importance in the future:

- *More changes*: Processes and applications will change even more frequently in the future, triggering a need for training and support among the employees using them
- *More collections of applications*: Instead of one large system installed and configured on premise, we will often see a collection of applications provided as a service out of the cloud. This asks for process guidance that works across applications and that can be configured and equipped with content by the user organization
- *Social networks will be used more at work*: We will also see more knowledge sharing and peer support using social network technologies at the workplace. Social BPG will provide users with access to the social network communication channels and will help to filter and display only messages that are relevant based on the process and application context of the user
- *Users will influence provision of content*: Statistics from software usage and user feedback will become an important source for content authors to provide

5.6.11　创建存储库的主要步骤

基于这些原则，创建相关的、有帮助的、最新的BPG学习内容存储库的典型方法如下。

（1）定义BPG系统支持的业务流程、应用程序和功能。

（2）确定个体学习小组（根据部门、职能等）。

（3）选择和收集学习内容（需要涵盖哪些主题以及在何种级别）。重点应该放在最有价值的流程，以及那些用户在与应用程序交互时遇到最多问题的流程上。

（4）将内容上传到BPG系统。

（5）定期分析用户对技术服务部门的要求和改进建议，并及时更新学习内容。

5.6.12　结论和展望

如图2所示，向组织引入应用程序的风险并不主要来自现有或新引入的技术。它们源于软件支持的流程的质量和员工正确使用技术的能力。

相同的BPG系统支持新应用程序的实现和相应业务流程的更新。它们将减少预先培训用户所需的工作量以及在引入阶段所需的技术支持部门的工作量。用户将学习更快地使用新的软件应用程序，从而在他们的流程工作中变得更有效率。他们还可以更好地了解整个流程以及不同应用程序在其中扮演的角色。这是确保按设计执行业务流程的重要部分。

预计通过BPG的实时学习在未来将变得越来越重要。

- 更多变化：流程和应用程序在未来将更频繁地发生变化，从而引发使用流程和应用程序的员工对培训和支持的需求。

- 更多应用程序集合：我们将经常看到作为云外服务提供的应用程序集合，而不是在本地安装和配置一个大型系统。这要求提供跨应用程序工作、用户组织可以配置和配备内容的过程指南。

- 社交网络将在工作中得到更多的使用：我们还将看到在工作场所使用社交网络技术进行更多的知识共享和同行支持。社交BPG将为用户提供访问社交网络通信渠道的权限，并将有助于根据用户的流程和应用上下文仅筛选和显示相关的消息。

- 用户将影响内容的提供：来自软件使用和用户反馈的统计信息将成为内容作者提供额外内容和改进BPG系统中现有支持内容的重要来源。

additional content and improve the existing support content in the BPG system

- *BPG will extend beyond the office*: Mobile devices will bring process guidance to new areas such as repair and maintenance of machines. First prototypes are built in research projects where information and work instructions will be displayed with augmented reality techniques on top of live pictures taken through the built-in camera. Users can call experts who support them directly, seeing the machine in real time through the camera.

BPG already is a good concept to support the introduction of new processes and applications. Its potential will grow in the future as it enables the organization.

End Notes

1. Drucker, Peter F., *The Age of Discontinuity Guidelines to our Changing Society*, 1969.

2. Rosen, E., *Every Worker Is a Knowledge Worker* (Business Week, 2011).

3. Toffler, A., *Powershift: Knowledge, Wealth and Violence at the Edge of the 21st Century*, 1990, ISBN 0-553-29,215-3..

4. Tapscott, Don; Williams, Anthony D., *How Mass Collaboration Changes Everything*, (Penguin, 2006) ISBN 1-59,184-138-0.

5. Güldenberg Stefan, "Wissensmanagement und Wissenscontrolling in lernenden Organisationen," (2003).

6. Mcdermott Michael, "Knowledge Workers: You can gauge their effectiveness," *Leadership Excellence* 22, no.10 (2005), ISSN: 8756–2308.

7. Overton Laura, "5 Practical Ideas for Embedding Learning into the Workflow," Available from towards Maturity Inc., published (July 2013) http://www.towardsmaturity.org/article/2013/07/04/5-practical-ideas-embedding-learning-workflow/.

8. Ibid.

9. Tapscott Don, Williams Anthony D., *How Mass Collaboration Changes Everything* (Penguin, 2006), ISBN 1-59184-138-0.

10. Hug T., *Micro Learning and Narration. Exploring possibilities of utilization of narrations and storytelling for the designing of "micro units" and didactical micro-learning arrangements* (MIT: Cambridge (MA), USA. 2005).

11. Raybould Barry, "An EPSS.Case Study," (1991).

12. Gery Gloria, "*Electronic Performance Support Systems: How and why to Remake the Workplace Through the Strategic Application of Technology*," (1991).

13. Milius Frank and Meiers Christina, "Performance Support für Mitarbeiter, Applikationen und Prozesse—Microlearning als methodischer Ansatz zur mitarbeiterorientierten Softwareschulung," *Information Management und Consulting* 26 (2011): 2.

- BPG将扩展到办公室以外：移动设备将为新领域带来流程指导，如机器的维修和维护。第一个原型是在研究项目中构建的，在通过内置摄像头拍摄的实况照片上，信息和工作说明将以增强现实技术显示。用户可以直接致电支持他们的专家，通过摄像头实时查看机器。

BPG已经是一个很好的概念来支持新流程和应用程序的引入。它的潜力在未来会随着它对组织的支持而增长。

参考文献

[1] Drucker, Peter F., The Age of Discontinuity Guidelines to our Changing Society, 1969.

[2] Rosen, E., Every Worker Is a Knowledge Worker (Business Week, 2011).

[3] Toffler, A., Powershift: Knowledge, Wealth and Violence at the Edge of the 21st Century, 1990, ISBN 0-553-29,215-3.

[4] Tapscott, Don; Williams, Anthony D., How Mass Collaboration Changes Everything, (Penguin, 2006) ISBN 1-59,184-138-0.

[5] Güldenberg Stefan, "Wissensmanagement und Wissenscontrolling in lernenden Organisationen," (2003).

[6] Mcdermott Michael, "Knowledge Workers: You can gauge their effectiveness," Leadership Excellence 22, no.10 (2005), ISSN: 8756−2308.

[7] Overton Laura, "5 Practical Ideas for Embedding Learning into the Workflow," Available from towards Maturity Inc., published (July 2013) http://www.towardsmaturity.org/article/2013/07/04/5-practical-ideas-embedding-learning-workflow/.

[8] Ibid.

[9] Tapscott Don, Williams Anthony D., How Mass Collaboration Changes Everything (Penguin, 2006), ISBN 1-59184-138-0.

[10] Hug T., Micro Learning and Narration. Exploring possibilities of utilization of narrations and storytelling for the designing of "micro units" and didactical micro-learning arrangements [MIT: Cambridge (MA), USA. 2005].

[11] Raybould Barry, "An EPSS.Case Study," (1991).

[12] Gery Gloria, "Electronic Performance Support Systems: How and why to Remake the Workplace Through the Strategic Application of Technology," (1991).

[13] Milius Frank and Meiers Christina, "Performance Support für Mitarbeiter, Applikationen und Prozesse—Microlearning als methodischer Ansatz zur mitarbeiterorientierten Softwareschulung," Information Management und Consulting 26 (2011): 2.

Business Process Management Alignment

Mona von Rosing, Henrik von Scheel, Justin Tomlinson, Victor Abele, Kenneth D. Teske, Michael D. Tisdel

INTRODUCTION

Alignment is a concept that dates back to the late 1990s, when it was described by Paul Strassmann[1]: "Alignment is the capacity to demonstrate a positive relationship between information technologies and the accepted financial measures of performance." Alignment of business process management (BPM) hence should follow a similar principle or pattern to be effective. The objective therefore is how this alignment to and between BPM can create value that is ultimately measurable as a favorable financial outcome for a commercial enterprise.

Business process management alignment, which is focused on both reusability and accelerating automation, requires that business managers have an understanding of what alignment is, how to develop an alignment competency, and what considerations should be made by organizations to ensure alignment is adequately adopted. This chapter discusses these aspects of alignment and gives credence to the development of aligned BPM.

BACKGROUND TO A NEW WAY OF LOOKING AT ALIGNMENT FOR BPM

The portfolio alignment-unity concept was developed for the United States Department of Defense (DOD), Department of Homeland Security (DHS), Department of Justice (DOJ), and Department of State (DOS) with the aim of

- Unifying common stakeholders, objectives, and size for common, complex, and critical missions and multidimensional warfare such as cyber war, combating weapon of mass destruction, combating transnational organized crime, and for security corporations.
- Achieving information sharing and unity of effort to meet national security objectives for the US DOD, DHS, DOJ, and the DOS.

US Government research involving the DHS, DOS, DOJ, and DOD initiated an alignment effort to:

1. Identify and specify common and repeatable patterns for business, application, and technology areas
2. Support analysis and stability operations planning efforts per JROCM 172-13
3. Change and update joint doctrine

The Complete Business Process Handbook. http://dx.doi.org/10.1016/B978-0-12-799959-3.00032-X

5.7　BPM的一致性

Mona von Rosing, Henrik von Scheel, Justin Tomlinson, Victor Abele, Kenneth D. Teske, Michael D. Tisdel

5.7.1　介绍

一致性是一个可以追溯到20世纪90年代后期的概念,当时P. Strassmann[1]描述了这一概念:"一致性能够证明信息技术与可接受的财务绩效衡量标准之间的正向关系。"因此,商业企业应该把BPM与财务指标之间的一致性作为衡量BPM是否创造价值的标准。

BPM一致性主要关注流程的复用性和对自动化的促进,它要求业务经理了解什么是一致性、如何具备开发一致性能力,以及组织应考虑哪些因素以确保一致性的充分性。本节将讨论这些方面,并为确保BPM一致性的开发提供依据。

5.7.2　了解BPM一致性的背景知识

美国国防部(the United States Department of Defense, DOD)、国土安全部(Department of Homeland Security, DHS)、司法部(Department of Justice, DOJ)和国务院(Department of State, DOS)制定了"一致性-统一"的概念,其目的是:

- 为共同、复杂和关键的任务和多维战争(如网络战、打击大规模杀伤性武器、打击跨国有组织犯罪和安全公司)统一共同利益相关者、目标和规模;
- 为美国国防部、国土安全部、司法部和国务院实现信息共享和共同努力,以实现国家安全目标。

美国政府的研究包含国土安全部、国务院、司法部和国防部发起了一项结盟致力于:

(1)为业务、应用程序和技术领域确定和指定通用的和可重复的模式。

(2)根据JROCM 172-13提供分析和稳定操作方法。

(3)改变和更新联合的原则。

4. Assess for use by the Executive Committee Joint Program Office (JPO) for Assignment of National Security and Emergency Preparedness Communications Functions per Executive Order 13,618
5. Benchmark, research and analyze, and identify alignment and unification patterns
6. Pilot first projects within US Government
7. Join and develop alignment and unity reference content that increases the level of reusability and replication within alignment and unity of stakeholders, portfolios, programs, and enterprise modeling, enterprise engineering, and enterprise architecture concepts
8. Extend with accelerators and templates, such as the:
    a. Alignment and Unity Stakeholder Map
    b. Alignment and Unity Quick Scan
    c. Alignment and Unity Maturity TCO-ROI evaluation
    d. Alignment and Unity Maturity Benchmark
    e. Alignment and Unity Development Path

The alignment-unity framework concept was such a success that DOD Stability Operations recommended the alignment-unity framework concept be used to support analysis and stability operation planning efforts per JROCM 172-13. A Unity of Effort Synchronization Framework Joint Knowledge online course was developed and over 600 DOD and other governmental personnel have taken and completed the course. In addition to this, Joint Doctrine Publication 3.0, 3.22, and others have adopted the alignment-unity framework and are incorporating it into the newest editions.

As of August 2014, the alignment-unity framework concept was being assessed for use by the Executive Committee JPO for Assignment of National Security and Emergency Preparedness Communications Functions. The framework has already been applied by US Special Operations Command to align its information technology portfolio as well as assist the J3-International division in finding commonality while building the Global Special Operations Forces Network (GSN) with its multinational mission partners.

Such a comprehensive alignment management concept uniquely recognizes that any organization, department, or even program, even if it has its own mission, vision, strategies, and critical success factors, is only one element of a larger delivery and service mechanism. In nearly all cases the success of strategy to execution depends on the ability to operate in alignment and therefore unity with the rest of the organizations with a common stake in the issues.

This truly encourages collaboration across areas, groups, portfolios, programs, and projects that will enable value creation and realization. However, realizing higher levels of alignment and unity requires identification of common objectives, initiatives, and standards or requirements.

Most organizations today face significant hurdles to ensure organizational alignment among goals, stakeholder, plans, programs, projects, and portfolios. Identification is the first step toward developing solutions or mitigation strategies. This US DOD Unity of Effort Framework project was developed with several organizational participants to identify important inhibitors to achieving unity of effort. Identifying the negatives is important, however; we learned that identifying the positives such

（4）根据13618号行政命令，评估执行委员会联合规划办公室（Joint Program Office，JPO）用于分配国家安全和应急准备通信功能方面的使用情况。

（5）标准测试、研究和分析，并识别一致性和统一模式。

（6）在美国政府内部试点新批项目。

（7）加入和开发一致性和统一性参考内容，在利益相关者、投资组合、程序和企业建模、企业工程和企业体系结构概念的一致性和统一性中提高可重用性和复制性的级别。

（8）使用加速器和模板进行扩展，例如：

a. 利益相关者关系的一致性；

b. 快速反应的一致性；

c. TCO-ROI（成本-收益）评估方法的一致性；

d. 成熟度标准的一致性；

e. 发展路径的一致性。

"一致性-统一"框架概念是如此成功，以至于国防部作战部门建议将"一致性-统一"框架概念用于支持JROCM 172-13的分析和规划工作，并开发了一个统一的工作同步框架联合知识的在线课程，共有超过600名国防部和其他政府人员参加并完成了该课程。除此之外，联合理论出版的3.0、3.22版本和其他已经采用了该框架，并将其纳入最新版本。

截至2014年8月，执行委员会JPO正在评估校准"一致性-统一"框架概念，以用于分配国家安全和应急准备通信职能。该框架已被美国特种作战司令部应用，以调整其信息技术组合，并协助"J3国际部"在与其多国任务伙伴建立全球特种作战部队网络（global special operations forces network，GSN）的同时寻求共性。

这种全面的一致性管理概念独特地认识到，任何组织、部门、甚至项目，即使它有自己的使命、愿景、战略和CSF，也只是更大的交付和服务机制的一个元素。在几乎所有的情况下，战略执行的成功取决于协调一致的运作能力，也取决于与在问题中有共同利益的其他组织的团结。

这确实鼓励跨领域、组织、项目组、项目的协作，从而创造价值。然而，实现更高层次的一致性和统一性需要确定共同的目标、计划和标准或需求。

如今，大多数组织都面临着在确保目标、利益相关者、计划、规程、项目和投资组合之间一致性的重大障碍。识别是制定解决方案或缓解策略的第一步。这个美国国防部统一工作框架（US DOD Unity of Effort Framework）项目是美国国防部与几个组织参与者共同开发的，旨在确定实现工作一致性过程中的重要阻碍因素。

as goals, areas of interest, and categories of effort applied by each of the organizations worked much better for gaining unity.

A lesson learned was that working together with a framework provided many more benefits than detractors on the way to improving unity of effort for complex governmental missions to include operational design, planning, and decision making about scarce resources. The framework also enables orchestrated development of planning to achieve regional and national objectives and is an enabler for building partnership capacity and security sector assistance.

The framework allowed for recommendations based on opportunities for strong organizational partnerships. Another lesson learned while working with stakeholders is that the framework allowed for identification of redundancies or overlaps, gaps in support requirements, seams in the operating environment, and shortfalls in resources.

We also learned that to develop true alignment, it requires representation, participation, and collection of information from stakeholder organizations. To facilitate this, an organization or group must be identified to manage the time and processes to complete a framework. In addition, some events must occur in person to allow time for stakeholders to validate and clarify collected information and participate through staffing activities.

The alignment as indicated in this chapter needs to be specific to business processes and their related objects and enterprise business elements. First, though, we need to define what alignment is in relation to BPM.

ALIGNMENT OF BPM

Most stakeholders across the enterprise landscape have some of the same external and internal forces and drivers influencing them, but different approaches. These stakeholders do not see what is common and hence they do not know how or why to work together. This indicates a lack of alignment maturity and results in enterprise strategy, management, and operations that are disjointed and do not provide the expected return on investment, representing an untapped potential of cost savings and operational excellence for both effectiveness and efficiency only based on the wide range of duplication of goals, competencies, services, process, functions, task, resources, roles, data, etc.

Alignment of BPM provides for the policy or strategy of the organization to drive the alignment of BPM portfolios, programs, and projects that require the relevant stakeholders (business process owners) to develop a common understanding of their business process so that there is a transformation of business process from the "as-is" through to the "to-be." The to-be business processes that have been aligned can then be used in enterprise transformation and innovation to enable improved financial measures of performance. This high level of BPM alignment is described in Figure 1.

ESTABLISHING ALIGNMENT TO BPM

One of the key tasks before even starting to establish alignment to BPM is to confirm that BPM alignment within an organization is even feasible. This requires two questions to be answered in the affirmative:

1. Is there a clear link with the organizations planning and budget commitments?
2. Does the organization have the level of competency required to carry out such a task?

我们意识到,识别诸如目标、感兴趣的领域和每个组织所应用的工作类别积极因素,对于获得一致性效果要好得多。然而,识别消极因素同样是很重要的。

一个经验教训是:在改善政府复杂的对稀缺资源的操作设计、规划和决策任务的统一性方面,该框架能够提供更多好处。此外,该框架在促进规划的协调发展、实现区域和国家目标、促进建立伙伴关系能力和安全部门援助等方面同样效果出色。

该框架基于强有力的组织伙伴关系提出建议,与利益相关方合作时学到的另一个教训是识别冗余或重叠的框架,包括识别出要求的差距、运营环境中的接缝以及资源的不足。

因此,我们了解到要发展真正的一致性,需要利益相关者组织的代表参与信息收集。为此,组织或组必须能够管理完成框架的时间和流程。此外,一些活动必须由利益相关者亲自进行,以便利益相关者有时间验证和澄清所收集的信息并参与人员配备活动。

本节中指出的对齐需要特定于业务流程及其相关对象和企业业务元素。第一步就是需要我们定义与BPM相关的对齐方式。

5.7.3　BPM一致性

企业环境中的大多数利益相关者都有一些相同的外部/内部力量和驱动因素影响他们,但是方法不同。这些利益相关者看不到什么是共同的,因此他们不知道如何或为什么要一起工作。这表明缺乏一致性的标准和结果与企业战略、管理和操作脱节。事实上,只有在目标、能力、服务、过程、职能、任务、资源、角色、数据等持续、重复的基础上,才能在效率和效益方面表现出尚未实现的节省成本和卓越运作的潜力。

BPM的一致性为组织的政策或战略提供了驱动BPM组合、计划和项目的一致性,这些项目需要相关的利益相关者(业务流程所有人)对其业务流程进行共同理解,从而使业务流程从"过去""as-is"转变为"未来""to-be"的未来业务流程,然后可用于企业转型和创新,以提高财务绩效指标。图1描述了这种高度的BPM一致性。

5.7.4　建立与BPM的一致性

在开始建立与BPM的一致性之前,关键任务之一就是确认组织中的BPM一致性是可行的。这需要两个问题得到肯定的回答:

(1)与本组织的规划和预算是否有明确的联系?

(2)组织是否具备完成此类任务所需的能力水平?

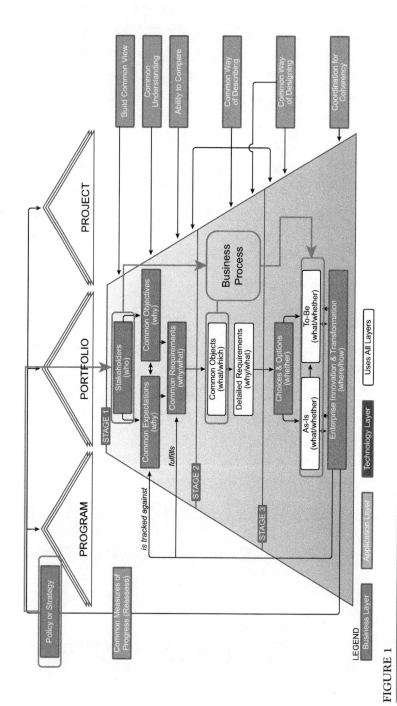

FIGURE 1

Business process management alignment from policy to enterprise innovation and transformation.[2]

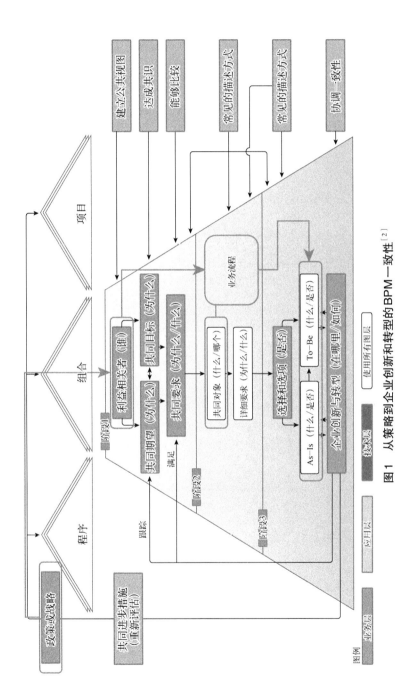

图 1 从策略到企业创新和转型的 BPM 一致性[2]

If the answer to these two questions is no or if it is uncertain, it is likely that the organization is suffering from one of the following symptoms in the area of BPM:

1. Stove pipes/silos (lack of process information sharing)
2. No visibility of BPM efforts and activities
3. Duplication of efforts and investments across the same set of business processes
4. Lack of planning resources to enable aligned BPM
5. No collective repository of process-centric information
6. Competing priorities among the stakeholders of a specific business process
7. Differing lexicon/taxonomy/language/vocabulary/semantics for BPM
8. Disparate activities across the organization relating to BPM

Overcoming these challenges requires the buy-in and leadership of senior executives. Decision makers and corporate governance requires a higher order of insight to effectively identify gaps and overlaps in its transformation and innovation plans. Furthermore, this is needed to identify opportunities to minimize costs and improve performance of its operational services, as well as to determine a road map for capital investments in corporate infrastructures. Finally, without a methodical way to link strategic business objectives across all layers of a corporate architecture, the organization runs the risk of disjointed execution and a diminished capacity to effectively control and assess the performance of service providers, both internal and external.

In both public and private sector organizations, the consequences translate to higher operating costs, disappointing returns on investment in transformation and innovation, and lost market opportunity.

Throughout business planning and project gating cycles, an organization needs to identify the portfolios, programs, and/or projects that are BPM related and identify aspects that are not aligned to their planning and budget commitments, correct misaligned and redundant efforts, and adjust where possible to an aligned state. Furthermore, the organization needs to identify what competencies are required to achieve the level of BPM alignment, which will bring improved efficiency and effectiveness and advance the organization's financial measures of performance. To methodically assess the potential for business process alignment, it is valuable to take an architectural view of which objects the business processes would relate to and therefore what templates could be used to facilitate improved alignment.

Table 1 describes objects that would be relevant to the overall business processes group object.

As indicated in Figure 1 the alignment starts at the highest level for BPM with the policy and the relevant stakeholders. They then need to ascertain to which business process meta objects can be related and aligned, bringing about synergy to the higher levels.

In Figure 2, alignment of the business process meta objects is highlighted through the relationships that can be made. The way this is achieved is through the development of maps, matrices, and models that cover from forces and drivers all the way to infrastructure high availability. This means that the templates that are relevant

如果这两个问题的答案是否定的,或者是不确定的,那么很可能组织在BPM领域正遭受以下症状之一:

(1)存在壁垒(缺乏过程信息共享);

(2)BPM工作和活动不可见;

(3)在同一组业务流程中重复工作和投入;

(4)缺乏支持一致BPM的规划资源;

(5)没有以流程为中心的信息的集中存储;

(6)特定业务流程之间没有优先级;

(7)BPM的词汇/分类/语言/语义存在不同;

(8)组织中存在与BPM相关的不同活动。

克服这些挑战需要高管的参与和领导。决策者和公司治理需要更高水平的洞察力,以有效识别其转型和创新计划中的差距和重叠。此外,还需要确定将其业务服务的成本降到最低并具备改善其业绩的机会,以及确定对公司基础设施进行资本投资的路线图。最后,如果没有系统的方法完成跨公司架构的所有层SBO的有效链接,组织将面临执行脱节和有效控制与评估服务提供者(包括内部和外部)的性能的能力下降的风险。

在公营和私营部门组织中,其后果将转化为更高的运营成本、转型和创新投资的失败以及丧失市场机会。

在整个业务规划和项目控制周期中,组织需要确定与BPM相关的投资组合、流程和项目,并确定与其规划和预算不一致的方面,纠正不一致和冗余的工作,并在可能的情况下调整到一致的状态。此外,组织需要确定实现BPM一致性所需要的能力,这将带来更高的效率和有效性,并推进组织的财务绩效度量。为了系统地评估业务流程一致性的潜力,有必要从体系结构的角度考虑业务流程将与哪些对象相关,以及可以使用哪些模板来促进改进的一致性。

表1描述了与整个业务流程组对象相关的对象。

如图1所示,BPM的对齐将从策略和相关工作最高级别开始。然后,他们需要确定哪些业务流程元对象可以关联和对齐,从而实现更高级别的协同。

在图2中,业务流程对象元素的对齐通过可以建立关系突出显示。实现这一点的方法是通过开发路径、矩阵和模型,这些模型涵盖了从驱动力量到基础设施高

Table 1 *Business Process Group and Its Related Objects Needed for BPM Alignment*

Meta Object	Description
Process area (categorization)	Highest level of an abstract categorization of processes
Process group (categorization)	Categorization and collection of processes into common groups
Business process	Set of structured activities or tasks with logical behavior that produce a specific service or product
Process step	Conceptual set of behaviors bound by the scope of a process, which, each time it is executed, leads to a single change of inputs (form or state) into a single specified output. Each process step is a unit of work normally performed within the constraints of a set of rules by one or more actors in a role who are engaged in changing the state of one or more resources or business objects to create a single desired output
Process activity	Part of the actual physical work system that specifies how to complete the change in the form or state of an input, oversee, or even achieve the completion of an interaction with others actors and which results in the making of a complex decision based on knowledge, judgment, experience, and instinct
Event	State change that recognizes the triggering or termination of processing
Gateway	Determines forking and merging of paths, depending on the conditions expressed
Process flow (including input/output)	Stream, sequence, course, succession, series, or progression, all based on the process input output states, where each process input/output defines the process flow that together executes a behavior
Process role	Specific set of prescribed set of expected behavior and rights (authority to act) meant to enable its holder to successfully carry out his or her responsibilities in the performance of work. Each role represents a set of allowable actions within the organization in terms of the rights required for the business to operate
Process rule	Statement that defines or constrains some aspect of work and always resolves to either true or false
Process measurement (process performance indicators)	Basis by which the enterprise evaluates or estimates the nature, quality, ability, or extent regarding whether a process or activity is performing as desired
Process owner	Role performed by an actor with the fitting rights, competencies, and capabilities to take decisions to ensure work is performed

表1　BPM对齐所需的业务流程组及其相关对象

对　　象	描　　述
流程区域(分类)	流程抽象分类的最高级别
流程组(分类)	将流程分类并归集到公共组中
业务流程	一组结构化的活动或任务,具有产生特定服务或产品的逻辑行为
流程步骤	受流程范围约束的行为的概念集,每次执行该行为时,都会将输入(形式或状态)更改为单个指定的输出。每个流程步骤都是一个工作单元,通常由角色中的一个或多个参与者在一组规则的约束下执行,这些参与者参与更改一个或多个资源或业务对象的状态以创建单个所需的输出
流程活动	实际工作系统的一部分,它规定了如何完成输入的形式或状态的变化,监督甚至完成与其他参与者的交互,从而根据知识、判断、经验和直觉做出复杂的决定
事　　件	识别出流程启动或终止的状态的变化
节　　点	根据表示的条件确定路径的分叉和合并
流程图(包括输入/输出)	流、序列、过程、连续或进程,都基于流程输入/输出状态,其中每个流程输入/输出定义一起执行行为的流程
流程角色	特定的一套规定行为和权利(行动授权),旨在使其持有人能够成功履行其在工作中的责任。每个角色代表组织内一组允许的行动,涉及业务运营所需的权限
流程规则	定义或约束工作某些方面并始终解析为"真"或"假"的语句
流程测量(PPI)	企业评估过程或活动是否按预期执行,以及执行的本意、质量、能力或程度
流程所有人	由具有适当权利和能力的参与者执行的角色,以作出决定以确保工作得到执行

LEAD Templates & LEAD Meta Object Relations	Forces & Drivers (FD)	Vision, Mission & Goals (VM	Requirement (Rq)	Stakeholder (ST)	Strategy (S)	Value (V)	Balanced Scorecard (BSC)	Performance (Pe)	Measurement & Reporting	Competency/Business Mod	Revenue (Rev)	Cost (Co)	Operating (Op)	Information (I)	Role (Ro)	Owner (O)	Organizational Chart (OC)	Object (Ob)	Workflow (WF)	Rule (Ru)	Channel (Ch)	Media (Me)
Business Process: Process Area (categorization)	1.2									2			1									
Process Group (categorization)	1.2									2			1									
Business Process			1.2							2		2	1									
Process Step												2										
Process Activity												2										
Events																						
Gateways																				2		
Object (Business & Information & Data)																		1.2				
Process Type (main/mgmt./support)													2									
Process Flow (incl. input/output)											3								1.2			
Process Roles															1.2							
Process Rules																				1.2		
Process Measurement (PPI)								1.2	1.2		2.3	2.3										
Process Owner				2					2				1.2			1.2						

LEAD Templates & LEAD Meta Object Relations	Process (P)	BPM Notations (BPMN)	Service (Se)	Application (A)	Application Service (AS)	Application Roles (ARo)	Application Rules (AR)	System Measurements/Rep	Application Interface (AI)	Application Screen (Asc)	Compliance (C)	Data (D)	Data Service (DS)	Data Rules (DR)	Platform (PL)	Platform Service (PLS)	Platform Rules (PLR)	Platform Distribution (PLD)	Infrastructure (IF)	Infrastructure Service (IFS)	Infrastructure Rules (IFR)	Virtualization (IFV)	High Availability (IFH)
Business Process: Process Area (categorization)	1.2	2																					
Process Group (categorization)	1.2	2																					
Business Process	1.2	2.3	2	2																			
Process Step	1.2	3	2	2																			
Process Activity	1.2	3	2	2																			
Events	1.2	3	2		3																		
Gateways	1.2	3	2		3																		
Object (Business & Information & Data)	2.3	2.3	2																				
Process Type (main/mgmt./support)	1																						
Process Flow (incl. input/output)	3	3	2		3				3	3													
Process Roles	2.3	2.3																					
Process Rules	2	2.3					2				2												
Process Measurement (PPI)	2	2.3						2															
Process Owner	1.2	2.3	2																				

FIGURE 2

Alignment across business process objects (1, maps, 2, matrices; 3, models).[3]

图 2　跨业务流程对象的一致性（1. 路径；2. 矩阵；3. 模型）[3]

LEAD模板和 LEAD元对象关系	新的通路分为 (FD)	需求，信息和目标 (VM)	需求 (Rq)	网络关系 (ST)	服务 (S)	价值 (V)	分数计分卡 (BSC)	测量 (Pe)	角色/业务描述	输入 (Rev)	版本 (Co)	容量 (Op)	事件 (I)	角色 (Ro)	序列者 (O)	相关 (OC)	报告 (Ob)	工作流 (WF)	规则 (Ru)	测量 (Ch)	材料 (Mo)	流程 (P)	BPMN符号 (BPMN)	服务 (Se)	应用 (A)	应用视图 (AS)	应用视图 (ARo)	重新测测/关系 (AR)	应用视图 (AI)	应用测量值 (Asc)	操作符 (C)	数据 (D)	数据服务 (DS)	数据视图 (DR)	分行 (PL)	分行视图 (PLS)	分行视图 (PLR)	分行存储 (PLD)	基础视图 (IP)	基础视图 (IPS)	基础视图 (IPR)	基础视图 (IPV)	喷射原格 (IPH)
流程域（分类）	1.2																					1.2	2	2	2																		
流程组（分类）		1.2																				1.2	2	2	2																	
业务流程			1.2							2	2	1										1.2	2.3	2	2	3																
流程步骤				2								1					1.2					1.2	3	2	2	3																
活动											2	1						1.2	2			1.2	3	2	2	3																
网关																						1.2	3	2	2																	
对象（商业与信息与数据）								1.2		2.3	2.3			1.2		1.2						1	2.3	3		3	3	2	2	3	2											
流程类型（主管/管理/支持）																						2.3	3	2																		
流程流（工业输入/输出）								1.2														2.3	2.3	2					3		2											
流程角色								1.2								1.2						2	2.3	2				2				2										
流程规则																				1.2																						
流程管理（PPT）				2																													2									
流程所有者																																										

LEAD模板和
LEAD元对象关系

流程分类示意图

to alignment within BPM and the strategic, tactical, and operational aspects are covered satisfactorily.

As an example of this alignment, the business processes (meta object) can be related to the requirements map and matrix, to the competency of an organization through a matrix, to cost through a matrix, to business process notations through a matrix and a model, and so on.

Each of the business process meta objects can be aligned in this way to the specific aspects required by an organization to fulfill its portfolio, program, and projects. It furthermore ensures that the business process alignment is applicable across the following layers of business and application.

Why is it important that the business processes be linked to the application layer? This is vital so that the process automation can be executed in line with the to-be business processes designed in the business layer.

BUSINESS SCENARIOS THAT WOULD REQUIRE BUSINESS PROCESS ALIGNMENT

The following section deals with some of the possible business scenarios that would require extensive review of the business processes and a transformation project to bring about alignment.

Stakeholder Alignment

Most stakeholders across the enterprise landscape have some of the same external and internal forces and drivers influencing them, but different approaches. These stakeholders do not see what is common and hence they do not know how to work together.

Alignment Portfolio, Program, and Project Management Challenges

Portfolio, program, and project management (PPPM) has a definite placeholder within the greater enterprise management organizational structure, as depicted in Figure 3. All three of these disciplines have been well documented and researched on their own and in combination. Within the enterprise structure the influence of their alignment is most noticeable and hence most influential. All organizations, whether larger or small, across all industry sectors will recognize that they need a combination of portfolio, program, and project management to delivery change initiatives that transform and or innovate their business. For PPPM the alignment context is multidirectional. Alignment needs to flow from both a top-down and bottom-up perspective. Aligning the portfolio at the strategic level through the programs at the tactical level to the projects at the operational level will enable smarter decisions. Alignment of PPPM is also influenced through the stakeholders who influence the enterprise structure at each of the organizational layers and all of the processes involved.

可用性的各个方面。这意味着与BPM内的一致性以及战略、战术和操作方面相关的模板均得到了令人满意的覆盖。

作为这种一致性的一个例子,业务流程(元对象)与流程的路径、矩阵以及通过矩阵形成的组织能力、成本、流程符号等相关。

每个业务流程元对象都可以通过这种方式与组织完成其产品组合、计划和项目所需的特定方面进行协调。此外,它还确保业务流程一致性适用于业务层和应用程序层。

为什么业务流程链接到应用程序层很重要?这是至关重要的,这样流程自动化就可以按照在业务层中设计的未来业务流程来执行。

5.7.5 需要业务流程一致性的业务场景

下面的部分讨论一些可能的业务场景,这些场景需要对业务流程和转换项目进行广泛的审查,以实现一致性。

1. 利益相关者的一致性

企业环境中的大多数利益相关者都有一些相同的外部和内部力量以及影响它们的驱动因素,但方法不同。这些利益相关者看不到共同点,因此他们不知道如何合作。

2. PPPM挑战

投资组合、计划和项目管理(PPPM)在大型企业管理组织结构中有明确的一席之地,如图3所示。所有这三个内容都有很好的文档记录,并且分别从各自和结合的角度进行了研究。在企业结构中,它们是最具影响力的,而它们的一致性的影响是最显著的。所有行业部门的所有组织都将认识到他们需要对项目、计划和项目管理来进行组合管理,包括交付、创新和业务的变更计划。对于PPPM,在流程环境中的一致性是多方向的。一致性需要同时从自顶向下和自底向上的角度进行。通过战术级别与操作级别的项目在战略级别上一致,将支持更明智的决策。PPPM的一致性还受到每个企业层级中的相关人员以及影响企业结构的股东的影响。

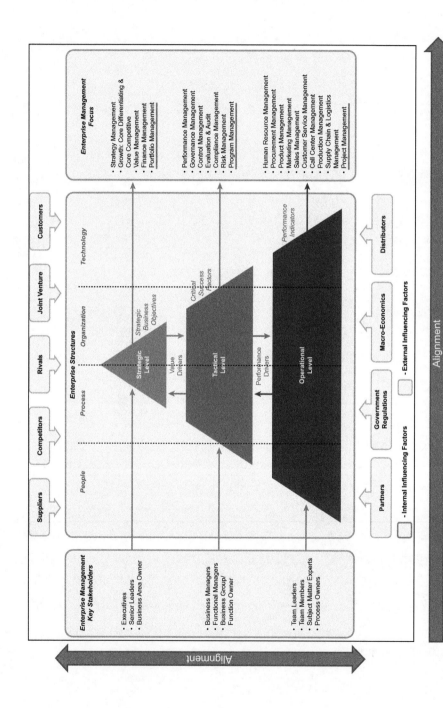

FIGURE 3

Alignment of PPPM across the enterprise structures.[4]

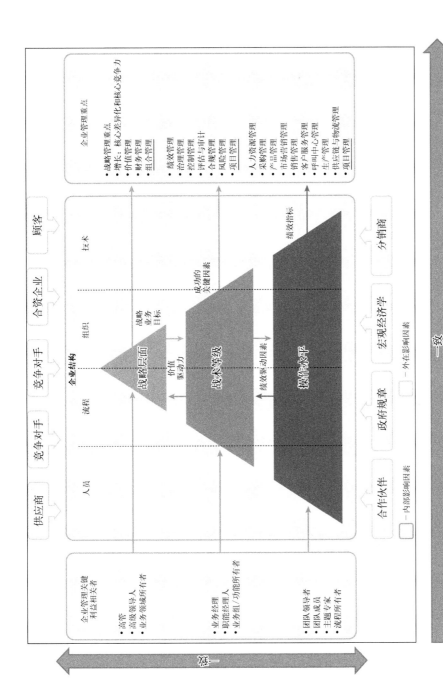

图 3 跨企业结构的 PPPM 对齐[4]

Merger and Acquisition

A typical example of when alignment of BPM would be necessary is when a company goes through a merger or when it acquires a new business entity through an acquisition. The merger or acquisition would require a transformation project that would focus its attention on identifying the common set of stakeholders who have a common set of business processes. The use of Figure 1 and Figure 3 is important to guide the flow from start to finish in terms of a transformation BPM alignment project.

Align BPM with Business Intelligence to Achieve Business Process Excellence

Today, many organizations implement BPM and business intelligence initiatives as separate programs. They are flooded with indicators—mostly process performance indicators and key performance indicators—but performance monitoring is carried out at too local a level in too isolated a way, and with too much focus on lagging indicators. Usually it is hard to see how the various factors measured contribute to different aspects of business value. Combining BPM and business intelligence to achieve closed-loop performance management makes it possible to relate all these indicators to each other. It thus becomes possible to analyze cause-and-effect relationships over different dimensions. As a result, management and staff can make better and more timely decisions and the organization becomes more efficient and effective. This is a crucial step for any enterprise with its sights set on intelligent business operations.

Align BPM with Master Data Management for Master Data Governance, Stewardship, and Enterprise Processes

When organizations align their master data management (MDM) and BPM (BPM) projects, they maximize the value of each solution. Analysts recommend that clients and vendors adopt a strategy that supports this aligned approach. An enterprise can gain differentiating value by aligning its MDM and BPM initiatives. Master data management provides data consistency to improve the integrity of business processes, making those processes smarter, more effective, and productive. Business process management is an agile process platform that can provide consistent visibility, collaboration, and governance. By aligning MDM and BPM initiatives, organizations can optimize their business performance through agile processes that empower decision makers with the trusted information that can provide a single version of truth.

Align BPM with SOA for a Business-Driven, Service-Oriented Enterprise

There is still a gap between business and information technology (IT), because until now the services provided by an Service Oriented Architecture (SOA) could not

3. 合并和收购

当公司进行合并或通过收购获得新的业务实体时 BPM 需要一致性。合并或收购时需要一个转换项目，该项目将把注意力集中在识别具有一组公共业务流程的公共利益相关者身上。图 1 和图 3 的使用对于指导从开始到结束的转换 BPM 项目对齐的流程非常重要。

4. 将 BPM 与商业智能（BI）结合起来以实现业务流程的卓越

今天，许多组织将 BPM 和 BI 活动作为单独的项目来实现。这个流程的指标大多是 PPI 和 KPI，但绩效监控大多以过于孤立的方式进行，并且过于注重滞后指标。通常很难看出所度量的各种因素如何对业务价值的不同方面做出贡献。结合 BPM 和 BI 实现闭环管理使所有这些指标能够相互关联。这样就有可能分析不同维度上的因果关系。因此，管理人员和员工可以做出更好、更及时的决策，组织也变得更有效率、更有效。对于任何着眼于 BI 的企业来说，这都是至关重要的一步。

5. 将 BPM 与 MDM 结合起来，用于主数据治理、管理和企业流程

当组织将它们的主数据管理（MDM）和 BPM 项目结合起来时，它们会将每个解决方案的价值最大化。分析人员建议客户和供应商采用支持这种一致方法的策略。企业可以通过调整 MDM 和 BPM 计划来获得不同的价值。MDM 提供数据一致性，以改进业务流程的完整性，使这些流程更智能、更有效和更高效。BPM 是一个敏捷流程平台，可以提供一致的可见性、协作和治理。通过调整 MDM 和 BPM 计划，组织可以通过敏捷流程优化其业务流程效率，该流程可向决策者提供单一版本的可信信息。

6. 业务驱动、面向服务的企业中确保 BPM 与 SOA 保持一致

业务和 IT 之间仍然存在差距，因为到目前为止，SOA 提供的服务不能立即支持业务流程。因此，将 SOA 和 BPM 项目结合、协调起来，会比单独启动这两个项

support the business processes immediately. Thus, combining and aligning SOA and BPM projects results in increased benefits that are achieved more quickly than when either is initiated alone, especially for larger initiatives, achieving a business-driven, service-oriented enterprise and with automated processes across business functions. Processes that need to execute across functions are often hampered by a lack of interoperability of underlying systems. Automate new processes with greater speed and change processes quickly in response to business needs. Avoid costly business errors and focus on improving business processes—not integrating systems. Align IT investments with business needs: With an SOA, it is straightforward to prioritize building services needed for key business processes and to establish service-level key performance indicators. That maximizes not only the alignment with business needs, but also the return on IT investments.

Align BPM with Cloud for Business Process as a Service

Many business processes, are good candidates for the cloud service. Alignment of BPM and cloud, called business process as a service, combines business processes and a cloud-based infrastructure enabling core computing resources best directed at the core business to be freed up. With goals of transparency and cost-efficiency in mind, it is logical to outsource many IT functions that are no longer cost-effective or when internal innovation is lacking. Increasing numbers of applications can be provided as a service with the right combination of technology and knowledge, from reporting and trade management to digital rights management and business analytics.

BENEFITS OF BPM ALIGNMENT

The strategic value of BPM alignment and the effect on organizational performance are significant, ranging from better processes produced to lower costs, higher revenues, motivated employees, and happier customers. The benefit checklist for the executive team includes the following:

1. Eliminates unnecessary process steps that are either regional- or system-driven
2. Standardizes and integrates the process across all geographies and business units for better benefit realization
3. Automates after elimination of unnecessary steps and standardizes the process
4. Enables process innovation using historical data from the BPM system once you have automated the process to transform it
5. Creates a repeatable pattern to align stakeholders with various portfolios, projects, and programs
6. Creates a consistent and institutionalized approach to align, plan, and resource programs and projects toward meeting common strategic objectives, expectations, and requirements
7. Improves alignment within planning, investments, and synchronization of effort across multiple portfolios, projects and programs, departments, interagencies, and resources

目更快地获得更多的好处,尤其是对于更大的计划,实现业务驱动的、面向服务的企业以及跨业务功能的自动化流程。需要跨职能执行的流程常常由于缺乏底层系统互操作性而受到阻碍。以更快的速度自动化新流程,并根据业务需求快速更改流程,可以避免出现专注于改进业务流程而不是集成系统等代价高昂的业务错误。使IT投资与业务需求保持一致的方法包括:使用SOA,可以直接确定构建关键业务流程所需服务的优先级,并建立服务级别关键性能指标。这不仅最大化了与业务需求的一致性,而且最大化了IT投资的回报。

7. 将BPM与云结合以实现BPaaS

许多业务流程都是云服务的良好候选者。BPM和云的结合可称为业务流程即服务(BPaaS),它将业务流程和基于云的基础设施组合在一起,从而释放最适合于核心业务的核心计算资源。考虑到透明度和成本效率的目标,将不再具有成本效益或缺乏内部创新的许多IT功能外包是个合理的做法。从报告和交易管理到数字版权管理和业务分析,越来越多的应用程序可以作为技术和知识组合管理的一种服务方式。

5.7.6　BPM一致性的好处

BPM一致性对战略价值和组织绩效非常重要,从产生更好的流程到更低的成本、更高的收入、有积极性的员工和更快乐的客户。这些好处包括:

(1)消除区域或系统驱动的不必要的流程步骤;

(2)标准化和集成跨所有地理区域和业务单元的流程,以实现更好的效益;

(3)在消除不必要的步骤后实现自动化,并使流程标准化;

(4)一旦您自动化了转换流程,就可以使用BPM系统中的历史数据进行流程创新;

(5)创建一个可重复的模式,使利益相关者与各种投资组合、项目和计划保持一致;

(6)创建一种一致的、制度化的方法,以协调、规划和为计划和项目提供资源,以实现共同的战略目标、期望和要求;

(7)在多个投资组合、项目和计划、部门、机构间和资源中改进计划、投资和工作同步;

8. Enables alignment for complex planning efforts
9. Reduces duplication of efforts across business, application, and or technology areas
10. Improves joint delivery and execution
11. Is a proven concept to reduce radical cost
12. Enables better transparency and traceability
13. Does not disturb existing efforts; rather, it provides a means to inform, integrate, synchronize, and control

CONCLUSIONS

In this chapter we have focused on BPM alignment and how it is a top priority for executives. We covered what BPM alignment is, why it is important, and how and where it can or should be applied. Business process management alignment establishes the basis for effective tactical planning and drives continuous improvement and change management. The effectiveness of BPM efforts can be predicted by the maturity of an organization's planning, alignment, and change management. We described this "how and where" to enable replication of the same success across projects, portfolios, and programs. Combined with Business Intelligence (BI), MDM, SOA, and or the cloud, BPM alignment offers significant potential to drive value and affect organizational performance.

End Notes

1. Paul A. Strassmann (1997), *The Squandered Computer*. Page 27–29 ISBN: 0-9620413-1-9.
2. LEADing Practice Alignment & Unity Reference Content #LEAD-ES60001AL.
3. Business Process Objects Relations from LEAD Template & LEAD Meta Object Relation, LEADing Practice.
4. LEADing Practice Alignment & Unity Reference Content #LEAD-ES60001AL.

（8）实现复杂计划工作的一致性；

（9）减少跨业务、应用程序和技术领域的工作重复；

（10）改善联合交付和执行；

（11）是一个行之有效的降低根本成本的概念；

（12）实现更好的透明度和可追溯性；

（13）不会干扰现有的工作，相反，它提供了一种通知、集成、同步和控制的方法。

5.7.7　结论

在本节中，我们重点讨论了BPM一致性以及它如何成为高管的首要任务。我们讨论了BPM一致性是什么、为什么它很重要，以及如何及应该在何处应用它。BPM一致性为有效的战术规划奠定了基础，并推动持续改进和变更管理。BPM工作的有效性可以通过组织的计划、一致性和变更管理的成熟度来预测。我们描述了"如何以及在何处"支持在项目、投资组合和程序之间复制相同的成功。结合BI、MDM、SOA、云这几方面，BPM一致性为驱动价值并影响组织绩效提供了巨大的潜力。

参考文献

［1］Paul A. Strassmann（1997），*The Squandered Computer*. Page 27-29 ISBN: 0-9620413-1-9.

［2］LEADing Practice Alignment & Unity Reference Content #LEAD-ES60001AL.

［3］Business Process Objects Relations from LEAD Template & LEAD Meta Object Relation, LEADing Practice.

［4］LEADing Practice Alignment & Unity Reference Content #LEAD-ES60001AL.

Business Process Outsourcing

Mark von Rosing, Gary Doucet, Gert O. Jansson, Gabriel von Scheel, Freek Stoffel, Bas Bach, Henk Kuil, Joshua Waters

INTRODUCTION

The great interest in outsourcing since the start of the 1980s had several causes and was influenced in numerous ways by process work. In this chapter, we will focus on what business process outsourcing is, why it is applied, and which aspects to consider before implementing business process outsourcing strategies in an organization.

Since the industrial revolution, companies have battled with how they can exploit their competitive advantage to increase their markets and their profits. The model for most of the twentieth century was a large integrated company that can own, manage, and directly control its assets. In the 1950s and 1960s, the rallying cry was diversification to broaden corporate bases and take advantage of economies of scale. By diversifying, companies expected to protect profits, even though expansion required multiple layers of management. Subsequently, organizations attempting to compete globally in the 1970s and 1980s were handicapped by a lack of agility that resulted from bloated management structures. To increase their flexibility and creativity, many large companies developed a new strategy of focusing on their core business, which required identifying critical processes and deciding which could be outsourced.[1]

Outsourcing was not formally identified as a business strategy until 1989 (Mullin, 1996). However, most organizations were not totally self-sufficient; they outsourced those functions for which they had no competency internally. Publishers, for example, have often purchased composition, printing, and fulfillment services. The use of external suppliers for these essential, but ancillary, services might be termed the baseline stage in the evolution of outsourcing. Outsourcing support services was the next stage. In the 1990s, as organizations began to focus more on cost-saving measures, they started to outsource functions necessary to run a company, but not related specifically to the core business. Managers contracted with emerging service companies to deliver accounting, human resources, data processing, internal mail distribution, security, plant maintenance and the likes as a matter of good housekeeping. Outsourcing components to affect cost savings in key functions was yet another stage as managers sought to improve their finances.

BUSINESS PROCESS OUTSOURCING: WHAT IS IT?

The short version is that business process outsourcing (BPO) is the contracting of a specific business task, such as payroll, to a third-party service provider. Business process outsourcing is a subset of outsourcing that involves contracting of operations and responsibilities of specific business functions (or processes) to a third-party

The Complete Business Process Handbook. http://dx.doi.org/10.1016/B978-0-12-799959-3.00033-1

5.8　BPO

Mark von Rosing, Gary Doucet, Gert O. Jansson, Gabriel von Scheel, Freek Stoffel, Bas Bach, Henk Kuil, Joshua Waters

5.8.1　介绍

自20世纪80年代初以来,人们对外包的极大兴趣有部分原因是在许多方面受到流程工作的影响。在这一节中,我们将关注什么是BPO、为什么要应用它,以及在组织中实现BPO策略之前要考虑哪些方面。

自工业革命以来,企业一直在为如何利用竞争优势扩大市场和利润而奋斗。在20世纪的大部分时间里,大型企业都是利用这种模式掌控、管理和直接控制自己的资产。20世纪50年代和60年代,企业的口号是多元化,并利用规模经济来扩大企业规模。尽管扩张需要多层管理,但企业通过多元化希望能够保护利润来源。到了20世纪70年代和80年代,由于管理结构臃肿而缺乏灵活性,企业在全球竞争中受到阻碍。为了提高灵活性和创造性,许多大公司制定了一种新的战略,把重点放在核心业务上,这就需要确定关键的流程,并决定哪些业务可以外包[1]。

直到1989年,外包这种方式才被正式确定为一种商业策略(Mullin, 1996年)。然而,大多数组织并不完全自给自足,他们外包了那些在公司内部没有能力胜任的职能。例如,出版商经常购买装订、打印和发行服务。使用外部供应商来提供这些基本辅助的服务,是外包的基本阶段。下一个阶段是外包支持服务。到了20世纪90年代,随着组织开始更多地关注节约成本的措施,他们开始与新兴服务公司签订合同,将财务、人力资源、数据处理、内部邮件分发、安全、工厂维护等与运营公司核心业务无关的职能外包,通过将这些职能外包逐渐成为管理人员改善其财务状况的手段。

5.8.2　BPO: 它是什么?

BPO是将特定的业务任务(如工资单)外包给第三方服务提供者。BPO是外包的一个子集,它涉及将特定业务功能(或流程)的操作和职责外包给第三方服务

service provider. Originally, this was associated with manufacturing firms such as Coca Cola that outsourced large segments of their supply chain.[2] Business process outsourcing is not considered only by large multinational organizations, however. Business process outsourcing is not a new field. Rochester, New York–based Paychex, for example, has been outsourcing payroll processing for small businesses since 1971. But the market is heating up these days thanks to companies' keen interest in cost cutting, their desire to improve business methods, and their growing comfort with outsourcing arrangements.[3] Business process outsourcing is something many organizations of any size apply. In this context, BPO is often divided into two categories[4]: (1) back office outsourcing, which includes internal business functions such as billing or purchasing; and (2) front office outsourcing, which includes customer-related services such as marketing or tech support. Business process outsourcing that is contracted outside a company's own country is sometimes called offshore outsourcing. Business process outsourcing that is contracted to a company's neighboring country is sometimes called near-shore outsourcing, and BPO that is contracted with the company's own county is sometimes called onshore outsourcing. Often the business processes are information technology (IT)-based, and are referred to as ITES-BPO, where ITES stands for information technology enabled service. Knowledge process outsourcing and legal process outsourcing are some of the sub-segments of the BPO industry.[5]

BUSINESS PROCESS OUTSOURCING VALUE CASE

The main advantage of BPO is the way in which it helps increase a company's flexibility. In the early 2000s, BPO was all about cost efficiency, which allowed a certain level of flexibility at the time. Owing to technological advances and changes in the industry (specifically, the move to more service-based rather than product-based contracts), companies who choose to outsource their back office increasingly looked for time flexibility and direct quality control.[6] Business process outsourcing enhances the flexibility of an organization in different ways:

- Most services provided by BPO vendors are offered on a fee-for-service basis, using business models such as remote in-sourcing or similar software development and outsourcing models.[7] This can help a company to become more flexible by transforming fixed into variable costs.[8]
- A variable cost structure helps a company respond to changes in the required capacity and does not require a company to invest in assets, thereby making the company more flexible.[9] Outsourcing may provide a firm with increased flexibility in its resource management and may reduce response times to major environmental changes.
- Another way in which BPO contributes to a company's flexibility is that a company is able to focus on its core competencies without being burdened by the demands of bureaucratic restraints.[10]
- Key employees are released from performing non-core or administrative processes and can invest more time and energy in building the firm's core

提供者。最初,这与可口可乐等将供应链的很大一部分外包出去的制造企业有关[2]。然而,BPO并不是一个新领域,也不仅仅被大型跨国组织所采用。例如,总部位于纽约罗切斯特的Paychex自1971年以来一直在为小企业外包薪资处理业务。但由于企业对削减成本的浓厚兴趣、改善业务方法的渴望,以及它们对外包服务越来越满意,如今市场正在升温[3]。因为BPO是任何规模的组织都可以应用的,所以在这种情况下,BPO通常分为两类[4]:① 后台业务外包,包括计费、采购等内部业务功能;② 前台部业务外包,包括市场营销、技术支持等与客户相关的服务。有些BPO发生在本国之外,此类外包被称为离岸外包。将业务流程外包给公司的邻国有时称为近岸外包,而将业务流程外包给公司所在的国家有时称为在岸外包。基于IT的外包通常被称为ITES-BPO,其中ITES代表支持信息技术的服务。此外,知识流程外包和法律流程外包是BPO行业的一些子细分市场[5]。

5.8.3 BPO价值的案例

BPO的主要优势在于它有助于提高公司经营的灵活性。在21世纪初,BPO完全是出于成本效率的考虑,这在当时允许一定程度的灵活性。由于技术的进步和行业的变化(具体地说,向更基于服务而非产品的合同转变),选择后台外包服务的公司越来越多地在寻求时间灵活性的同时更加重视质量控制[6]。因此,BPO以不同的方式增强了组织的应变能力。

- 大多数BPO供应商使用诸如远程内包或软件开发外包模型按收取服务费的方式提供外部服务[7],这可以通过将固定成本转换为可变成本来帮助公司变得更加灵活[8]。
- 成本结构可变性避免公司过多投资于资产,能够帮助公司应对各种变化,从而使公司更加灵活[9]。因此,外包可以增加企业资源管理的灵活性,减少对重大环境变化的响应时间。
- BPO为公司的灵活性做出贡献的另一种方式是,公司能够专注于其核心竞争力,而不会受到官僚限制要求的负担[10]。
- 核心员工能够从非核心或行政流程中解放出来,可以投入更多的时间和精

businesses.[11] The key lies in knowing which of the main value drivers to focus on: customer intimacy, product leadership, or operational excellence. Focusing more on one of these drivers may help a company create a competitive edge.[12]

- Business process outsourcing increases organizational flexibility by increasing the speed of business processes. Supply chain management with the effective use of supply chain partners and BPO increases the speed of several business processes, such as the throughput in the case of a manufacturing company.[13]
- Flexibility is seen as a stage in the organizational life cycle: A company can maintain growth goals while avoiding standard business bottlenecks.[14] Business process outsourcing therefore allows firms to retain their entrepreneurial speed and agility, which they would otherwise sacrifice to become efficient as they expanded. It avoids a premature internal transition from its informal entrepreneurial phase to a more bureaucratic mode of operation.[15]

A company may be able to grow at a faster pace because it will be less constrained by large capital expenditures for people or equipment that may take years to amortize, may become outdated, or may turn out to be a poor match for the company over time. The economic benefits of BPO are clear: the CNET News report[16] states some of these benefits in numbers: Whereas IT outsourcing, such as farming out control of a data center, can cut costs by 10–15%, outsourcing a business process may shave 40–60% off the bottom line, Pool said. "What you're providing on the BPO side is much more valuable to the client."

The business case[17] for traditional finance and administration BPO is well proven; typically the service can be delivered as effectively, i.e., with no worsening in quality of service, and more efficiently in a lower cost location, typically with an improved system of controls. Most of the business case is a combination of efficiency, centralization, standardization, process improvement, and automation coupled with labor arbitrage benefits of low-cost locations. The combination of these levers can often bring benefits of 40–50% on original cost. However, the business case for procurement is about sourcing and compliance savings, i.e., effectiveness savings (doing it better or, in the case of procurement, buying the same or better for less), which can dwarf the efficiency savings (doing it quicker or cheaper). Also, the efficiency savings in procurement are often less (perhaps 20–30%) because the resources will need to be located across regional locations to support language requirements, not just in far eastern low-cost locations; hence, average savings will be less than the 40–50% above (Figure 1).

I therefore believe that the largest portion of the benefit from outsourcing can come from sourcing savings and in particular through better compliance management.

However, it is critical to approach the design of the outsourcing in a way that maximizes the efficiency (process) savings and maximizes the (sourcing) effectiveness savings. You can construct a business case for procurement BPO on efficiency savings, but you will sell it to the business on effectiveness savings.

力来构建公司的核心业务[11]，例如，将工作重点放在：客户关系、产品竞争力、更加卓越的运营等。更多地关注这些驱动力能够帮助公司创造有竞争力的优势[12]。

- BPO通过提高业务流程的速度来提高组织的灵活性。有效利用供应链合作伙伴和BPO的供应链管理可以提高多个业务流程的速度，如加快制造公司的吞吐量[13]。

- 灵活性被视为组织生命周期的一个阶段：可以让公司保持增长，同时又避免业务瓶颈的出现[14]。因此，BPO允许公司保持其发展速度和敏捷性，否则，随着企业的扩张，它们将牺牲这些速度和敏捷性来提高效率。它避免了内部过早地从非正式的创业阶段过渡到更官僚的运作模式[15]。

一家公司之所以能够以更快的速度增长，是因为它较少受到人力或设备方面的巨额资本支出的限制，这些支出可能需要数年才能摊销。这些投入可能会过时，或者随着时间的推移，这些支出可能与公司不匹配。BPO的经济效益是显而易见的，CNET新闻报道[16]以数字形式阐述了其中的一些好处：尽管IT外包（如将数据中心的控制权外包出去）可能会使利润减少40% ~ 60%，但也可以将成本削减10% ~ 15%。Pool认为："您在业务流程外包以外的方面提供的服务对客户来说更有价值。"

传统金融和行政BPO的商业案例[17]已得到充分证明：通常在服务质量没有恶化的前提下，这些服务可以有效地交付，而且在成本较低的地区在控制系统得到改善的情况下产生的效率更高。所以，大多数业务案例是效率、集中化、标准化、流程改进和自动化以及低成本地区的劳动力优势的组合。这些组合通常可以在原始成本的基础上带来40% ~ 50%的收益。然而，商业案例在采购方面则体现在关于采购成本和资源的节省（做得更好，或者在同等采购标准的情况下，购买相同或更好的花费更少），这可以使效率更高（做得更快或便宜）。此外，采购成本通常会降低（可能是20% ~ 30%）。资源存在异域性，不仅只存在于远东的低成本地区，语言沟通也存在困难，因此，平均节省量将低于40% ~ 50%（图1）。

因此，我认为外包带来的最大好处可能来自采购节约，特别是实现更好的合规管理。

然而，以一种最大化效率（流程）节约和（资源）节约的方式来设计外包是至关重要的。您可以为采购BPO构建一个关于效率节约的典型案例，前提是您将其应用到业务部门实践之中。

Building the business case for BPO Procurement

Note: Client with $500 million in indirect source-able spend

FIGURE 1

Example of BPO procurement business case.

THE BPO MARKET

In 2014–2015, the growth of BPO was 4% and surpassed $950 billion; it is expected to average a 5% clip each year through 2018.[18] The global BPO and related IT service market size in 2014 (in billion US dollars) by segment were the following[19] (see Figure 2):

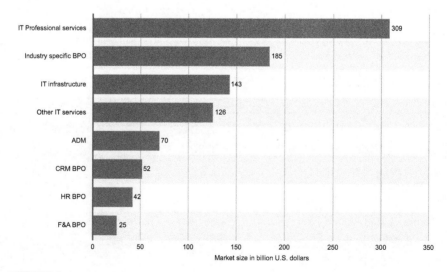

FIGURE 2

The Global BPO and the related IT service market size in 2013–2014.

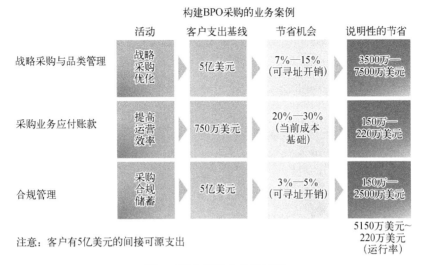

图1　BPO采购业务的示例

5.8.4　BPO的市场规模

2014～2015年,BPO的增长率为4%,超过了9亿～500亿美元,预计到2018年,每年平均增长5%。2014年全球BPO和相关IT服务市场规模(按照10亿美元)分段如下[19](图2)。

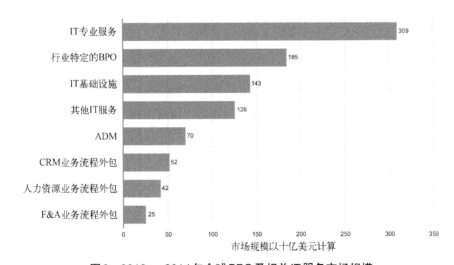

图2　2013～2014年全球BPO及相关IT服务市场规模

BUSINESS PROCESS OUTSOURCING: POSSIBLE PITFALLS

Although the BPO value case arguments favor the view that BPO increases the flexibility of organizations, management needs to be careful with implementing it because there are issues that work against these advantages. In this section we will try to illustrate them.

These days, outsourcing is often an expression of the facility management way of thinking: that is, how I can minimize my risks and secure the best relevant price/performance, for example, in accounts departments, IT operations, staff restaurants. The challenge for all of these lies in the interfaces. What are the expectations regarding upstream and downstream to have successful outsourcing? This may vary, but it is interesting that 70% of all outsourcings are reported as failures, especially those that are primarily aimed at cutting cost.[20] This is alarming. As Chief Executive Officer Peter Bendor-Samuel of the Everest Group pointed out,[21] part of the reason is that many of BPO concepts are large contracts with durations of 3–10 years. Many of these contracts include substantial capital for assets such as people, servers, networks, and capitalized transformational costs. These contracts are notoriously inflexible, driven by a combination of factors including the need to predefine service-level agreements and scope over a long time, pricing that has to anticipate changes in volume and technology, and the substantial capital cost that must be retired over the life of the contract. As you can see from the chart, these contracts delay the profits to the service provider and deliver only modest profitability late in the contract term (Figure 3).

The combination of unrealized earnings and un-depreciated assets has the potential to create substantial stranded costs if the contracts are terminated early or significantly renegotiated mid-term. These stranded costs have been the bane of

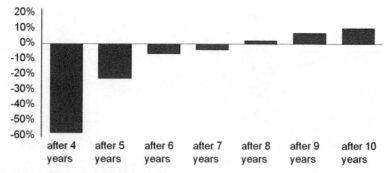

Rate of Return for a typical traditional IT outsourcing contract

Source: Bernstein Research analysis

FIGURE 3

Rate of return within outsourcing.

Source: *Bernstein Research analysis.*

5.8.5　BPO：可能的陷阱

虽然BPO价值案例支持BPO增加组织灵活性这样的观点，但是管理者在实现它时需要格外小心，因为存在不利于这些优势的问题。在本节中，我们将尝试说明它们。

如今，大家往往将外包的思维方式等同为一种设施管理，即如何在会计部门、IT运营部门、员工餐厅等方面将风险降到最低，并确保最佳的相关价格/性能，所有这些的挑战都在于接口。对上游和下游成功外包的期望是什么？这可能有所不同，但有趣的是，70%的外包项目被定为失败，尤其是那些主要是为了削减成本而进行的外包项目[20]，这是令人担忧的。Everest Group首席执行官Peter Bendor-Samuel指出，原因之一是许多BPO项目都是期限为3～10年的大型合同，其中许多合同包括大量的资本投入，如人员、服务器、网络和巨额的转换成本。众所周知，这些合同缺乏灵活性，其驱动因素包括长期需要预先定义服务水平协议和范围、价格必须能够预测数量和技术的变化，以及合同期限内必须收回的大量成本。从图表中可以看出，这些合同压低了服务提供商的利润，或者在合同期限的后期只取得了少量的利润（图3）。

如果外包合同在履行过程中提前终止或重新协商，则未实现收益和未折旧资产可能会产生大量搁置成本。这些搁置的成本一直是该行业的祸根，是出现大量

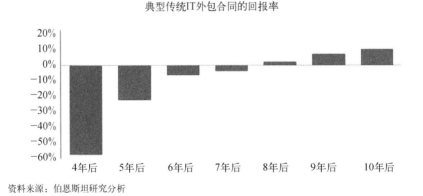

典型传统IT外包合同的回报率

资料来源：伯恩斯坦研究分析

图3　外包内的收益率

the industry, creating a steady stream of blow-up deals, some of which consistently further suppress earnings in the sector. Sometimes a decision to transfer assets to a service provider was driven by an artificial increase in return on capital that the buyer would show after assets were moved from its books to those of a provider. Nonetheless, even this benefit eventually disappeared owing to changes in assets accounting in an outsourcing transaction.

Among other problems that arise in practice are a failure to meet service levels, unclear contractual issues, changing requirements and unforeseen charges, and a dependence on the BPO, which reduces flexibility. Consequently, these challenges need to be considered before a company decides to engage in business process outsourcing.[22] A further issue is that in many cases there is little that differentiates the BPO providers other than size. They often provide similar services, have similar geographic footprints, leverage similar technology stacks, and have similar quality improvement approaches.[23]

Possible pitfalls and risks are major drawbacks with BPO. Outsourcing an information system, for example, can cause security pitfalls and risks factors from both a communication and privacy perspective. For example, security of North American or European company data is more difficult to maintain when accessed or controlled in the Indian subcontinent. From a knowledge perspective, a changing attitude in employees, underestimation of running costs, and the major risk of losing independence, outsourcing leads to a different relationship between an organization and its contractor.[24] Pitfalls, risks, and threats of outsourcing must therefore be managed to achieve any benefits. To manage outsourcing in a structured way, maximizing positive outcome, minimizing risks, and avoiding threats, a business continuity management model is set up. Business continuity management consists of a set of steps to successfully identify, manage, and control business processes that are or can be outsourced.[25]

BUSINESS PROCESS OUTSOURCING: HOW TO GO ABOUT IT

As mentioned earlier, BPO has to link to the aspects of the core businesses.[26] Therefore, the link between the organization's competencies and process execution provides the means of identifying ways to appropriately define which areas to outsource and thereby increase flexibility and/or innovation to support value creation and revenue growth, or where to improve the effectiveness and efficiency of operations or reduce cost. Without this context there is no means to judge the goodness of a particular process outsourcing strategy. For example, if it is not possible to detect that a process contributes to the unique value creation of the business and helps it differentiate, it is best not to outsource to cut cost; however, if you have business areas and thereby processes that are non-core and do not compete or differentiate in a business context, it should be done in the cheapest way possible.

Figure 4 shows a summary of the concepts for categorizing the six domains of business models, the competencies that enable the business models, and the type of practice standards that correspond to the different competencies.

For non-core competencies, it makes sense to do BPO with standard BPO best practices, in an effort to optimize operations and minimize cost. Similarly, industry

毀约行为的原因,其中一些进一步抑制了该项目的盈利。有时,外包方将资产转移到承包方的行为可以使外包方的资本回报率人为地增加,但由于外包交易中资产会计的变化,这种好处也会最终消失。

在实践中出现的其他问题包括:服务水平未能满足需求、合同项目不明确、需求发生变化和产生未预见的费用以及对BPO的依赖降低了灵活性。因此,在公司决定从事BPO之前,需要考虑这些挑战[22]。另一个问题是,除规模以外,在许多情况下BPO供应商之间几乎没有什么区别。他们提供相似的服务、处于相似的地理位置、利用相似的技术体系,并具有相似的质量改进方法[23]。

潜在的陷阱和风险是BPO的主要劣势。例如,外包信息系统可能在通信和隐私方面出现安全隐患和风险因素。例如,当在印度访问或控制北美或欧洲公司数据时安全性更难保证。从知识的角度来看,员工态度的变化、对运营成本的低估以及失去独立性的主要风险,均会导致外包商与其承包商之间的关系出现差异[24]。因此,为确保利益,必须对外包的陷阱、风险和威胁进行管理。为了系统化地管理外包、最大限度地实现收益、最大限度地降低风险、避免威胁,有必要建立业务连续性管理模型。业务连续性管理包括成功地识别、管理和控制外包或可以外包的业务流程等一系列步骤[25]。

5.8.6 BPO:如何实现目标

如前所述,BPO必须与核心业务的各个方面联系起来[26]。因此,组织能力与流程执行之间的联系提供了一种方法,该方法可以确定外包领域,从而提高灵活性和/或创新能力,支持价值创造和收入增长或在某处提高运营效率或降低成本。相反,如果没有这种背景,就没有办法判断特定流程外包战略的优劣。例如,如果无法检测到流程外包有助于业务价值,则最好不要通过外包以降低成本,但是,如果您拥有非核心业务领域,并在业务环境中不具有竞争性或差异性,那么应该以尽可能低的成本完成。

图4显示了对业务模型的六个领域、支持业务模型的能力以及与不同能力相对应的实践标准类型进行分类的概念的摘要。

对于非核心竞争力,使用标准的BPO最佳实践来进行BPO是有意义的,可以

FIGURE 4

When to apply BPO practices and services.[27]

BPO best practice may be adopted for core-competitive competencies because the business only aims to compete effectively with its competitors and maximize its performance. However, to drive growth in revenue and value, new products and services have to be developed to give the business a competitive advantage. By its nature, an advantage requires something that is not offered elsewhere; thus, the business strives toward developing and nurturing core-differentiating competencies. However, applying such differentiating competencies in a standardized BPO offering will at best result in high performance, but not differentiating value.

As shown in Figure 4, true differentiating competencies typically comprise a small portion of the business (only 5%), although it may be a much larger percentage in truly innovative enterprises. Although it could be argued that there are relatively few cases in which a BPO organization could offer unique or even market-leading products and services in a field where you are differentiating, it is crucial for the business to find partners that can support you in your process of innovation and differentiation in the market. Until recently, it had been axiomatic that no organization would outsource core competencies, those functions that give the company a strategic advantage or make it unique. Often, for core-differentiating competencies and all processes involved, some organizations that do not know which competencies are core-differentiating just define it as any function that gets close to customers. In the early 2000s, when BPO matured in the market, this concept of only outsourcing non-core competency areas and the various related processes started to change with the realization that outsourcing some core functions may be good strategy, not anathema, that threatens nether the business model nor the operating model. For example, some organizations outsource customer service, precisely because it is so important.[28]

The key lies in knowing which of the main value drivers and related competencies to focus on, and which produce customer intimacy, product leadership, or operational excellence. Focusing on the drivers that build uniqueness will help a company create a competitive edge (Figure 5).[29]

Before consideration is given to whether to undertake a BPO program, a diagnostic of a company's expenditure must first be undertaken.

图4　何时应用BPO实践和服务[27]

优化操作和最小化成本。类似地,行业BPO的最佳实践也可以被用于核心竞争力,因为BPO的唯一目标是与竞争对手有效地竞争并最大化其绩效。然而,为了推动收入和价值的增长,必须开发新产品和服务,使企业获得竞争优势。从本质上讲,优势是指其他主体不具备的东西,因此,企业应该努力发展和培养核心差异化能力。然而,在标准化的BPO产品中应用这样的差异化能力,最多只能获得高绩效,而不会产生差异化价值。

如图4所示,真正的差异化能力通常只占企业的一小部分(只有5%),但是在真正的创新型企业中,这一比例可能要大得多。BPO组织可以在一个领域内提供独特的甚至是市场领先的产品和服务,尽管这种情况较少,但对于企业来说,找到能够在市场创新和差异化过程中为您提供支持的合作伙伴至关重要。直到最近,没有一个组织会将给公司带来战略优势或使公司与众不同的核心能力外包,这是一个不言自明的事实。通常,对于无法分清哪些是核心差异化能力和哪些是所涉及的一般流程,这些组织只是将其定义为获取客户的一般功能。21世纪初,当BPO在市场上逐渐成熟时,只外包非核心能力领域和各种相关流程的观念开始发生变化,人们意识到将客户服务等某些核心功能外包可能是一种好的战略,而不是一种威胁,既不会威胁到业务模式,也不会威胁到运营模式[28]。

关键在于要有能力了解哪些主要价值驱动因素和相关能力需要关注,哪些能够产生客户亲密度、产品领导力或卓越运营。因此,专注于建立独特性的驱动因素将有助于公司创造竞争优势(图5)[29]。

所以,在考虑是否实施BPO计划之前,必须首先对公司的支出情况进行判断。

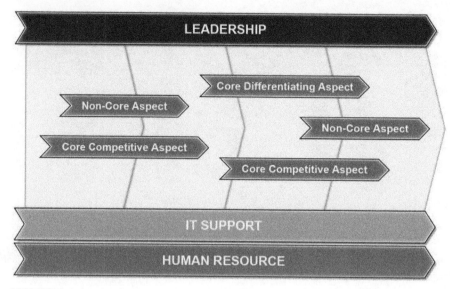

FIGURE 5

Specific processes are within the organization's non-core, core competitive, and core-differentiating aspects across business units.[30]

- Divide the total corporate expenditure into categories that relate to the supplier markets, and then further divide expenditure categories by business units or locations to identify each supplier.
- This initial diagnostic is required to be only around 80% accurate to have relevance and to offer valuable insight.
- If and when BPO teams need to refine spend data for each category, the supplier may be a more accurate source (this is discussed further below).

Diagnostics can provide a expenditure map by category. These expenditure categories should be classified according to competitiveness in the supplier marketplace compared with how important they are to the organization. This results in an expenditure category matrix that will help direct the team toward a potential sourcing strategy for each category.

The Kraljic portfolio purchasing model plots categories as strategic (low supply market competitiveness, high business impact), leverage (high supply market competitiveness, high business impact), bottleneck (low supply market competitiveness, low business impact), and routine (high supply market competitiveness, low business impact) (Figure 6).

Once diagnostics are complete, the business must decide which categories to address immediately and which to delay until internal or external conditions are better. For categories that need immediate attention, the seven-step process begins.

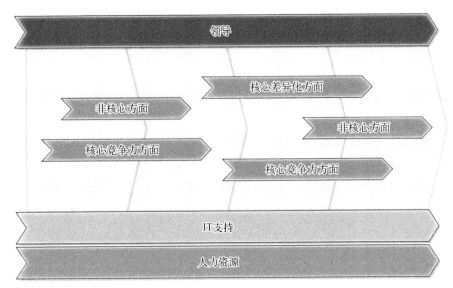

图5 特定流程位于组织的非核心、核心竞争和跨业务单元的核心差异化方面[30]

- 将公司总支出按照供应商市场相关类别进行划分,然后按业务单位或地点进一步划分支出类别,以确定每个供应商。
- 最初诊断的准确度要求只有80%左右,以便具有相关性并提供有价值的见解。
- 当BPO团队需要细化每个类别的支出数据时,供应商是最准确的来源(下文将进一步讨论)。

诊断可以按类别得到支出图。这些支出类别应根据供应商市场的竞争力和它们对组织的重要性进行分类。这将产生一个支出类别矩阵,该矩阵将有助于指导团队针对每个类别制定潜在的采购策略。

Kraljic组合采购模型(卡拉杰克模型,Kraljic portfolio purchasing model)将类别划分为战略(低供应市场竞争力、高业务影响)、杠杆(高供应市场竞争力、高业务影响)、瓶颈(低供应市场竞争力、低业务影响)和常规(高供应市场竞争力、低业务影响)四类(图6)。

一旦诊断完成,企业必须决定立即处理哪些类别,以及推迟处理哪些类别,直到内部或外部条件好转。对于需要立即注意的类别,将开始七个步骤的过程。

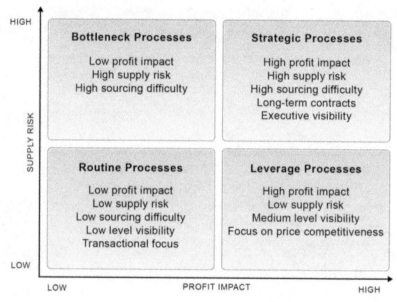

FIGURE 6

Kraljic's purchasing model.

Step 1: Fully Understand the Expenditure Category

This step, along with the next two, is conducted by the business process sourcing team. At this stage, the team needs to ensure it understands everything about the expenditure category itself. For example, if the category is corrugated packaging at a consumer products company, the team will need to understand the definition of the category and usage patterns and why the particular types and grades were specified. Stakeholders at all operating units and physical locations would need to be identified: for example, logistics, which may need to know about shipping specifications, or marketing, which may need to understand certain quality or environmental characteristics, where applicable.

The five key areas of analysis are:

- Total historic expenditure and number of processes
- Expenditure categorised by main, management, and supporting processes
- Expenditure by division, department, or user
- Expenditure by supplier
- Future demand projections or budgets

Step 2: Supplier Market Assessment

Concurrently run supplier market assessment for seeking alternative suppliers to existing incumbents. Understand the key supplier marketplace dynamics and current trends. Prepare should-cost information from the major components of the

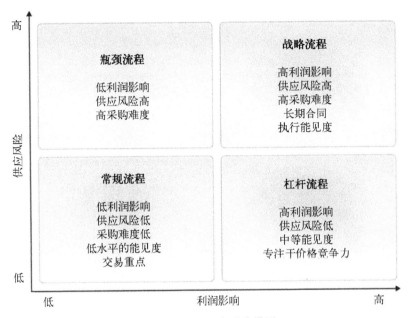

图6   Kraljic组合采购模型

步骤1：充分了解支出类别

此步骤以及接下来的两个步骤由业务流程寻源团队执行。在这个阶段，团队需要确保了解支出类别本身的所有信息。例如，如果采购类别是消费品公司的瓦楞纸包装，团队将需要了解类别和使用模式的定义、类型及等级等原因。所有运营单位和物理位置的利益相关者都需要确定，例如，可能需要了解"物流"运输规范，或需要了解"营销"的某些质量或环境特征。

分析的五个关键领域是：

- 总历史支出和流程数量；
- 按主要、管理和支持流程分类的支出；
- 按部门、小组或用户划分的支出；
- 供应商支出；
- 未来需求预测或预算。

步骤2：供应商市场评估

同时进行供应商市场评估，寻找现有供应商的替代者。了解关键供应商市场动态和当前趋势。准备关键产品的主要组成部分的成本信息、查看关键供应商的

key products. Take a view on the key suppliers' sub-tier marketplace and analyze for any risks as well as opportunities. Should-cost analysis is not appropriate for every item. In many cases, traditional strategic sourcing techniques work well. However, in cases where strategic sourcing cannot be applied, should-cost analysis provides a valuable tool that can drive cost reductions and supplier continuous improvement efforts.

Step 3: Prepare a Supplier Survey

Next, develop a supplier survey for both incumbent and potential alternative suppliers. This survey will help evaluate the supplier capabilities. At this point, consider verifying spend information using data that incumbent suppliers have from their sales systems. The survey is to assess the capability and capacity of the market to meet your requirements. It enables you to assess at an early stage whether your proposed project is feasible and can be delivered by the identified supply base. It also provides an early warning of your requirements to the market, and enables suppliers to think about how they will respond. The key aim here is to encourage the right suppliers with the right structure to respond to you.

Look to gather knowledge in these key areas:

- Feasibility
- Capability
- Maturity
- Capacity

Step 4: Building the Strategy

This step involves developing the business process sourcing strategy. The combination of the first three steps provides the essential ingredients for the sourcing strategy. However, for each area or category it will depend on:

1. How competitive the supplier marketplace is: Armed with the supplier information, you can build the competitive landscape in the supply marketplace. This can help demonstrate the size of the prize to alternative suppliers and communicates the seriousness of a potential sourcing exercise to incumbent suppliers.
2. How supportive your organization's users are to testing incumbent supplier relationships: A business process sourcing team has two sets of internal stakeholders: people who use the services that are bought and executives who manage overall costs. The people who consume the expenditure category will accept cost reductions as long as the process is started in another department, does not mean a change in suppliers, and does not jeopardize a good relationship with the supply base, generate complaints, or affect issues such as delivery reliability, service, or payments. For executives, cost and service competitiveness is a key objective, but they, too, are users of various corporate services, and so are often caught between the pursuit of cost improvement and a user mentality

其他市场、分析全面的风险和机遇。那么,成本分析是否不适用于每个项目呢? 在许多情况下,传统的战略寻源技术工作得很好。然而,在不能应用战略采购的情况下,成本分析应提供一个有价值的工具,它可以推动成本降低和促使供应商持续改进。

步骤3:准备供应商调查

接下来,对现有和潜在的替代供应商开展供应商调查。此调查将有助于评估供应商的供应能力。此时,调查人员可以从其销售系统获得现有供应商数据来验证潜在供应商未来的信息。该调查旨在评估市场满足您要求的能力。它使您能够在早期阶段评估您提议的项目是否可行,并且可以由确定的供应基地交付。它还向市场提供您的需求的早期警告,并使供应商能够考虑他们将如何响应。这里的关键目标是鼓励具有正确结构的正确供应商响应您。

在这些关键领域搜集知识:

- 可行性;
- 胜任能力;
- 成熟;
- 潜在能力。

步骤4:制定战略

此步骤涉及开发业务流程寻源策略。前三个步骤的结合为采购战略提供了基本要素。但是,对于每个区域或类别,它将取决于以下几点。

(1)供应商的市场竞争力:有了供应商信息,您可以在供应市场中构建竞争格局。这有助于向其他供应商展示优惠的规模,并向现有供应商传达潜在采购活动的严重性。

(2)您的用户对测试现有供应商的支持程度:业务流程采购团队有两方面的内部利益相关者,分别是使用所购买服务的人员和管理总体成本的管理人员。只要流程在另一个部门启动,支出类别的人员将接受降低成本,但这并不意味着供应商的变化会危及与原有供应商的良好关系,避免产生投诉或影响诸如交付可靠性、服务或付款的情况。对于管理人员而言,成本和服务竞争力是一个关键目标,但他们也是各种公司服务的用户,因此往往陷入追求成本改善和抵制变革的用户心态

of resisting change. To mobilize users' and executives' support for the category strategy, it is vital to communicate all benefits and overcome any potential risks.

3. What alternatives exist to competitive assessment: If the supply base is competitive, you can harness those forces to leverage better pricing or terms owing to increased number of a streamlined product or service specification. Once the result of the competitive sourcing effort is determined, it will be useful to set up a collaborative program that will run until the next competitive sourcing event takes place. If a competitive approach to sourcing is not a viable option, it is worth considering what the alternatives are, such as collaborating with suppliers:

   a. To reduce complexity and in turn increase productivity
   b. To create corroborative process improvements that reduce the cost of doing business
   c. To change the way the relationship is structured. For example, firms may invest in supplier operations to guarantee access to supply, new technology, or process improvements.

These alternatives are pursued typically when a buying company has little leverage over its supply base. They will be relying on good faith that suppliers will share the benefits of a new approach. The sourcing strategy is an accumulation of all the drivers thus far mentioned.

Step 5: RFx Request for ...

Where a competitive approach is used, which is the general case for most expenditure categories, a request for proposal (RFP) or bid will need to be prepared (request for quotation (RFQ), electronic request for quotation (eRFQ), etc.). This will define and make clear the requirements to all prequalified suppliers. It should include product or service specifications, delivery and service requirements, evaluation criteria, pricing structure, and financial terms and conditions. A communication plan should also be implemented at this stage to attract maximum supplier interest. Ensure that every supplier is aware they it is competing on a level playing field.

Once the RFP is sent out to all suppliers, make sure they are given enough time to respond. Follow-up messages should also be sent out to encourage a greater response.

Step 6: Selection

This is about selecting and negotiating with suppliers. The sourcing team should apply its evaluation criteria to the supplier responses. If extra information beyond the RFP response is required, do not be afraid to ask for it. If carried out manually, the negotiation process is conducted first with a larger set of suppliers, then narrowed to a few finalists. If the sourcing team uses an electronic negotiation tool, a greater number of suppliers may be kept in the process for longer, giving more diverse suppliers a better chance at winning the business.

之间。为了动员用户和管理人员对此类战略的支持,沟通所有利益相关者并克服任何潜在风险至关重要。

(3)评估有哪些替代方案:如果供应商具有竞争力,您可以利用这些优势优化产品或服务规范,以便产生更好的定价或条款。一旦确定了寻源结果的竞争力,建立一个合作计划将非常有用,该计划将一直运行到下一个竞争性寻源活动发生。如果竞争性的采购方案不是可行的选择,那么有必要考虑替代方案,如与供应商合作。

a. 降低复杂性,进而提高生产率。

b. 建立可行的改进流程,以降低开展业务的成本。

c. 改变关系结构。例如,企业可以投资于供应商运营,以确保获得改进的新技术或新工艺。

这些替代方案通常是在采购商对其供应商几乎没有影响力的情况下采用的。他们将依靠供应商分享新方法的好处的诚意。采购策略是迄今为止提到的所有驱动因素的累积。

步骤5:提出要求

如果采用通常的竞争性的采购方法,则需要编制招标邀请书(request for proposal, RFP)或报价邀请书(request for quotation, RFQ)、电子报价请求书(electronic request for quotation, eRFQ)等,这将规定并明确所有供应商资格预审的要求。它应包括产品或服务规范、交付和服务要求、评估标准、定价结构以及财务条款和条件。在这一阶段还应与供应商实施有效沟通,以最大限度地吸引供应商的兴趣,确保每个供应商都知道他们在公平竞争。

一旦将招标邀请书发送给所有供应商,确保他们有足够的时间来响应,还应发出后续信息,鼓励作出更大的反应。

步骤6:选择

这是关于选择和与供应商谈判。采购团队应将其评估标准应用于已经响应的供应商。如果需要招标文件以外的额外信息,不要害怕要求。如果是人工进行的,谈判过程首先是与一组较大的供应商进行,然后缩小到几个最终确定者。如果采购团队使用电子谈判工具,更多的供应商可能会在这一过程中停留更长时间,从而使更多的多样化供应商有更好的机会赢得业务。

Compare outcomes in terms of total value or implementation cost differences. Departments directly affected can be brought into the final selection process. Senior executives should be briefed on the final selection, to gain their approval and also be given the rationale behind the decision, to prepare them for any calls they receive from disappointed suppliers.

Step 7: Communicate with Your New Suppliers

Once the winning supplier(s) are notified, they should be invited to participate in implementing recommendations. Implementation plans vary depending on the degree of supplier switches. For incumbents, there will be a communication plan that will include any changes in specifications, improvements in delivery, and service or pricing models. These ought to be communicated to users as well. Because the company may have significantly benefited from this entire process, it is important that this be recognized by both company and supplier.

For new suppliers, a communication plan has to be developed that manages the transition from old to new at every point in the process that is touched by the spend category. Department, finance, and customer service are affected by this change, and their risk antennae will be particularly sensitive during this period. It is particularly important to measure closely the new supplier's performance during the first weeks of engagement.

Being able to demonstrate that performance matches, or is superior to, that of the former supplier will be vital during this sensitive time. It is also important to capture the intellectual capital your sourcing team has developed during the seven-step process so it can be used the next time that category is sourced.

CONCLUSIONS

In this section we have gone through what BPO is, and how and why it is applied. We have furthermore detailed where it can or should be applied. We have also elaborated on the current stage in the evolution of outsourcing and the development of strategic partnerships, including the steps and the to-do's such a partnership as well as BPO tendering would include. To gain maximum benefit, a BPO program should go through a formal close-down. There is no point in arguing lost causes once irrevocable decisions have been taken. Staff and companies alike need to accept the new situation and move forward. However, there will be a lot of information generated during the life of the program, and this will have been stored with varying degrees of formality by the team members. This information needs to be formally filed away for future reference.

In this light, there are no simple criteria to conduct an outsourcing versus in-house analysis. The benefits associated with outsourcing are numerous, and one should consider each project on its individual merits. Ongoing operational costs that may be avoided by outsourcing are also a consideration. In a nutshell, outsourcing allows organizations to be more efficient, flexible, and effective, while often reducing costs.

When considering a BPO program, a few main factors influence successful outsourcing. Critical areas for a successful outsourcing program as identified are:

根据总报价或实施成本差异比较结果。可靠的供应商可以进入最终选择流程，采购部门应向高级管理人员简要介绍最终选择，以获得他们的批准，并给出决策的理由，为他们接到落选供应商的任何电话做好准备。

步骤7：与您的新供应商沟通

一旦通知中标供应商，应邀请他们参与实施建议。实施计划因供应商更换的程度而异。对于现有企业，应有一个沟通计划，其中将包括规范中的任何更改、交付中的改进以及服务或定价模型。这些也应该传达给用户。由于公司可能从整个过程中获得了巨大的利益，公司和供应商都必须认识到这一点。

对于新的供应商，必须制定一个沟通计划，在支出类别涉及的流程中的每一点管理从旧到新的过渡。部门、财务和客户服务都会受到这种变化的影响，在此期间，它们的风险意识将特别敏感。特别重要的是，在新供应商参与的前几周要密切衡量其表现。

能够证明性能匹配或优于之前供应商产品的性能对新供应商来说在这一敏感时期至关重要。获取采购团队在七步流程中开发的知识资本也很重要，以便下次采购该类别时使用。

5.8.7 总结

在本节中，我们已经介绍了什么是BPO以及它的应用方式和原因。我们已经进一步详细说明了它的应用领域和范围，还阐述了外包和战略伙伴关系发展的当前阶段，包括BPO招投标步骤和事项，为了获得最大的收益，BPO项目应该有始有终。

一旦做出了不可撤销的决定，就没有必要为失败的原因争论。员工和公司都需要接受新形势并向前发展。但是，在项目的生命周期中会产生大量的信息，并且团队成员将以不同程度的形式存储这些信息。这些信息需要正式归档以备将来参考。

因此，没有简单的标准来进行外包与内部分析。与外包相关的好处是很多的，我们应该根据每个项目的特点来考虑每个项目。外包可能避免的持续运营成本也是一个考虑因素。简而言之，外包可以使组织更高效、灵活和有效，同时通常能够降低成本。

在考虑BPO项目时，有几个主要因素会影响成功的外包。成功外包计划的关键领域如下：

- Understanding company goals and objectives
- A strategic vision and plan
- Identify a business model and operating model in which are core differentiating, core competitive, and non-core competencies
- Map the relevant processes for the chosen business competencies and business functions that are desired to be outsources
- Selecting the right vendor
- Ongoing management of relationships
- Properly structured contract
- Select value and performance measures that are the basis for the service-level agreement
- Open communication with affected individual/groups
- Senior executive support and involvement
- Careful attention to personnel issues
- Short-term financial justification (value case and business case)

Some of the top advantages brought about by outsourcing include the following:

- Staffing flexibility
- Acceleration of projects and quicker time to market
- High-caliber professionals who hit the ground running
- Ability to tap into best practices
- Knowledge transfer to permanent staff
- Cost-effective and predictable expenditures
- Access to the flexibility and creativity of experienced problem solvers
- Resource and core competency focus

End Notes

1. Robert Handfield, *A Brief History of Outsourcing, 2006*.
2. Tas J and Sunder S, 2004, "Financial Services Business Process Outsourcing", *Communications of the ACM* 47, no. 5.
3. C.N.E.T. News, *IT Firms Expand from PCs to Payroll*, ed. Frauenheim.
4. http://searchcio.techtarget.com/definition/business-process-outsourcing.
5. Nellis J. G. and David Parker, "Principles of Business Economics," Financial Times Prentice Hall (2006): 213, ISBN 978-0-273-69306-2.
6. Sagoo A, "How IT is Reinvigorating Business Process Outsourcing" CIO (6 September 2012), Retrieved 25 March 2013.
7. B.P.M. Watch, In-Sourcing Remotely: A Closer Look at an Emerging Outsourcing Trend, http://www.bpmwatch.com/columns/in-sourcing-a-closer-look-at-an-emerging-outsourcing-trend/.
8. Willcocks L. Hindle J. Feeny D and Lacity M, "IT and Business Process Outsourcing: The Knowledge Potential," *Information Systems Management* 21, (2004): 7–15.
9. Gilley K. M and Rasheed A, "Making More by Doing Less: An Analysis of Outsourcing and its Effects on Firm Performance," *Journal of Management* 26, no. 4 (2000): 763–790.

- 了解公司目标；
- 战略愿景和规划；
- 确定业务模式和运营模式，其中包括核心差异化、核心竞争力和非核心竞争力；
- 为所选业务能力和期望外包的业务职能制订相关流程；
- 选择合适的供应商；
- 对现有关系持续管理；
- 签订合理的合同；
- 以服务水平为指标订立价值和绩效评价方法；
- 与个人/团体进行开放沟通；
- 获得高管人员的支持和参与；
- 注意人事问题；
- 短期财务识别（价值案例和业务案例）。

外包带来的一些主要优势包括：

- 人员配置灵活性；
- 加速项目并加快上市时间；
- 加速培养高水平的专业人员；
- 检验最佳实践经验；
- 将知识传授给普通员工；
- 具有成本效益和可预测的支出；
- 让有经验的问题解决者得到灵活性和创造性；
- 聚焦于资源和核心能力。

参考文献

[1] Robert Handfield, *A Brief History of Outsourcing, 2006.*

[2] Tas J and Sunder S, 2004, "Financial Services Business Process Outsourcing", *Communications of the ACM* 47, no. 5.

[3] C.N.E.T. News, *IT Firms Expand from PCs to Payroll*, ed. Frauenheim.

[4] http://searchcio.techtarget.com/definition/business-process-outsourcing.

[5] Nellis J. G. and David Parker, "Principles of Business Economics," Financial Times Prentice Hall (2006): 213, ISBN 978-0-273-69306-2.

[6] Sagoo A, "How IT is Reinvigorating Business Process Outsourcing" CIO (6 September 2012), Retrieved 25 March 2013.

[7] B.P.M. Watch, In-Sourcing Remotely: A Closer Look at an Emerging Outsourcing Trend, http://www.bpmwatch.com/columns/in-sourcing-a-closer-look-at-an-emerging-outsourcing-trend/.

[8] Willcocks L. Hindle J. Feeny D and Lacity M, "IT and Business Process Outsourcing: The Knowledge Potential," *Information Systems Management* 21, (2004): 7−15.

[9] Gilley K. M and Rasheed A, "Making More by Doing Less: An Analysis of Outsourcing and its Effects on Firm Performance," *Journal of Management* 26, no. 4 (2000): 763−790.

10. Kakabadse A and Kakabadse N, "Trends in Outsourcing: Contrasting USA and Europe," *European Management Journal* 20, no. 2 (2002): 189–198.

11. Weerakkody, Vishanth , Currie L. Wendy and Ekanayake, Y, "Re-engineering business processes through application service providers - challenges, issues and complexities," *Business Process Management Journal* 9, no. 6 (2003): 776–794.

12. Leavy B, "Outsourcing strategies: Opportunities and Risk," *Strategy and Leadership* 32, no. 6 (2004): 20–25.

13. Tas J and Sunder S, "Financial Services Business Process Outsourcing," *Communications of the ACM* 47, no. 5 (2004).

14. Fischer L. M, 2001, From vertical to Virtual; How Nortel's Supplier Alliances Extend the enterprise [online], Strategy+Business.

15. Leavy, 2004, 20–25.

16. See the note 3 above.

17. Capgemini Consulting BPO Research.

18. http://www.horsesforsources.com/hfs-index-q12013_02221#sthash.crBfQZEL.dpuf.

19. http://www.statista.com/statistics/298574/bpo-and-it-services-market-breakdown-worldwide/.

20. Christian Schuh et al.,*The Purchasing Chessboard: 64 Methods to Reduce Cost and Increase Value with Suppliers*, Pages 11–33, (Springer: Berlin Heidelberg, 2009), ISBN 978-3-540-88724-9.

21. Code Red – Stat!, Sherpas in Blue Shirts, Peter Bendor-Samuel, Chief Executive Officer, Everest Group and Ross Tisnovsky, Senior Vice President, Everest Group, (June 2, 2011).

22. Michel V and Fitzgerald G, "The IT outsourcing market place: vendors and their selection," *Journal of Information Technology* 12, (1997): 223–237.

23. Adsit, D, (2009) Will a Toyota Emerge from the Pack of Me-Too BPO's?, In Queue.

24. Bunmi Cynthia Adeleye, Fenio Annansingh and Miguel Baptista Nunes, "Risk management practices in IS outsourcing: an investigation into commercial banks in Nigeria," *International Journal of Information Management* 24 (2004): 167–180.

25. F. Gibb and S. Buchanan, "A framework for business continuity management," *International Journal of Information Management* 26, no.2 (2006): 128–141.

26. See the note 11 above.

27. BPO Practices & Services Model, LEADing Practice Competency Modelling Reference Content (LEAD-ES20013BC).

28. See the note 1 above .

29. See the note 12 above.

30. See the note 27 above.

［10］ Kakabadse A and Kakabadse N, "Trends in Outsourcing: Contrasting USA and Europe," *European Management Journal* 20, no. 2 (2002): 189−198.

［11］ Weerakkody, Vishanth, Currie L. Wendy and Ekanayake, Y, "Re-engineering business processes through application service providers—challenges, issues and complexities," *Business Process Management Journal* 9, no. 6 (2003): 776−794.

［12］ Leavy B, "Outsourcing strategies: Opportunities and Risk," *Strategy and Leadership* 32, no. 6 (2004): 20−25.

［13］ Tas J and Sunder S, "Financial Services Business Process Outsourcing," *Communications of the ACM* 47, no. 5 (2004).

［14］ Fischer L. M, 2001, From vertical to Virtual; How Nortel's Supplier Alliances Extend the enterprise ［online］, Strategy+Business.

［15］ Leavy, 2004, 20−25.

［16］ See the note 3 above.

［17］ Capgemini Consulting BPO Research.

［18］ http://www.horsesforsources.com/hfs-index-q12013_02221#sthash.crBfQZEL.dpuf.

［19］ http://www.statista.com/statistics/298574/bpo-and-it-services-market-breakdown-worldwide/.

［20］ Christian Schuh et al.,*The Purchasing Chessboard: 64 Methods to Reduce Cost and Increase Value with Suppliers*, Pages 11−33, (Springer: Berlin Heidelberg, 2009), ISBN 978-3-540-88724-9.

［21］ Code Red　Stat!, Sherpas in Blue Shirts, Peter Bendor-Samuel, Chief Executive Officer, Everest Group and Ross Tisnovsky, Senior Vice President, Everest Group, (June 2, 2011).

［22］ Michel V and Fitzgerald G, "The IT outsourcing market place: vendors and their selection," *Journal of Information Technology* 12, (1997): 223−237.

［23］ Adsit, D, (2009) Will a Toyota Emerge from the Pack of Me-Too BPO's?, In Queue.

［24］ Bunmi Cynthia Adeleye, Fenio Annansingh and Miguel Baptista Nunes, "Risk management practices in IS outsourcing: an investigation into commercial banks in Nigeria," *International Journal of Information Management* 24 (2004): 167−180.

［25］ F. Gibb and S. Buchanan, "A framework for business continuity management," *International Journal of Information Management* 26, no.2 (2006): 128−141.

［26］ See the note 11 above.

［27］ BPO Practices & Services Model, LEADing Practice Competency Modelling Reference Content (LEAD-ES20013BC).

［28］ See the note 1 above .

［29］ See the note 12 above.

［30］ See the note 27 above.

The Business Process Management Way of Training and Coaching

Mark von Rosing, Henrik von Scheel, August-Wilhelm Scheer

INTRODUCTION

The high demand for skilled and experienced process professionals to both lead and participate in business process management (BPM) initiatives, has created a major skills gap that traditional BPM and BPM notation bodies of knowledge and certifications do not meet. In this chapter, we elaborate on the way of process training. The way of training focuses on the most effective way to build and mature cross-disciplinary competencies across the breadth of the ways of thinking, working, modeling, implementing, and governance. We advocate an approach that uniquely combines training and coaching with hands-on experience by using all of these disciplines in context with real-life process-oriented projects so as to address the existing knowledge gap.

In this chapter we explore a standardized and common way to train process professionals with a detailed career path for process experts, process architects, and process engineers. The result is that we turn traditional process education into performance-based project coaching that provides a different set of cost–benefit value ratios, representing the most effective way for organizations to build their process skills.

The Complete Business Process Handbook. http://dx.doi.org/10.1016/B978-0-12-799959-3.00034-3

第六部分

6.1 业务流程管理方式的培训和辅导

Mark von Rosing, Henrik von Scheel, August-Wilhelm Scheer

介绍

对技术和经验丰富的流程专业人员的高要求,无论是领导还是参与BPM计划,都会产生一个主要的技能差距。传统的BPM和BPM符号知识体系和认证都无法满足这些差距。在这一部分,我们详细阐述流程培训的方式。培训方式侧重于在思维、工作、建模、实施和治理方法的各个方面建立和发展跨学科能力的最有效方式。我们倡导一种独特的方法,将培训和指导与实践经验相结合,在现实的面向过程的项目中使用所有这些学科,以解决现有的知识差距。

在这一部分中,我们探索一种标准化和通用的方法来培训流程专业人员,为流程专家、流程架构师和流程工程师提供详细的职业发展道路。结果是,我们将传统流程教育转变为基于绩效的项目辅导,提供不同的成本−收益价值比率,代表了组织建立流程技能的最有效方式。

The Need for a Standardized and Common Way of Process Training

Joshua von Scheel, Mark von Rosing, Marianne Fonseca, Ulrik Foldager

INTRODUCTION

The need for skilled and experienced personnel to lead and participate in business process management (BPM) activities is obvious and in high demand in today's global market. Professional certification can be found in many industries and professions, and BPM is no exception. The market offers a variety of vendor-, technology-, and methodology-driven certifications. Certification in BPM, as described in this chapter, does not refer to certification in well-known methodologies used in BPM such as Six Sigma, Lean, or the IT Infrastructure Library (ITIL), nor in any vendor-specific tool. Instead, we refer to and focus on more generic, broadly scoped training in BPM as a discipline, including a much needed new way of thinking, working, modeling, implementing, and governing process modeling, process architecture, and process engineering principles (see Figure 1). There is rapidly growing interest in this type of certification, and a number of organizations have already established their own distinct approaches to curricula, exams, assessments, and certifications for BPM.

SKILLS REQUIREMENTS

As BPM rapidly matures, the need for a standardized and common way of training for process professionals is also evident. In this chapter, we will outline a fully standardized and common way of training for process professionals, focusing on what skills are required to succeed in BPM today, what career path a process professional should follow to meet the market demand, and in particular what the most effective way is to build on existing process competencies as well as develop new ones.

A recent global business process training and certification research based on a survey of 1765 organizations representing all major countries across both public and private sectors examined which kinds of skills are required to succeed with process modeling, architecture, and engineering in BPM projects. The research identified that:

- 72% of available business process certification programs do not meet the skills required for a process project today
- 93.4% of business process-certified practitioners do not know how to apply the knowledge gained from a classroom setting in a real-life project
- Business process certification programs are not up-to-date. Although businesses matured, applying processes to improve their performance, the certification

The Complete Business Process Handbook. http://dx.doi.org/10.1016/B978-0-12-799959-3.00035-5

6.2　对流程标准化和常用方法进行培训的需求

Joshua von Scheel, Mark von Rosing, Marianne Fonseca, Ulrik Foldager

6.2.1　介绍

在当今的全球市场上,领导和参与BPM活动需要更加熟练和经验丰富的人员。专业认证在很多行业都有,BPM也不例外。市场提供各种供应商、技术和方法驱动的认证。如本章所述,BPM中的认证并未涉及BPM中使用的众所周知的方法(如六西格玛、精益或ITIL)以及任何特定于供应商的工具中的认证。相反,我们参考并关注BPM中作为一门学科的更通用、范围广泛的培训,包括一种急需的新思维方式、工作方法、建模流程、实施和管理流程建模、流程架构和流程工程原理(图1)。人们对这种认证的兴趣正在迅速增长,许多组织已经为BPM的课程、考试、评估和认证建立了自己独特的方法。

6.2.2　技能要求

随着BPM的迅速发展,对流程专业人员在标准化和常规方法方面的培训需求也很明显。在本节中,我们将论述流程专业人员在标准化和常规方法方面的培训,重点关注今天在BPM中取得成功所需的技能、流程专业人员应该遵循哪些专业知识来满足市场需求,事实上,最有效的方法是具备在现有流程能力基础上建立、开发新流程的能力。

最近的一项全球业务流程培训和认证研究是基于对1765个组织的调查,这些组织来自主要国家的国有和私营部门,通过研究BPM项目中成功进行流程建模、体系结构和工程所需的技能发现以下几点。

- 72%的现有业务流程认证项目不符合流程项目所需的技能。
- 93.4%的业务流程认证从业人员不知道如何将从课堂环境中获得的知识应用到实际项目中。
- 业务流程认证项目不是最新的。尽管通过流程来提高绩效已经很成熟,但

programs never really followed along. With this significant skills gap, both practitioners and organizations were left on their own to reinvent and piece together the skills required.

- Organizations require cross-disciplinary business process practitioners; this includes skills within process modeling, architecture, and engineering (see Figure 1), although no existing training vendor or organization actually offers cross-disciplinary process certification programs.

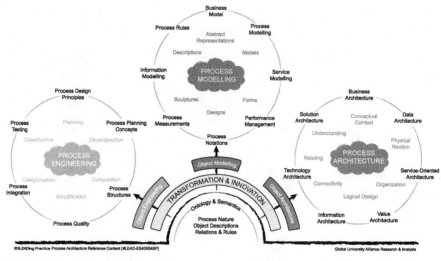

FIGURE 1

Example of cross-disciplinary business process skills: process engineering, process modeling, and process architecture.

Based on the key findings of the research, we can conclude that traditional BPM certification has a high learning versus forgetting curve and it is not up-to-date to meet the skills needed on the market today. There simply is a critical need for cross-disciplinary programs that share common aspects, and existing classroom training programs inhibit and restrain the strong transfer of knowledge from theory to practice in real-life projects.

LEARNING VERSUS FORGETTING CURVE

Academic studies by Profs. P. Quinn[1] and Thailheimer[2] on the most effective way to learn (learning versus forgetting curve) concluded that:

1. Most theoretical content ought to be converted to an e-learning or online setting
2. Traditional classroom training is improved by 58% when it is combined with coaching to convert the new knowledge gain into practical skills acquisition

是认证程序从未真正遵循。由于此项差距，从业者和组织都只能靠自己来重新定义和拼凑所需的技能。

- 组织需要跨学科的业务流程从业者，这包括流程建模、体系结构和工程中的技能（图1），虽然有现成的培训供应商或组织，但实际上没有提供跨学科的流程认证计划。

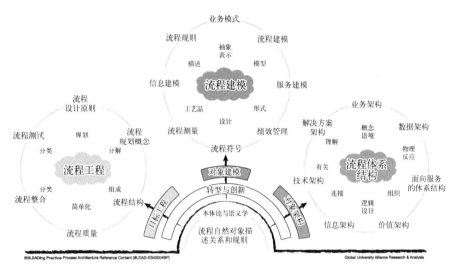

图1　跨学科业务流程技能示例：流程工程、流程建模和流程体系结构

根据研究的主要结果，我们可以得出结论：传统的BPM认证具有较高的学习与遗忘曲线，并且它不是最新满足当今市场所需的技能。简单地说，我们迫切需要跨学科的培训项目，而现有的课堂培训项目抑制和限制了知识在现实项目中从理论到实践的转移。

6.2.3　学习与遗忘曲线

根据P. Quinn[1]和Thailheimer[2]教授的学术研究，可得到最有效的学习方式（学习与遗忘曲线），结论为：

（1）大多数理论内容应该转换为在线学习；

（2）传统课堂培训与教练相结合，将获取新知识转化为实际技能的比率提高了58%；

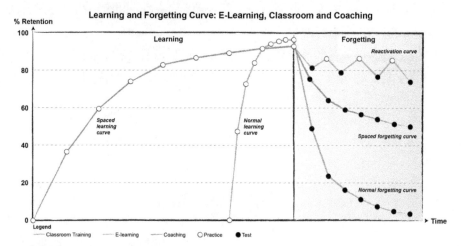

FIGURE 2

Learning versus forgetting curve for e-learning, classroom, and coaching.

3. Combining e-learning with training and coaching improves understanding and decreases the loss of knowledge (forgetting curve) by 79%[3,4] (see Figure 2).

This academic research shows that the most effective way to build process skills is a combination of online training (theoretical), classroom training (theoretical with practical examples), and coaching with hands-on experience using the different disciplines and program content in the context of real-life process-oriented projects. This requires turning traditional education into a performance-based project coaching exercise, instead—something that simultaneously provides a new whole new cost–benefit ratio and, through this new way of training, is achievable for both the individual practitioner and the entire project teams.

STANDARDIZED WAY OF TRAINING FOR PROCESS PROFESSIONALS

To face the skills gap that we encounter today, it is evident that a standardized and common way of training is required to build and develop the skills that process experts, process architects, and process engineers need. In our effort to standardize a common way of training for process professionals, we have:

1. Mapped the skills requirement based on the BPM ontology and the skill sets needed
2. Structured the entire learning process and educational material into a way of thinking, working, modeling, and governing (see Figure 3), and
3. Integrated the most effective way to build skills.
4. Support or exchange physical education with online training.

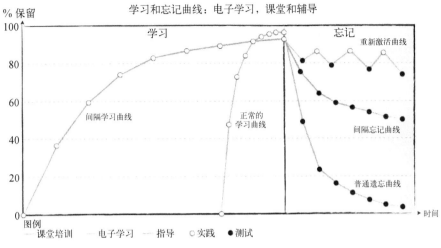

图2 在线学习、课堂和辅导的学习与遗忘曲线

（3）将在线学习与培训和辅导相结合，可以提高理解能力并将知识损失（遗忘曲线）减少79%[3,4]（图2）。

这一学术研究表明，构建流程技能最有效的方法是将在线培训（理论）、课堂培训（理论与实践案例）和指导与实践经验相结合，在面向流程的实际项目中使用不同学科和项目内容。这就需要将传统的教育转变成一种基于绩效的项目指导实践，并且通过这种新的培训方式，对个体从业者和整个项目团队来说都是可以实现的。

6.2.4 针对流程专业人员的标准化培训

为了应对我们今天遇到的技能差距，我们需要采用标准化和常规的培训方式来建立和发展流程专家、流程架构专家和流程工程师所需的技能。在我们为流程专业人员提供标准化培训方式的努力中，我们有：

（1）基于BPM本体和所需的技能映射技能需求；

（2）将整个学习过程和教育材料组织成一种思考、工作、建模和管理的方式（图3）；

（3）整合最有效的技能培养方式；

（4）以线上学习的方式支持或交流学习内容。

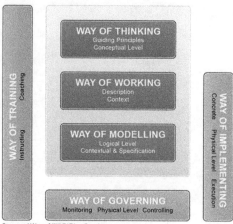

FIGURE 3

Structured way of thinking, working, modeling, implementing, governing, and training with processes.

End Notes

1. Xiao, W.S., Quinn, P.C., Pascalis, O., & Lee, K. "Own- and other-race face scanning in infants: Implications for perceptual narrowing." *Developmental Psychobiology*, (Special Issue on Perceptual Narrowing), 56, (2014): 262–273.
2. Spacing Learning Events Over Time: What the Research Says A Work-Learning Research, Inc. *Publication.* 2006 by Will Thalheimer.
3. Ritter, F.E., & Schooler, L.J. "The learning curve." in *International Encyclopedia of the Social and Behavioral Sciences* (2002), 8602–8605. Amsterdam: Pergamon.
4. Meek, C., Thiesson, Bo, Heckerman, David (Summer 2002). "The Learning-Curve Sampling Method Applied to Model-Based Clustering." *Journal of Machine Learning Research* 2(3): 397.

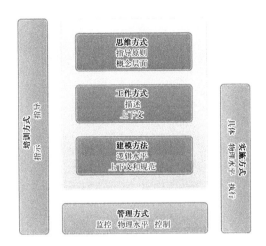

图3   结构化的思维方式、工作方式、建模方式、实现方式、治理方式和流程培训方式

参考文献

［ 1 ］ Xiao, W.S., Quinn, P.C., Pascalis, O., & Lee, K. "Own- and other-race face scanning in infants: Implications for perceptual narrowing." *Developmental Psychobiology*, (Special Issue on Perceptual Narrowing), 56, (2014): 262−273.

［ 2 ］ Spacing Learning Events Over Time: What the Research Says A Work-Learning Research, Inc. *Publication.* 2006 by Will Thalheimer.

［ 3 ］ Ritter, F.E., & Schooler, L.J. "The learning curve." in *International Encyclopedia of the Social and Behavioral Sciences* (2002), 8602−8605. Amsterdam: Pergamon.

［ 4 ］ Meek, C., Thiesson, Bo, Heckerman, David (Summer 2002). "The Learning-Curve Sampling Method Applied to Model-Based Clustering." *Journal of Machine Learning Research* 2(3): 397.

Process Expert Training

Program duration: Five working days which is supported and or can be exchanged by online training.

Target audience: This certification program has been designed for professionals with 3 or more years of experience:

- *Specialists*: process experts, business analysts, process specialists, process method specialists (*Business Process Reengineering* (BPR), Six Sigma, *Total Quality Management* (TQM), and/or Lean practitioner) and quality/production/manufacturing specialists
- *Consultants*: process consultants, service consultants, business consultants and transformation consultants
- *Architects*: process architects, enterprise architects, technology architects, solution architects, application architects, business architects, service architects, information/data architects, and value architects
- *Managers*: process managers, service managers, business managers, and project managers
- *Directors*: process owners, business owners, *Line of Business* (LoB) directors, and LoB owners

Program type: The ideal program type would be a classroom (physical location) or online, depending on the size and location of the process team and its willingness to accept online training as a form of skills building.

The process expert certification program has been structured to build on the existing competencies of the practitioner, but most importantly, with a main focus on developing new skills through the use of our unique modeling principles. The aim is also to infuse the practitioner with an entirely new way of thinking, working, and modeling with business processes.

This is done through 5 days of intensive classroom or online training, in-depth tutoring, and coaching coupled with hands-on project experience, in which the practitioner applies the acquired process modeling (Figure 1) techniques and its related disciplines to the practitioner's own company projects.

The Complete Business Process Handbook. http://dx.doi.org/10.1016/B978-0-12-799959-3.00036-7

6.3 流程专家培训

课程持续时间：5个工作日现场培训，或者使用在线培训的方式代替。

目标受众：本认证计划专为拥有3年或以上经验的专业人士设计。

- 专家团队：流程专家、业务分析人员、流程专家、流程方法专家（BPR、六西格玛专家、TQM专家、精益实践者）以及质量/生产/制造专家。
- 顾问团队：流程顾问、服务顾问、业务顾问和变革顾问。
- 架构团队：流程架构专家、企业组织架构专家、技术架构专家、解决方案架构专家、应用专家、业务架构专家、服务架构专家、信息/数据架构专家和价值架构专家。
- 管理团队：流程管理专家、服务支持管理专家、业务管理专家和项目管理专家。
- 主管团队：负责流程人员、负责业务人员、业务线（line of business，LOB）主管和负责人。

课程类型：理想的课程类型是现场教学（工作现场）或在线培训，这取决于流程团队的规模和场所，以及将在线培训作为一种技能培养方式的意愿。

流程专家认证课程的构建基于从业者的现有能力，但最重要的是，应该聚焦在通过使用我们独特的建模原则而得到的新技能方面。这样做的目的在于为参与者注入一种全新的思维、工作和业务流程建模方法。

这主要是通过为期5天的密集现场教学或在线培训，以及手把手式的教练式经验传授来完成，其中流程专家学习模型（图1）技术及其相关质量将应用于本公司项目。

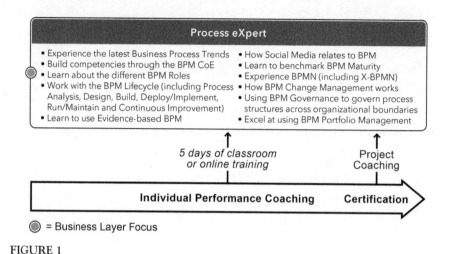

FIGURE 1

Process expert learning model.

CONTENT OF THE PROGRAM

The process expert certification program provides the practitioner with an extensive tool kit and profound knowledge of the Layered Enterprise Architecture Development (LEAD) enterprise standards and industry standards as well as highly detailed and descriptive process reference content that links and connects directly to hot topics such as:

- Business process trends
- How to build business process management (BPM) competencies: the BPM center of excellence
- The various BPM roles
- Working with the BPM life cycle
  1. Process analysis (project preparation and blueprinting)
  2. Process design (project realization)
  3. Process build (final preparations)
  4. Process deployment/implementation (process release and deployment management)
  5. Process run and maintenance (process governance and monitoring)
  6. Continuous process improvement (optimization and improvements)
- Evidence-based BPM
- Social media and BPM
- Business process management maturity
- Business process management notation (BPMN) (including X-BPMN, the extended BPMN discipline)
- Business process management change management
- Business process management governance
- Business process management portfolio management

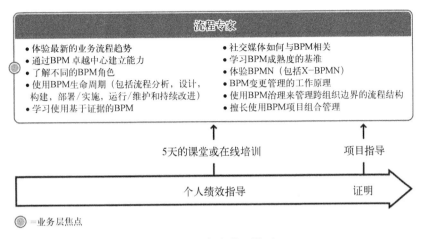

图1　流程专家学习模型

6.3.1　课程内容

流程专家认证课程为参与者提供了实用的工具包,以及与非常详细和描述性的流程参考内容一样的对企业层级架构(LEAD)中企业标准和行业标准的深刻认识,如以下方面。

- 业务流程趋势。

- 如何构建BPM能力:BPM卓越中心。

- 各种BPM角色。

- 使用BPM生命周期:

(1)流程分析(项目准备及蓝图);

(2)流程设计(项目实现);

(3)流程构建(最终准备);

(4)流程部署/实现(流程发布和部署管理);

(5)流程运行和维护(流程治理和监控);

(6)流程持续改进(优化和改进)。

- 基于证据的BPM。

- 社交媒体和BPM。

- BPM成熟度。

- 业务流程建模标识法(BPMN)(包括X-BPMN,扩展的BPMN规程)。

- BPM变革管理。

- BPM治理。

- BPM组合管理。

These topics have already been covered extensively throughout this handbook, and the process expert certification program aims to deliver actual, practical, and hands-on experience by using all of these disciplines in context with real-life process-oriented projects. This ensures the highest level of knowledge transfer and skills building to meet today's organizations cross-disciplinary competency requirements for professionals involved with process-oriented subjects.

PROCESS EXPERT LEARNING MODEL

The process expert learning model (Figure 1) is based on an intensive training module supported by in-depth individual performance coaching on a selected project. The hands-on experience ensures that the BPM and process modeling skills are applied to the following disciplines (Figure 1):

The needed skill for abstraction level for a process expert is:

- *Concrete*: Tangible, existing, and actual
- *Descriptive and specification*: Explanation, depiction/sketch, and portrayal, often using a process map, matrix, and/or model
- *Design*: Plan, intend, and aim

WHAT THE PRACTITIONER GETS

This certification program includes:

- Five working days of classroom and online training
- Training material with practical and usable reference content (i.e., the BPM life cycle, process templates)
- Individual performance coaching during the course
- Three 1.5-hour digital prerecorded sessions
- Process expert certificate

这些主题已经在本手册中进行了详细的介绍,流程专家认证课程旨在通过这些亲身体验、专业知识为工作实践提供切合实际的专业经验。这能够保证最高水平的知识继承和技能建设,以满足组织中流程人员具备跨学科能力的要求。

6.3.2　流程专家学习模型

流程专家学习模型(图1)建立在对选定项目的深入、高强度的培训基础之上,实践经验确保BPM和流程建模技术应用于以下学科(图1)。

流程专家所需的技能包括以下方面。

- 具体的技能:有形的、现有的和实际的。
- 描述和规范能力:解释、描述/示意图和画像,通常使用流程图、矩阵和/或模型。
- 设计能力:计划、意图和目标。

6.3.3　参与者获得了什么

该认证课程包括:

- 五个工作日的课堂学习和在线培训;
- 具有实用和可用参考内容的培训材料(即BPM生命周期、流程模板);
- 课程期间的个人教练指导;
- 三个1.5小时的光盘课程;
- 流程专家认证证书。

Process Architect Training

Program duration: Ten working days.

Target audience: This certification program has been designed for professionals with 3–10 years of experience:

- *Specialists*: senior process experts, senior business analysts, and Enterprise Architecture (EA) method specialists (The Open Group Architecture Framework (TOGAF), Zachman, and *Ministry of Defence Architecture Framework* (MODAF) practitioners)
- *Consultants*: senior process consultants, senior business consultants, and senior transformation consultants
- *Architects*: process architects, enterprise architects, solution architects, application architects, business architects, service architects, information/data architects, and value architects
- *Managers*: process managers if they have a process ownership

Program type: The ideal program type is to combine classroom training (physical location) with online training, depending on the size and location of the process team. The first week is an optional choice (classroom or online); the second week is always classroom.

The process architect certification program has been structured to build on the existing competencies of the practitioner, but most importantly, with a main focus on developing new skills through the use of our unique architecture principles. The aim is also to infuse the practitioner with an entirely new way of thinking, working, and modeling with business processes.

This is done through 10 days of intensive classroom and/or online training, in-depth tutoring, and coaching coupled with hands-on project experience, in which the practitioner applies the acquired process architecture (Figure 1) techniques and its related disciplines to the practitioner's own company projects.

The Complete Business Process Handbook. http://dx.doi.org/10.1016/B978-0-12-799959-3.00037-9

6.4　流程架构师培训

课程持续时间：10个工作日。

目标受众：本认证计划专为拥有3 ~ 10年经验的专业人士设计。

- 专家团队：高级流程专家、高级业务分析专家和企业架构（EA）方法方面的专家［开放组架构框架（TOGAF）、Zachman框架和国防部结构框架（Ministry of Defence architecture framework，MODAF）专家］。
- 顾问团队：高级流程顾问、高级业务顾问、高级变革顾问。
- 架构团队：流程架构专家、企业组织架构专家、解决方案架构专家、技术架构专家、应用专家、业务架构专家、服务架构专家、信息/数据架构专家和价值架构专家。
- 管理团队：流程管理人员。

课程类型：课程类型是现场教学（工作现场）或在线培训，这取决于流程团队的规模和场所。第一周通常为选修课（课堂或网上），第二个周可在教室里现场教学。

流程架构认证课程的构建基于从业者的现有能力，但最重要的是，应该聚焦在通过使用我们独特的建模原则而得到的新技能方面。这样做的目的在于为参与者注入一种全新的思维、工作和业务流程建模方法。

这主要是通过为期10天的密集现场教学或在线培训，以及手把手式的教练式经验传授来完成，其中流程架构师学习模型（图1）技术及其相关质量将应用于本公司项目。

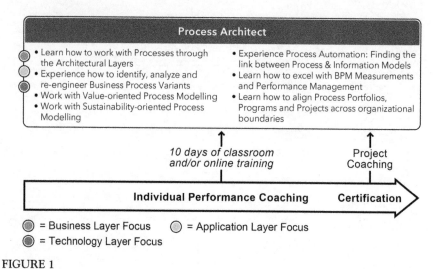

FIGURE 1

Process architect learning model.

CONTENT OF THE PROGRAM

The process architect certification program provides the practitioner with an extensive tool kit and profound knowledge of the Layered Enterprise Architecture Development (LEAD) enterprise standards and industry standards as well as highly detailed and descriptive process and architecture reference content that links and connects directly to hot topics such as:

- How to work with processes in architectural layers
- How to work with and model business process variations
- Value-oriented process modeling
- Sustainability-oriented process modeling
- Process automation: link between process models and information models
- Business process management measurements and performance management
- Business process management alignment

These topics have already been covered extensively throughout this handbook, and the process architect certification program aims to deliver actual, practical, and hands-on experience by using all of these disciplines in the context of real-life process-oriented projects. This ensures the highest level of knowledge transfer and skills building to meet today's organizations cross-disciplinary competency requirements for professionals involved with process-oriented subjects.

PROCESS ARCHITECT LEARNING MODEL

The process architect learning model (Figure 1) is based on an intensive training module supported by in-depth individual performance mentoring on a selected project. The

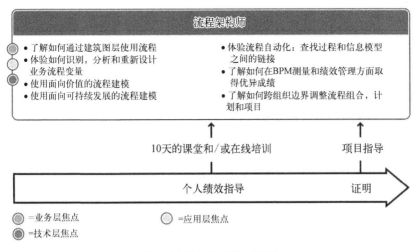

图1 流程架构师学习模型

6.4.1 该计划的内容

流程架构认证课程为参与者提供了实用的工具包,以及与非常详细和描述性的流程参考内容一样的对企业层级架构(LEAD)中企业标准和行业标准的深刻认识,如以下方面。

- 架构层级体系中的流程的运转方式。
- 如何使用和建模业务流程变更。
- 以价值为导向的流程建模。
- 面向可持续发展的流程建模。
- 流程自动化:流程模型和信息模型之间的链接。
- BPM测量和绩效管理。
- BPM一致性。

这些主题已经在本手册中进行了详细的介绍,流程架构认证课程旨在通过这些亲身体验、专业知识为工作实践提供切合实际的专业经验。这能够保证最高水平的知识继承和技能建设,以满足组织中流程人员具备跨学科能力的要求。

6.4.2 流程架构师学习模型

流程架构师学习模型(图1)建立在对选定项目的深入、高强度的培训基础

hands-on learning experience ensures that the business process management (BPM) and process architecture skills are applied in the following disciplines (Figure 1):

The needed skill for abstraction level for a process architect:

- *Conceptual*: Theoretical, abstract, and intangible; the high-level description of the logical
- *Context*: Situation, milieu/environment, and perspective
- *Concrete*: Tangible, existing, and actual
- *Descriptive and specification*: Explanation, depiction/sketch, and portrayal, often using a map, matrix, and/or model
- *Design*: Plan, intend, and aim
- *Execution*: Completing, performing, and realization

WHAT YOU GET

This certification program includes:

- Ten working days of classroom and online training
- Six 1.5-hour digital online sessions
- Individual performance coaching for project and individual development during the course
- Training material with practical and usable reference content (i.e., the BPM life cycle, process templates)
- Process architect certificate

之上。实践经验确保BPM和流程建模技术应用于以下学科(图1)。

流程架构师所需的技能如下。

- 概念抽象能力:理论,抽象和无形;高水平的逻辑描述。
- 背景认知能力:情境,环境/背景和观点。
- 总结能力:有形的、现有的和实际的。
- 描述性和规范:解释、草图和描绘,通常使用地图、矩阵和/或模型。
- 设计能力:计划、打算和目标。
- 执行能力:完成、执行和实现。

6.4.3 您能得到什么

该认证课程包括:

- 十个工作日的课堂和在线培训;
- 六个1.5小时的数字化课程;
- 课程期间为项目和个人发展提供个人表现指导;
- 具有实用和可用参考内容的培训材料(即BPM生命周期、流程模板);
- 流程架构师认证证书。

Process Engineer Training

Program Duration: Ten working days.

Target Audience: This certification program has been designed for professionals with 3–10 years of experience:

- *Specialists:* The engineering specialist who needs process knowledge, such as a systems engineer, quality engineer, or software engineer
- *Consultants:* Engineering consultants (depending on whether it is an engineering-centric organization)
- *Architects:* Process architects (depending on whether it is an engineering-centric organization)
- *Managers:* Process managers (depending on whether it is an engineering-centric organization)

Program type: The ideal program type is to combine classroom training (physical location) with online training, depending on the size and location of the process team. The first week is an optional choice (classroom or online); the second week is always classroom.

The process engineer certification program has been structured to build on the existing competencies of the practitioner, but most importantly, with a main focus on developing new skills through the use of our unique engineering principles. Process engineers focus on the design, operation, control, and optimization of business, application, and technology processes. They use specific engineering principles to enable better enterprise-related processes. Process engineering training therefore focuses on the daily job of the process engineer:

- Construct and maintain the endeavor-specific process from the process landscape
- Evaluate process tools for consistency with the organizational process landscape and process life cycle and/or endeavor-specific process
- Ensure that the endeavor-specific process is constructed based on endeavor-specific needs before process tool selection rather than being driving by early selection of a potentially inappropriate process tool
- Provide input to the environments team regarding required process tool support
- Provide local guidance and mentoring in the proper adoption and use of the endeavor process
- Identify and document the enterprise's own leading practices; disseminate and evangelize industry best practices and common best practices
- Work to support strategic process initiatives including recommending improvements to the organizational process framework
- Support multiple endeavors within a local region
- Staff regional process help desks

The Complete Business Process Handbook. http://dx.doi.org/10.1016/B978-0-12-799959-3.00038-0

6.5　流程工程师培训

课程持续时间：10个工作日现场培训。

目标受众：本认证计划专为拥有3 ~ 10年经验的专业人士设计。

- 专家团队：知识工程专家，如系统工程专家、质量工程专家或软件工程专家。
- 顾问团队：工程顾问（取决于它是否是以工程为中心的组织）。
- 架构团队：流程架构师（取决于它是否是以工程为中心的组织）。
- 管理团队：流程管理人员（取决于它是否是以工程为中心的组织）。

课程类型：理想的计划类型是将课堂培训（实际位置）与在线培训相结合，具体取决于流程团队的规模和位置。第一周通常为选修课（课堂或网上），第二个周可在教室里现场教学。

流程工程师认证课程的构建基于从业者的现有能力，但最重要的是，应该聚焦在通过使用我们独特的建模原则而得到的新技能方面。流程工程师专注于业务、应用过程和技术流程的设计、操作、控制和优化。他们使用特定的工程原理来实现更好地与企业相关的流程。因此，流程工程培训侧重于流程工程师的日常工作：

- 从流程使用背景角度构建和维护工作流程；
- 从流程背景和流程生命周期以及行为过程角度确定流程评估工具；
- 确保在选择流程工具之前根据特定需求构建个性化的流程，避免产生不适当的流程；
- 向环境团队提供有关所需流程工具支持的信息；
- 在流程确定过程中提供本地指导；
- 识别并记录企业自身的优势；传播和宣传行业和本领域最佳实践经验；
- 致力于支持战略流程计划，包括建议改进组织流程框架；
- 支持本地化的多项尝试；
- 员工帮助平台；

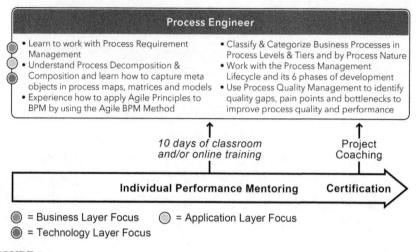

FIGURE 1

Process engineer learning model.

- Present local training on process-related topics
- Research advances in process engineering (e.g., new software development methods)

The aim is also to infuse the practitioner with a supporting process way of thinking, working, and modeling with business processes. This is done through 10 days of intensive classroom and/or online training, in-depth tutoring, and coaching, coupled with hands-on project experience in which the practitioner applies the acquired process engineering (Figure 1) techniques and its related disciplines to the practitioner's own company projects.

CONTENT OF THE PROGRAM

The process engineer certification program provides the practitioner with an extensive tool kit and profound knowledge of the LEAD enterprise standards and industry standards as well as highly detailed and descriptive process and engineering reference content that links and connects directly to topics such as:

- Process analysis techniques
- Process requirement management
- Process design practices and concepts
- Process decomposition and capturing in process templates/models
- Process composition and capturing in process templates/models
- Process categorization and classification
  - Process levels
  - Process tiers
  - Process nature

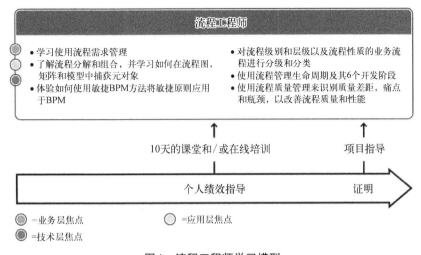

图1　流程工程师学习模型

- 提供有关流程相关主题的本地培训；
- 流程工程的研究进展（如新的软件开发方法）。

目的是通过为期10天的课堂学习、在线培训经验辅导的方式完成，为从业者提供一种对业务流程进行思考、工作、建模的流程知识方式。结合实践项目经验，从业者可将获得的流程工程师学习模型（图1）技术及其相关学科应用于公司项目。

6.5.1　该计划的内容

流程工程师认证项目为从业者提供了广泛的工具包，以及与非常详细和描述性的流程参考内容一样的对企业层级架构（LEAD）中企业标准和行业标准的深刻认识，如以下方面。

- 流程分析技术。
- 流程需求管理。
- 流程设计实践和概念。
- 流程模板/模型中的流程分解和捕获。
- 流程组合和流程模板/模型中的捕获。
- 流程分集和分类：
- 流程级别；
- 流程系列；
- 流程性质。

- Process life cycle management
- Process quality management
- Agile business process management (BPM)

These topics have already been covered extensively throughout this handbook, and the process engineer certification program aims to deliver actual, practical, and hands-on experience by using all of these disciplines in the context of real-life process-oriented projects. This ensures the highest level of knowledge transfer and skills building to meet today's organizations' cross-disciplinary competency requirements for professionals involved with process-oriented subjects.

PROCESS ENGINEER LEARNING MODEL

The process engineer learning model (Figure 1) is based on an intensive training module supported by in-depth individual performance mentoring on a selected project. The hands-on learning experience ensures that the business process management and process engineering skills are applied within the following disciplines (Figure 1):

Needed skills at an abstraction level for a process engineer are:

- *Conceptual*: theoretical, abstract, and intangible—high-level description of the logical
- *Context*: situation, milieu/environment, and perspective
- *Concrete*: tangible, existing, and actual
- *Descriptive and specification*: explanation, depiction/sketch, and portrayal, often using a map, matrix, and/or model
- *Design*: plan, intend, and aim
- *Execution*: completing, performing, and realization

WHAT YOU GET

This certification program includes:

- Ten working days of classroom and online training
- Six 1.5-hour digital online sessions
- Individual performance coaching for project and individual development during the course
- Training material with practical and usable reference content (i.e., the BPM life cycle, process templates)
- Process engineer certificate

CONCLUSIONS

In this chapter we have outlined a standardized and common way of training process professionals with a detailed career path for the process expert, process architect, and process engineer. We have explained which skills are required for the way of

- 流程生命周期管理。
- 流程质量管理。
- 敏捷BPM。

这些主题已经在本手册中进行了详细的介绍,流程工程师认证课程旨在通过这些亲身体验、专业知识为工作实践提供切合实际的专业经验。这能够保证最高水平的知识继承和技能建设,以满足组织中流程人员具备跨学科能力的要求。

6.5.2 流程工程师学习模型

流程工程师学习模型(图1)建立在对选定项目的深入、高强度的培训基础之上,实践经验确保BPM和流程建模技术应用于以下学科(图1)。

流程工程师所需的技能如下。

- 概念抽象能力:理论,抽象和无形;高水平的逻辑描述。
- 背景认知能力:情境,背景/环境和观点。
- 总结能力:有形的、现有的和实际的。
- 描述性和规范:解释、草图和描绘,通常使用地图、矩阵和/或模型。
- 设计能力:计划、打算和目标。
- 执行能力:完成、执行和实现。

6.5.3 您能得到什么

本认证计划包括:

- 十个工作日的课堂和在线培训;
- 六个1.5小时的数字化课程;
- 课程期间为项目和个人发展提供个人表现指导;
- 具有实用和可用参考内容的培训材料(即BPM生命周期、流程模板);
- 流程工程师认证证书。

6.5.4 总结

在本节中,我们概述了一种标准化和通用的培训流程专业人员的方法,以实现为流程专家、流程架构师和流程工程师提供详细的职业发展道路的目标。我们已经解释了思维方式、工作方式、建模和管理方式所需的技能。我们的经验表明,建

thinking, working, modeling, and governing. Our experience has shown that the most effective way to build process skills is a combination of online (theoretical) and classroom training (theoretical with practical examples) and coaching, and by using all of the described disciplines in the context of real-life process-oriented projects, turning traditional education into performance-based project coaching that provides a whole new cost–benefit ratio for the practitioner.

立流程技能的最有效方式是在线（理论）、课堂培训（理论与实践范例）和辅导相结合，并通过在现实生活过程中面向流程的项目，将传统教育转变为基于绩效的项目指导，为从业者提供全新的成本效益比。

Process Owner Training

Program duration: 0.5 to 2 days.

Target audience: This certification program has been designed for professionals with business experience who want to build process ownership skills:

- **Directors**: Business executives, business owners, Line of Business (LoB) directors, and LoB owners
- **Managers**: Business managers, service managers, operational managers, project managers, and business process managers
- **Specialists**: Business experts, operational process owners, business analysts, process specialists, process method specialists (BPR, Six Sigma, TQM, and/or Lean practitioner), and quality/production/manufacturing specialists

PROGRAM TYPE

The speed of business changes is increasing; because of this, the work of the process owner has become more complex and demanding. In most organizations the process owner is responsible for the governance of processes and their performance as well as continuous improvement. For most organizations this includes defining the process mission, vision, goals, and objectives; relating them to the various business key performance indicators (KPIs) and measures that are a part of multiple reports; and aligning them with the organization's decision making and strategies. Process owners monitor and report process performance against these KPIs and on the health of execution versus established plans. Furthermore, they are involved in synchronizing process improvement plans with other process owners within the value chain and other interconnected processes. Their process aim is to continuously increase the maturity of the processes and sustain each level of maturity.

Because of the nature of the work of process owners, it is not easy to take half a day or 1 or 2 days of work out of an already tight schedule, so training has to be available for whenever it is needed. The ideal program type for process owners and their on-demand model is therefore online training. Depending on the size and location of the process team, however, the program can be taught as classroom training at a physical location. Any process owner training program should be designed to build knowledge for business people who have process ownership and focus on specialized skills in this direction.

The process owner certification program is structured so as to build on the existing competencies of the practitioner, but more important, with a main focus on developing new process-focused skills through the use of our unique strategic, managerial, and operational process principles. The overall goal is to infuse the process owner with an entirely new way of thinking, working, modeling, and governing with business processes.

The Complete Business Process Handbook. http://dx.doi.org/10.1016/B978-0-12-799959-3.00039-2

6.6　流程所有人培训

课程持续时间：0.5 ~ 2个工作日。

目标受众：这个认证项目是为希望获得构建流程全部技能的具有业务经验的专业人士设计的。

- 主管团队：业务主管、业务负责人、LOB总监和负责人。
- 管理团队：业务经理、服务经理、运营经理、项目经理和业务流程经理。
- 专家团队：业务专家、运营流程所有人、业务分析专家、流程专家、流程方法专家(BPR、六西格玛、TQM和/或精益实践者)以及质量/生产/制造专家。

6.6.1　流程类型

业务变化的速度正在增加。因此，流程所有人的工作变得更加复杂和苛刻。在大多数组织中，流程所有人负责流程及其性能的治理以及持续改进。这包括定义流程使命、愿景、目的和目标，将它们与各种业务KPI和措施联系起来，并使它们与组织的决策和策略保持一致。流程所有人根据这些KPI以及执行状况确定流程绩效。此外，他们还参与了与价值链中的其他流程所有人和其他相互关联的流程同步流程改进计划。他们的流程目标是不断提高流程的成熟度并维持每个成熟度。

由于流程所有人的工作性质，在已经很紧的工作时间表上花费额外的半天、1天或2天并不容易，因此必须在需要时进行相应的培训，而在线培训是最理想的培训方式。但是，根据流程团队的规模和位置，可以将该计划作为实际位置的课堂培训进行教学。任何流程所有人培训计划都应该旨在为拥有流程所有权并专注于此方向的专业技能的业务人员构建知识。

对流程所有人进行认证的前提是建立在从业者的现有能力基础之上，但更重要的是，让其通过使用我们灵活的战略、管理和运营流程规则来开发新的以流程为中心的能力。总体目标是为流程所有人提供一种全新的思维、工作、建模和管理业务流程的方式。

This is done through intensive online training (classrooms are also an option), in-depth tutoring, and coaching coupled with hands-on project experience. The learning route of the process owner program has been designed in three areas:

- Strategic process owner (executive/director focus), which requires 0.5–1 day of training
- Tactical process owner (management focus), which requires 1 day of training
- Operational process owner (execution focus), which requires 2 days of training

The process owner certification program is split into three kinds of training components, but it is possible to train and focus on one, two, or all select areas: for instance, only the strategic or tactical and operational aspects.

CONTENT OF THE PROGRAM

The process owner role is as a specialization of the business owner role, and his or her responsibilities focus almost exclusively on defining and organizational planning of process strategies and goals, governing process execution, and evaluating service delivery through value and performance measurements, as well as setting up and initiating process transformation and innovation through the business process management (change management lifecycle).

The process owner certification program provides the practitioner with an extensive tool kit and profound knowledge of the LEAD enterprise standards and industry standards as well as highly detailed and descriptive process reference content. The process owner certification program consists of three areas that focus on different kinds of subject matter content:

- *Strategic process owner*
  - Establishment of process mission and vision
  - Strategy definition and planning (Strategic Business Objectives (SBOs))
  - Process planning aspects
  - Goals and objectives setting (Critical Success Factors (CSFs))
  - Process innovation thinking
  - Value-oriented process thinking
  - Budgeting and forecasting
- *Tactical process owner*
  - Strategic advisory and guidance
  - Performance design and monitoring (Key Performance Indicators (KPIs), PPIs, and Service Performance Indicators (SPIs))
  - Link to business policies and procedures
  - Defining process policies and procedures
  - Performance management
  - Link to business innovation and transformation enabling
  - Risk management and communication
  - Continuous improvement

要实现这个目标,密集的在线培训(现场课堂)、深入的辅导、教练技术以及项目实践经验是必需的。流程所有人项目的学习路线设计主要集中在如下三个方面:

- 战略流程所有人(重点是执行/指导),需要0.5 ~ 1天的培训;
- 战术流程所有人(重点是管理),需要1天的培训;
- 运营流程所有人(重点是执行),需要2天的培训。

流程所有人认证项目可分为三种培训组件,但可以培训其中一个。例如,仅进行战略或战术或运营方面的培训,也可以选择两个或所有项目。

6.6.2　该计划的内容

流程所有人角色是业务负责人角色的专业化,他或她的职责几乎专注于流程策略和目标的指定与规划、管理流程执行过程,以及通过价值和绩效测量评估技术,实现对流程生命周期的建立、启动、变革、创新等方面的管理。

流程所有人认证项目为流程实践者提供了广泛的工具包,对领先企业和行业标准有深刻理解,具备高度详细和描述性的过程参考。流程所有人认证项目包括三个领域,重点关注不同的主题。

- 战略流程所有人:
  - 建立流程使命和愿景;
  - 战略定义和规划(SBO);
  - 流程规划;
  - 目标和目标设定(CSF);
  - 流程创新思维;
  - 以价值为导向的流程思维;
  - 预算和预测。
- 战术流程所有人:
  - 战略咨询和指导;
  - 性能设计和监控(KPI、PPI和SPI);
  - 链接到业务策略和流程;
  - 流程定义的原则和程序;
  - 绩效管理;
  - 与业务创新和转型支持相关联;
  - 风险管理和沟通;
  - 持续改进。

- *Operational process owner*
  - Process governance
  - Operational planning and processing
  - Process maturity and process improvement
  - Operational advisory and guidance
  - Performance measurements and reporting
  - Issue management
  - Communication and collaboration

These topics have already been covered extensively throughout this handbook. The process owner certification program aims to deliver actual, practical, hands-on experience by using all of these disciplines in the context of real-life process-oriented projects. This ensures the highest level of knowledge transfer and skill building to meet today's organizations' cross-disciplinary competency requirements for professionals involved in process-oriented initiatives.

PROCESS OWNER LEARNING MODEL

The process owner learning model is based on an intensive training module supported by in-depth individual performance coaching on a selected project. The hands-on experience ensures that strategic management, tactical administration, and operational execution skills are applied for the following disciplines (Figure 1).

Needed Skill for Abstraction Level for a Process Owner

- **Conceptual**: Theoretical, abstract, and intangible—the high-level description of the logical way of working
- **Context**: Situation, milieu/environment, and perspective regarding the organization's business model

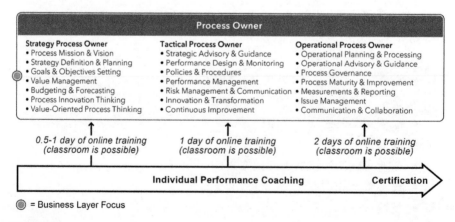

FIGURE 1

Process owner learning model.

- 运营流程所有人：
  - 流程治理；
  - 运营规划和处理；
  - 流程成熟度和流程改进；
  - 业务咨询和指导；
  - 绩效衡量和报告；
  - 问题管理；
  - 沟通和协作。

这些主题已在本手册中广泛涉及。流程所有人认证项目旨在通过在现实生活中面向流程的项目中使用所有这些学科来提供实际、实用的实践经验。这确保了最高水平的知识转移和技能培养，以满足当今组织对参与过程导向计划的专业人员的跨学科能力要求。

6.6.3　流程所有人学习模型

流程所有人学习模型基于一个密集的培训模块，支持对选定的项目进行深入的个人绩效指导。实践经验表明，战略管理、战术管理和操作执行技能应用于以下领域（图1）。

流程所有人业务抽象能力

- 概念性：理论性、抽象性和无形性——对逻辑工作方式的高层次描述。
- 环境：关于组织商业模式的情况、背景/环境和观点。

图1　流程所有人学习模型

- **Concrete**: Tangible, existing, and actual to the organization's operating model
- **Descriptive and specification**: Explanation, depiction/sketch, and portrayal often using a process map, matrix, and/or model
- **Design**: Plan, intention, and aim

WHAT THE PRACTITIONER GETS

This certification program includes:

- 0.5–4 days of on-demand online training (classroom training is also possible)
- Training material with practical and usable reference content (i.e., process workflows, process templates, etc.)
- Individual performance coaching during the course
- Process owner certificate

- 具体：组织的运营模型是有形的、现有的和实际的。
- 描述和规范：解释、描述/示意图和描述,通常使用流程图、矩阵和/或模型。
- 设计：计划、意图和目标。

6.6.4　实践者获得了什么

该认证项目包括：

- 0.5 ~ 4天的按需在线培训(也可以进行课堂培训);
- 具有实用和可用参考内容的培训材料(即工作流、流程模板等);
- 课程期间的个人表现指导;
- 流程所有人证书。

Conclusion

As it is so often, things have an ending, every story has an ending, and every book has an ending, however, in process modeling the end of something is just the start of something else. In this body of knowledge, we have shared with you concepts, detailed guidelines, methods, and approaches with integrated and standardized templates/artifacts. As a part of your journey we have the following additional recommendations:

For You as an Individual, it is a career opportunity. Our recommendations to succeed in your journey are as follows:

Stay Informed—push the bar and build the expertise required to work with process—do not let the vendors or consultants set the agenda. Follow the websites, webinars, and blogs of thought leaders and pioneers.

Join a Community—get involved in agnostic and vendor-neutral process community and learn from your peers, share insight, best practices, and leading practices.

Get Certified—we see too many process specialists who do not get formal training, missing out on the ability to be able to have any documentation i.e., certification on their skills. So our clear recommendation, get your vendor neutral certification as a Process eXpert, Process Architect, and/or Process Engineer with focus on cross-disciplines, e.g., Business Process Principles (BPR, Six Sigma, TQM, Lean, etc.), BPMN 2.0, eXtended BPMN (X-BPMN), Process Monitoring, Value-Oriented Process Modelling, Continuous Improvement, and Layered Architecture Modelling (Business, Application, Technology).

For You as a Business, it is a core discipline of any organization to understand the "as is" situation and design the "to be" state in order to model, engineer, and architect how the organization creates and realizes value. With our experience with leading organizations (i.e., outperformers), our recommendations are as follows:

Innovation & Transformation Enablement—Get executive coaching or project coaching on Business Innovation & Transformation Enablement. Due to lack of ability to execute what was defined, 70% of strategies fail to be implemented. The ability to identify what needs to be changed, why it needs to be changed, and how, is a very important aspect for process modeling. Without this link the cross-road to link strategy with operational execution is missing.

Do not Reinvent the Wheel—learn from others and apply business process best practices (reference content) to standardize the noncore competencies of your organization and thereby focus on the Cost and Operating Model.

Thrive on Accelerators—within your own industry, apply industry best practices and accelerators (industry capability models, industry performance indicators, industry measures, and other industry reference content) that will improve competitive parity and standardize core competitive competencies with focus on Performance Model and Service Model.

The Complete Business Process Handbook. http://dx.doi.org/10.1016/B978-0-12-799959-3.00040-9

第七部分

结论

通常来讲,任何事情有一个结局,每个故事都有一个结局,每本书都有一个结尾,然而,在流程建模中,一个阶段的结束只是另一个阶段的开始。在这个知识体系中,我们与您分享了集成和标准化模板/组件的概念、详细指南和方法。作为工作的一部分,我们为您提供以下额外建议。

对于您个人而言,这是一个职业机会。为确保您在今后工作中取得成功,给您建议如下。

保持知情:这是推动标准并建立处理流程所需的专业知识,不要让供应商或顾问设定议程。关注本领域的专家领袖、前沿网站、网络研讨会和博客。

加入一个社区:参与到探索未知领域以及具有客观中立观点的流程社区中,向同行学习、分享见解以及最佳和前沿实践。

获得认证:我们看到太多没有接受过正式培训的流程专家,他们错过了获得技能认证的机会。我们的明确建议是,在流程架构、流程工程等方面取得认证,重点关注跨学科的知识,如业务流程原则(BPR、六西格玛、TQM、精益等)、BPMN 2.0、扩展BPMN(X-BPMN)、过程监控、价值导向的流程建模、持续改进和分层架构建模(业务、应用、技术)。

为了您的企业:了解"现状"是任何组织的核心任务,设计"未来"状态,以便组织进行创造和实现价值,进行建模、设计和架构。凭借我们在领先组织(即表现优异的组织)方面的经验,我们的建议如下。

创新与变革推动者:获得关于业务创新与变革推动者的高管培训或项目培训。由于缺乏执行所定义内容的能力,70%的策略无法实现。识别需要更改什么、为什么需要更改以及如何更改的能力是流程建模的一个非常重要的方面。如果没有此链接,则缺少将战略与运营执行联系起来的纽带。

不要做重复性工作:向他人学习并应用业务流程最佳实践(参考内容)来标准化组织的非核心竞争力,从而专注于成本和运营模式。

Become a Leader—learn from leading organizations, the outperformers. Using leading practices will strengthen your competitive advantage, innovation, and efficiency in the core differentiating competencies with a focus on the Revenue Model and the Value Model. Our experience is that only few can do it, linking innovation aspects to BPM, especially since it is done in an executive closed-door session and it takes strategic insight and detailed enterprise modeling, engineering, and architecture knowledge.

We wish you luck in your journey of business process modeling and we realize that your business processes are a big deal, not only because they are in the center of business performance, but also because value creation and realization is always related to business processes. We know it can be a very big opportunity for you and your business.

Finally, if you need any help, we are always up for a challenge.

Prof. Mark von Rosing
Henrik von Scheel

　　站在成功者的肩膀上：利用您所在行业内的最佳实践和参考内容（行业能力模型、行业绩效指标、行业指标和其他行业参考内容），通过关注性能模型和服务模型来提高竞争力的平等性和标准化核心竞争力。

　　成为领导者：向领先的组织学习，向表现优异的个人学习。通过关注收益模型和价值模型，使用领先一步的实践经验加强您在核心差异化能力方面的竞争优势、创新和效率。我们的经验是：只有少数人能够做到将创新方面与BPM联系起来，尤其是当它在一个闭门环境中完成的时候，所以它需要战略洞察力和详细的企业建模、工程和体系结构知识。

　　我们祝您在业务流程建模的旅途中取得好成绩，我们意识到您的业务流程非常重要，不仅因为它们是业务绩效的中心，而且因为价值创造与实现总是与业务流程相关。我们知道这对您和您的业务来说是一个非常大的机会。

　　最后，如果您需要帮助，我们随时准备迎接挑战。

Prof. Mark von Rosing

Henrik von Scheel